CHROMATOGRAPHIC SCIENCE

A Series of Monographs

VOLUME 1

Dynamics of Chromatography (in three parts)

by J. Calvin Giddings

Additional volumes in preparation

DYNAMICS OF CHROMATOGRAPHY

PART I

Principles and Theory

DYNAMICS OF CHROMATOGRAPHY

J. CALVIN GIDDINGS

DEPARTMENT OF CHEMISTRY
UNIVERSITY OF UTAH

PART I
Principles and Theory

MARCEL DEKKER, Inc., New York

1965

Library of Congress Catalog Card Number 65-24914

Printed in the United States of America

FOREWORD

The dynamic processes of chromatography are as varied as they are important. Here flow, diffusion and other processes interact with one another, determining the evolution and, eventually, the separation of component zones. The importance of dynamics to both theory and practice are discussed in *Dynamics of Chromatography*.

When the idea of this work was first conceived in 1957, it was hoped that a comprehensive treatise on chromatographic principles could be bound under one cover. The subsequent explosion of knowledge has forced an expansion to three parts and a restriction of subject matter to the dynamic aspects of chromatography. Even for this subject, the coverage is not complete; e.g., nonlinear chromatography has received little more than incidental attention.

The underlying mechanism of all forms of chromatography is discussed in Part I; gas chromatography and liquid chromatography will be discussed in a more specific manner in Part II and Part III, respectively. Although a wide chasm exists between many of the uses and users of different techniques, they are all so similar as dynamic processes that it would be almost total duplication to write the fundamentals for each separately. Even those with interest in a limited area would find perhaps 90 per cent or more of Part I applicable to their particular studies.

Any effort such as this must proceed from a viewpoint that is sometimes a matter of personal choice. It is perhaps natural then that this approach is flavored by the author's direction of thinking evolved over years of research and by a continual effort to reduce complex chromatographic phenomena, insofar as possible, to a simple language. In fact, simplification is one of the main objects here. *Dynamics of Chromatography* is not primarily a theoretical work designed to cast ideas into mathematics; it is more an interpretive work whose object is to make physical sense of theoretical developments in chromatography. Presently, there is no other book available that attempts, as this one does, to satisfy the pressing need of most analysts to reach an understanding of chromatography with realistic physical concepts. The word "realistic" unfortunately forces us into a world beyond the primers, for, at their best, chromatographic dynamics are complex and require much work on the part of the reader.

Finally, I cannot resist an attempt to destroy the blind, artificial barrier inserted by some between theory and practice. Theory (when correct) and experiment (if carefully executed) describe the same truths. The science of chromatography requires both approaches if it is to grow in proportion to the demands made on it. For some reason, this rather obvious fact has been resisted by a staunch minority. On several occasions I have heard pronouncements comparable to this public assertion: "Despite what the theoreticians say, we have found it best to run at flow velocities higher than optimum." If the speaker had bothered to look, he would have found his theoretical explanation in the abundant literature on the increase of practical analysis speed. Now it is obviously impossible to acquaint oneself with the entire literature of chromatography. Yet, if some attempt is not made by the field's active workers to correlate the two, the study of chromatography will be in danger of becoming, on one side, an unrelated array of tens of thousands of separate facts and observations and, on the other side, a meaningless set of mathematics. It is hoped that this work, despite certain mathematical excursions, will serve to show in some small way the common ground of theory and practice, and how the two may be combined with enormous benefits to both.

I am indebted to a large number of people for their help with this work. Many stimulating discussions and useful comments have been offered by Drs. John Knox, Stephen Dal Nogare, Donald DeFord, Paul Schettler and Stephen Hawkes. Dr. N. C. Saha, Margo Eikelberger, Ed Fuller, and Marcus Myers, along with some of those above, made valuable suggestions during our research seminars. (I hope the general reader is as generous with his comments, so that future editions will be more complete.) The talent of Alexis Kelner is apparent in the drawings. The credit for the difficult task of typing goes to Miss Margaret McMullen and Mrs. Athene Hunt. Finally, this work would not have been possible without patience and encouragement from my wife, Jennifer.

JUNE, 1965
SALT LAKE CITY, UTAH

J. Calvin Giddings

PREFACE

While general in scope, Part I of *Dynamics of Chromatography* is not limited to generalities. Frequent reference is made to practical forms of gas and liquid chromatography, and many numerical examples are used.

A good deal of the material found here has not appeared before in the chromatographic literature; for example, much of the material on the relationship of fluid flow and chromatography, the extensive random walk treatment, and many other items.

Part I can be broken into roughly four sections. Chapters One and Two introduce the concepts of migration and zone spreading. Here the random walk model is presented in detail because it gives perhaps the clearest picture of column processes.

Chapters Three and Four deal with nonequilibrium concepts. Here, to progress, one must become rather mathematical. The difficult parts of Chapter Four (and, occasionally, elsewhere in the book) are enclosed in brackets; these parts may be omitted without loss of continuity. My excuse for writing Chapter Four, the most difficult in the book, is that it lies at the heart of chromatographic theory and yet has never been covered except by specialized research papers.

Chapters Five and Six are concerned with the physical events that underlie chromatographic processes—flow, diffusion, and kinetics. Since these are classical studies, well documented, I have mainly summarized and simplified them for the student of chromatography.

Chapter Seven is concerned with separations. It is in part a distillation of the practical ideas scattered throughout the earlier chapters. I have been particularly interested in comparing the potentials of gas and liquid chromatography, a subject too long ignored.

Throughout the book schematic figures are used as a pictorial guide to concepts. These are not faithful mirrors of detail, as the chromatographic medium is one of the most complex three-dimensional structures now under intensive study.

A table of symbols appears at the end of the book. Even with this I have repeated definitions profusely for immediate clarity and to make the text more suitable for reference purposes.

CONTENTS

Chapter One

Introduction and Dynamics of Zone Migration

1.1 A Viewpoint

Chromatography has been defined in numerous and sometimes contradictory ways. We wish to avoid the tedious differentiation between the chromatographic and the nonchromatographic. We prefer here to approach the word "chromatography" by association with the laboratory technique bearing that name. We should like to approach the laboratory technique as would a surgeon, dissecting it to discover the underlying features and the processes related thereto. We may look, first, at those events which occur over the dimensions of a single granular particle; then, step by step, change our magnification to the level of the small pores and finally to the molecular world itself. We should then like to synthesize a whole picture of chromatography from the discoveries thus made.

It is inevitable that we shall partially fail in this task, for the detailed processes are too complex and varied, and empirical work has not yet offered the type of precise, discriminatory data needed. This is merely one way of saying that the science of chromatography is not complete. We must still invent models, average out details, compartmentalize the interrelated, rely on inadequate data, ignore the tenuous, and otherwise simplify our thinking to a comprehensible level. At the end of all this we find, fortunately, that our kaleidoscopic picture is in accord with the major findings of experimental chromatography. We thus have at our disposal a means of correlating, predicting, and, most importantly, understanding the phenomenon of chromatography. This is the real justification for the present task.

1.2 Terminology

For the most part we have used terms and expressions commonly appearing in the chromatographic literature. In a few cases it has been necessary to draw gas and liquid chromatography together by compromise. In gas liquid chromatography, for instance, it has been conventional to affix subscript g to gas phase terms (e.g., C_g) and subscript l to the liquid stationary phase terms. While this distinction is appropriate for Part II, here we must (except in specific cases) use subscript m for the mobile phase—liquid or gas—and s for the stationary phase.

Frequent use is made of expressions such as "column migration" or "processes within the column." The use of the word "column" merely identifies the location of the relevant process, and is not intended to exclude noncolumn techniques such as paper and thin layer chromatography. We could use the general term "chromatographic bed," but this has been done only on occasion. By the same token the word "particle" = "fiber" = "bead" in the relevant cases.

The terms "mobile phase" and "stationary phase" are, of course, standard in chromatography. We have been rather free with the former, sometimes referring to it by the obvious alternates, "mobile fluid" and "moving fluid." The stationary phase divides itself into two categories, depending on whether molecular attachment occurs by adsorption (adsorption chromatography) or absorption (partition chromatography). The latter may be distinguished by the term "bulk stationary phase." The substance being chromatographed is the "solute" or the "sample."

The source of migration in chromatography is the downstream motion of the mobile phase. We must clearly understand the measure of this motion in order to make quantitative sense out of migration. The appropriate measure is the flow velocity, and it comes in several forms. At any particular microscopic point there is a local velocity v'. This varies widely within and between the particles. Its average over 5–10 particle diameters is the *regional velocity*, v. If, furthermore, there is a serious variation of this across the column (as in many preparative columns), the average over the entire cross section is denoted by $\tilde{v}$. The latter variations can usually be ignored and the observed velocity identified with v.

The best measure of the mobile phase velocity v is the distance/time ratio of some inert material which does not attach to the stationary phase. Such material will occasionally be swept at great speed through the open flow passages between particles, and sometimes held back in stagnant pockets and pores. As an average, however, its velocity best represents the overall motion of mobile phase. On occasion we are interested only in the mean interparticle velocity (thus excluding mobile phase held stagnant within pores that might exist), which can be denoted by v_0.

Due to retention, chromatographic zones move along at only a fraction R of the mobile phase velocity. Thus the zone velocity (i.e., the velocity of the zone's center of gravity) is Rv. The quantity R is the basic retention ratio characterizing migration rates. The quantities R and R_f, while closely related, are sometimes mistakenly equated.[1,2] The latter differs slightly (about 15%) from R by being the ratio of the zone velocity to the velocity of the solvent (mobile phase) front in paper or thin layer chromatography. Since the velocity of the front is greater than that of the following mobile phase, $R > R_f$.

In gas chromatography the retention parameter most commonly used is the capacity factor k. This is related to R by $R = 1/(1 + k)$. It is entirely valid to characterize retention by either parameter. Despite the discord with much gas chromatographic literature, the use of R is more desirable from several points of view. First, it and closely related parameters (e.g., R_f) have been used in liquid chromatography for years. It is unfortunate that a departure from this sound terminology was ever made, as it has tended to widen the gulf between related techniques. (The use of R for "resolution" has compounded the difficulties; Rs is used here.) Second, R is proportional to the migration rate, and migration in differential form is the basis of separation. Third, and perhaps most important, nearly every mathematical expression containing R is more complex when written in terms of k. Of the more than 150 numbered equations in this book containing R (excepting those characterizing R), over 90% are simpler (fewer type characters) as a consequence of not using k. Most of the well-known equations of chromatography follow this trend. The simplicity associated with R is related to the fact that it is a direct migration velocity parameter.

Numerous other terms and parameters appear throughout the volume. These are explained along the way. And a section, Symbols, at the end of the book summarizes the principal mathematical quantities.

1.3 Dynamics of Migration and Equilibrium

Separation is achieved through differential migration. Hence the migration process is a cornerstone of chromatography. This section introduces the migration concept from a dynamic viewpoint.

As usually observed, a zone migrates along smoothly at some fraction R of the mobile phase velocity. At the molecular level, by contrast to outward appearances, migration occurs in a state of chaos, each molecule progressing in a stop-and-go sequence independent of any other molecule. Each time a molecule affixes itself to the stationary phase, its migration is interrupted while the zone as a whole passes over. One molecule may pass ahead of another that is immobilized, only to be overtaken by the same molecule at a subsequent point. The process might best be described as a game of uncoordinated leapfrog. This analogy gives a picture of how the zone can move continually ahead although only a fraction of the zone's molecules are moving at any one time.

Since each molecule follows a random, jerk-like path through the column, there is a statistical dispersion of the zone's molecules, that is, the zone broadens out. The zone spreading which accompanies migration in all normal forms of chromatography is thus a reflection of the molecular basis of migration. Although any given molecule is jumping rapidly back and forth between mobile and stationary phases, the probability at any instant (after an initial transient period) that it is in a given phase becomes a rigidly fixed physical parameter. Let the probability that the molecule is in the mobile phase be R. As more and more random jumps are made and the relative degree of statistical fluctuations restrained (as always) by the large number of such transfers, the time fraction spent in the mobile phase will approach this fraction R. Since net downstream motion is achieved solely during residence in the mobile phase,[3] the molecule moves only with a fraction $\sim R$ relative to the velocity v of the mobile phase. Thus the R value may be defined identically as a relative migration rate (as earlier), as the probability that a solute molecule is in the mobile phase or as the limiting fraction of time a molecule spends in the mobile phase. Furthermore, if we have a large number of such molecules (e.g., all those in a zone), the fraction in the mobile phase at equilibrium must equal the same parameter R. This can be used essentially as a fourth definition of R.

The latter definition leads us to the viewpoint that zone migration, despite its chaotic molecular origin, is related to equilibrium parameters. All equilibrium has a kinetic origin, so this dual viewpoint is not unusual. Since the equilibrium assumption is a basic part of much chromatographic work, particularly the identification of components through R values, it merits further comment.

If we have a large number of molecules in a chromatographic zone, then the *fraction* in the mobile phase at any point and time is identical to the *probability* that any given molecule is in the mobile phase at the same point and time. Let us follow this simple probability concept to establish where and when equilibrium is valid. A value of exactly R denotes complete equilibrium.

In many systems the initial zone is carried onto the head of a column in a stream of mobile fluid, and migration commences without hesitation. A given molecule, thrust suddenly onto a column, remains in the mobile phase a finite time until its first capture by the stationary phase. The probability that the molecule is in the mobile phase at the instant of contact with the column is thus one. The probability decays exponentially (in the simplest cases) to R as sorption commences. The time constant of the exponential is usually about equal to the time required for a single sorption step. Since in most runs a large number (about 10^3–10^4) of sorptions and desorptions take place, the probability rapidly approaches R with great precision. Thus the zone as a whole may be presumed at equilibrium except perhaps for about one-thousandth of its lifetime right at the beginning. Hence the average distance migrated can usually be based on equilibrium considerations to within one part in a thousand.

On occasion, where there is a heterogeneous or uneven sorptive material with certain parts resisting fast equilibration, the probability, after decaying rapidly at first, may only slowly make its final approach to R. If this slowness is enough to disturb the equilibrium approximation, it will usually manifest itself in nonsymmetrical zones, usually with tailing[4] (see Sections 2.13 and 6.5).

The above tells us that the migration of a zone as a whole is usually governed by equilibrium. But it does not tell us whether equilibrium is valid throughout the entire width of the zone. Recall that R is an average time fraction and that zone spreading occurs mainly as a result of individual molecules fluctuating around this value.

If, after considerable migration, we select a given molecule near the front of a zone, this molecule will naturally be one which has reached its advanced position by spending a fraction of time slightly larger than R (say, $R + \Delta R$) in the mobile phase. At the particular time t (or at any time before t) when we look at this molecule, its probability of being in the mobile phase is just this value $R + \Delta R$.† Hence the fraction (= probability) of molecules in the mobile phase near the front of a zone exceeds the equilibrium value R. This same argument shows that the probability—hence the number fraction at the rear of a zone—is less than R. In fact the precise value R is applicable only at the zone center; that is, true equilibrium occurs at the zone center and nowhere else. This will be shown by a different kind of argument in Chapter 3. It suffices here to say that if we are to relate chromatographic migration to the equilibrium distribution of molecules between phases, we must refer to the center of the zone, and we must have a system with a fairly rapid molecular exchange between phases.

1.4 Forms of the Migration Equation

The R value, being now established as an equilibrium parameter when evaluated at the center of normal zones, can be expressed in several useful ways.

† This makes use of the concept of conditional probabilities. The given molecule departs from the norm only because we have purposely selected a rapidly advancing molecule. The nonaverage probability, $> R$, is conditional on the selection of just such a molecule.

If R is the equilibrium fraction of solute in the mobile phase, then $1 - R$ is the solute equilibrium fraction in the stationary phase. The term $R/(1 - R)$ is thus the ratio of the solute in the mobile phase to that in the stationary phase. Along with most equilibrium ratios, this one is the concentration-volume product $c_m'V_m$, applicable to the numerator (mobile phase) divided by the corresponding term $c_s'V_s$ of the denominator (stationary phase); that is,

$$\frac{R}{1 - R} = \frac{c_m'V_m}{c_s'V_s} \tag{1.4-1}$$

The equilibrium ratio of concentration in the two phases $(c_s'/c_m')_{eq}$ is the distribution coefficient† or partition coefficient K (sometimes denoted as α). Thus

$$\frac{R}{1 - R} = \frac{V_m}{KV_s} \tag{1.4-2}$$

The quotient V_m/V_s is simply the mobile phase to stationary phase volume in the various forms of partition chromatography (cross-sectional areas, frequently used in place of volumes, are equally correct). For adsorption chromatography V_s can be replaced by the surface area.

On solving the last equation for R, we have an equivalent of the classical expression of Martin and Synge[6]

$$R = \frac{V_m}{V_m + KV_s} \tag{1.4-3}$$

If retention is due to several simultaneous mechanisms (e.g., partial adsorption in partition chromatography), the above arguments can be easily extended to yield the generalization[7]

$$R = \frac{V_m}{V_m + \sum K_i V_{si}} \tag{1.4-4}$$

In any case R can always be related to the equilibrium term (or terms) K and to the amount of each phase present.

In much of chromatographic practice it is impossible and unnecessary to deal with the local concentrations c_s' and c_m', applicable to microscopic regions within the phases. Given a unit volume of the column material, it is often only necessary to know the following; the total amount c of solute in that volume, the amount c_s associated with the stationary phase, and the amount c_m in the mobile phase. We have, of course, $c_s + c_m = c$.

† This ratio only approximates the true thermodynamic distribution coefficient, except in ideal solutions and gases, since the latter is a ratio of activities (rather than concentrations).[5]

The c terms as defined here are overall concentrations, taking no account of the microscopic distribution of solute within the phases (the latter is measured by the local concentration units discussed above, always denoted by a prime, e.g., c_s').

We have already seen that R is the fraction of solute in the mobile phase at equilibrium. Thus $R = (c_m/c)_{\text{eq}}$. Henceforth equilibrium concentrations will be denoted by an asterisk so that R is now given by the simple expression†

$$R = c_m{}^*/c \tag{1.4-5}$$

In a like manner the solute content of the mobile phase over that of the stationary phase at equilibrium is

$$\frac{R}{1-R} = \frac{c_m{}^*}{c_s{}^*} \tag{1.4-6}$$

Although both $c_m{}^*$ and $c_s{}^*$ change from point to point in the zone, their ratio is constant. Equations using overall concentration terms, like the above two, will be employed extensively later.

Taking another approach, we have already noted that a molecule left long enough in the column will spend a fraction closely approaching R of its total time in the mobile phase. In any finite interval R will not exactly equal this fraction due to the molecule's fluctuations, but if we take the average time fraction for all solute molecules, some fluctuating in one direction and some in another, R may be identified precisely. Thus

$$R = t_m/(t_m + t_s) = t_m/t \tag{1.4-7}$$

where in total time t an average time t_m is spent in the mobile phase and t_s in the stationary phase. This equation is due to LeRosen.[8] As with the earlier cases we may write this in other equivalent forms; for example,

$$\frac{R}{1-R} = \frac{t_m}{t_s} \tag{1.4-8}$$

The above equations can be related more directly to the dynamics of sorption and desorption as follows. During its migration down the column a given molecule will sorb and desorb many times in rapid succession. Since each sorption step must be followed by a desorption step, the same number of each must occur during migration (give or take one for end effects). Let this number be r. Now if a molecule spends a total time t_s with the stationary phase, and desorbs r times, the average stay must be t_s/r. This is denoted by t_d since it is the average time required to get the

† The total concentration c is not starred because it does not change as solute transfers between phases. We have $c_s + c_m = c_s{}^* + c_m{}^* = c$.

molecule desorbed. Similarly, the average stay in the mobile phase before an adsorption can occur is $t_a = t_m/r$. On writing these expressions as $t_s = t_d r$ and $t_m = t_a r$, equations (1.4-7) and (1.4-8) become

$$R = t_a/(t_a + t_d) \tag{1.4-9}$$

$$\frac{R}{1 - R} = \frac{t_a}{t_d} \tag{1.4-10}$$

thus relating R to the dynamic parameters t_a and t_d. As will be shown in Section 2.7, t_a and t_d are equal to the reciprocal of the (apparent) rate constants of adsorption and desorption.

In summarizing this section, we should draw attention to the fact that R is based on equilibrium considerations but that true equilibrium even in the most ideal cases is applicable at only one point (the center) in the zone. The reference of R to strictly equilibrium conditions, out of keeping with nearly all the solute in the zone, is reflected by certain (unrealistic) restraints applicable in most of the equations for R in this section. Thus each time R is written as a concentration ratio, we must take the equilibrium ratio rather than the real value found at some random point in the zone. And each time R is taken as a time ratio (in mobile and stationary phases), it must be the ratio applicable to the hypothetical average molecule or to a molecule spending infinite time in the column rather than the fluctuating value of real molecules.

1.5 Chromatographic Methods

It is unnecessary to describe in full the numerous modifications of chromatography. These have been cataloged in many books on the subject. We should like, however, to indicate very briefly the dynamic similarities and differences of the main categories.

Depending on the characteristics of interest, chromatography may be split in numerous, arbitrary ways. The list below is by no means complete, but will illustrate the enormous scope of the subject.

(*1*) *Linear-Nonlinear*. In linear chromatography all solute molecules behave independently; in nonlinear chromatography there is some measurable degree of interaction between them. Nonlinear chromatography is usually thought of as involving only nonlinear sorption isotherms. It may, however, also involve nonlinear dynamic processes such as diffusion and kinetics. Nonlinearity always yields to linearity with sufficiently small samples of solute. The dynamics of interacting molecules, as found in nonlinear chromatography, is so complex that it has never been treated with real satisfaction, and therefore constitutes only a minor part of this book. Equilibrium treatments of nonlinear chromatography, by contrast to dynamic treatments, abound in the classical papers† on the subject.

(*2*) *Liquid-Gas.* As far as experimental work goes, there are two main camps divided according to whether a liquid of a gas is used for the mobile phase. As shown in Chapter 7, there is little basic difference in the nature of the dynamics involved. Thus the work of this volume is applicable almost in total to both. But the numerical differences in certain physical properties (e.g., diffusivity, viscosity) are so enormous as to cause a very real inequality in practical performance (speed, resolution, etc.). The specific and detailed nature of gas chromatographic processes will be discussed in Part II, while liquid chromatography will be reserved for Part III.

(*3*) *Partition-Adsorption.* This is perhaps the most basic division between techniques because there is a fundamental difference in the retention mechanism. Each form is frequently used in both gas and liquid chromatography. Adsorption chromatography is based on the retention of solute by surface adsorption, whereas partition chromatography is based on the adsorption of solute by bulk‡ stationary phase usually held in place by an inert scaffold of solid particles. Both methods, along with certain "mixed" cases, are considered at length in this volume.

(*4*) *Zone-Frontal.* Solute may be introduced as a discrete zone or as a step function (i.e., as a continuous train of solute beginning with a sharp front). Resolution is best when the discrete zone is introduced as narrow as possible (approaching a δ-function) and, in the frontal case, when the front begins as a sharp, right-angle profile. A δ-function profile (infinitely thin zone) introduced into a linear column which has a rapid exchange between phases will develop on the column into a Gaussian ("normal error curve") zone. The sharp front will develop into an S-shaped "error function." The mathematical treatment of zone shape in linear chromatography automatically describes fronts because the latter is the integral of the former (Section 2.14). Since zone chromatography is the common departure point in the laboratory our discussion will largely center on this method. The terms δ-function and Gaussian will consequently appear often.

(*5*) *Elution-Nonelution.* Solute may be washed clear through the column for analysis or it may be stopped at some point within the chromatographic bed. The latter is especially useful in paper and thin layer techniques. The dynamics of the two cases are essentially identical except that in one case the process is terminated earlier. Both cases are covered in this volume.

† For example, references 9 and 10.

‡ The word "bulk" in this book means three dimensional, not necessarily that it has the properties of large masses.

(*6*) *Column-Noncolumn.* Most chromatographic processes are confined by the walls of a column. Paper and thin layer chromatography are not physically confined. The latter retains its mobile phase by surface tension and because capillary forces substitute for pressure drop. Hence the noncolumn methods are restricted to liquid chromatography.

(*7*) *Packed-Capillary.* Although most chromatography results from percolation through a granular material, capillary (or open tube) chromatography involves flow through a narrow open tube with the stationary phase confined to the wall. Capillary methods have been confined almost entirely to gas chromatography, but extensions to liquid chromatography have been made. Dynamic processes in capillary columns are simpler than in packed columns, and may usually be regarded as a special case of the latter.

(*8*) *Nongradient-Gradient.* The composition of the incoming mobile phase may be varied with time to change the speed of elution. Such gradient methods are confined almost solely to liquid chromatography. In theory they could be applied to gas chromatography using high pressures and dense carrier gases, but ordinarily gases are so inert that they have only a small effect on elution. The gradient method will be discussed in Part III.

(*9*) *Isothermal-Programmed Temperature.* In gas chromatography the isothermal technique has in many cases yielded to programmed temperature methods. By increasing column temperature with time a more even spacing of peaks is obtained. This method can be applied advantageously also to liquid chromatography. Because of its current prevalence in gas chromatography, discussion will be reserved for Part II.

(*10*) *Other Categories.* There are, of course, many more categories than those listed above. We might include the displacement methods, but in this rough ordering displacement must be considered as a special case of nonlinear chromatography. We might also consider tapered columns, but these have such a minor role in the broad scheme of things that inclusion as a major category would be misleading. Furthermore, we could start a grouping based on solutes, as, for example, inorganic and organic analyses.

Virtually any combination of the above may be imagined to exist in one technique. Thus we may have *liquid* chromatography in an *isothermal*, *packed column*, based on a *linear*, *partition* mechanism, with solute introduced as a *zone* and developed to the point of *elution* without *gradients*. There are $2^9 = 512$ such combinations. In practice many are excluded—for example, those combinations having both gas and noncolumn characteristics. Many more are useless in the present state of things—the use of mobile phase gradients in gas chromatography. Other combinations show great promise but have not reached the laboratory level.

1.6 References

1. J. C. Giddings, G. H. Stewart, and A. L. Ruoff, *J. Chromatog.*, **3,** 239 (1960).
2. K. L. Mallik and J. C. Giddings, *Anal. Chem.*, **34,** 760 (1962).
3. A. L. LeRosen, *J. Am. Chem. Soc.*, **67,** 1683 (1945).
4. J. C. Giddings, *Anal. Chem.*, **35,** 1999 (1963).
5. M. R. James, J. C. Giddings, and R. A. Keller, *J. Gas Chromatog.*, in press.
6. A. J. P. Martin and R. L. M. Synge, *Biochem. J.*, **35,** 91 (1941).
7. R. A. Keller and G. H. Stewart, *J. Chromatog.*, **9,** 1 (1962).
8. A. L. LeRosen, *J. Am. Chem. Soc.*, **67,** 1683 (1945).
9. J. N. Wilson, *J. Am. Chem. Soc.*, **62,** 1583 (1940).
10. D. DeVault, *J. Am. Chem. Soc.*, **65,** 532 (1943).

Chapter Two

Dynamics of Zone Spreading

2.1 Importance of Zone Spreading

The effectiveness of separation in chromatography depends on two important requirements. First and most obvious, a disengagement of zone centers must be obtained through a difference in migration rates of individual solutes. The control of migration rates is related to thermodynamic equilibrium, and a thorough understanding is beyond the scope of this volume. Second, the zone must be kept narrow and compact to avoid overlap. The subject of zone spreading and its control will be discussed in this and subsequent chapters.

It is difficult to generalize regarding the relative importance of controlling migration rate differences and zone spreading. They are of identical importance in the sense that the separability of a given pair is equally enhanced (using any reasonable criterion) by either doubling the migration rate difference or halving the spread of the peaks. In practice the relative importance of these two depends entirely on individual circumstances. In some cases specific interactions can be employed to yield widely different migration rates, and the control of zone spreading is of secondary importance. In other cases, especially where similar structural features exist, it is difficult to enhance the difference in migration rates. Here zone spreading must be carefully controlled. Spectacular advances have been made in both areas, leading to the tremendous versatility of chromatography as an analytical tool.

There is one important class of separations in which the control of zone spreading may be regarded as the paramount requirement. To separate a large number of solutes simultaneously, those in a petroleum fraction, for instance, the requirement for narrow zones is critical. This is illustrated in Figure 2.1-1 by a gas chromatographic run of premium grade gasoline.[1] If the relative migration rates were changed, we would merely scramble the peak locations, improving some separations and hindering others. If each peak or zone were reduced in width, each and every peak would be more completely isolated from its neighbors. With many component solutes the basic difficulty lies in the fact that there is simply not sufficient space in a chromatogram into which can be crowded numerous zones of moderate width. A clean separation can be obtained only if the zones are kept narrow.

Some of the physical processes leading to zone spreading are present in all forms of chromatography. These will be considered in detail later in the chapter. An especially sharp distinction is evident between chromatographic runs in which the sorption isotherm is essentially linear throughout the procedure and those in which the isotherm differs significantly

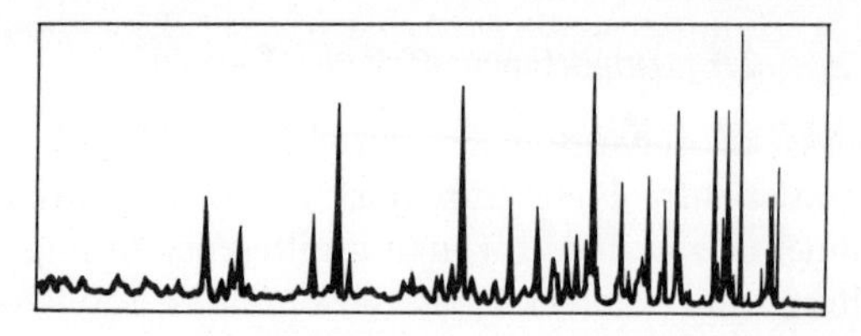

Figure 2.1-1 This analysis of premium gasoline by gas chromatography shows the necessity for keeping peak width minimal in order to accommodate the large number of peaks present.

from linearity. The linear case is far more important in a practical sense, far more tractable theoretically, and will thus occupy the chief attention in this book. For linear (or nearly linear, since exact linearity is rarely obtained) chromatography a number of diffusion-like processes are responsible for a continual enlargement of zones. If we were to postulate certain "ideal" conditions—instantaneous mass transfer, zero axial diffusion, and flow paths of exact equivalence—the diffusion-like processes would cease to operate and each zone would maintain its original shape as it migrated through the column. This is not the case for nonlinear isotherms. Even under such ideal conditions a change in shape and width would continually occur with migration. The diffusion-like processes, always encountered in practice, would simply be superimposed on the normal evolution of the zone profile. For all practical purposes the addition of the two zone spreading mechanisms in nonlinear chromatography makes the zone spreading excessive and the technique less useful.†

2.2 The Evolution of Zone Spreadiug Concepts

The development of concepts and mathematical expressions for zone spreading has been one of the major theoretical and practical challenges of chromatography. It cannot yet be said that the final chapter has been

† Exceptions are pointed out in reference 2.

added to this development. Remarkable advances have been made, however, in the face of the enormous geometrical complexities of the packing materials and the complications of the flow, diffusion, and kinetic processes occurring therein. In fact, in the slightly more than two decades in which theoretical studies have been made, chromatography has advanced from complete empiricism to a science in which zone structure and spreading can in some cases be successfully predicted before a single chromatographic measurement is made. The evolution of these important concepts and the scope of their practical applications are well worth describing.

The theory of chromatography originated in 1940 in a paper entitled "A Theory of Chromatography" by J. N. Wilson.[3] The important historical role of this work has been largely neglected, partly because the quantitative theory did not allow for diffusion nor the lack of equilibrium between the two phases. Aside from the quantitative theory, however, Wilson's paper appears to have been the first to offer a sound qualitative description of nonequilibrium and its important place in chromatography. The role of longitudinal diffusion was also explained. In describing the role of nonequilibrium, Wilson states that "the width of a band may increase . . . because the leading edge of the band migrates too rapidly on account of a low rate of adsorption, or because the trailing edge of the band migrates too slowly on account of a low rate of desorption." Although the reversible relationship between adsorption and desorption is perhaps overlooked here, the description is obviously a forerunner of our present concept of the role of nonequilibrium (see Section 3.1). A further concept, still of paramount importance, was established in the following statement: "The effects arising from low rates of adsorption and desorption can be diminished by decreasing the rate at which liquid flows through the column, but as the rate of flow decreases the importance of diffusion effects increases." Thus it is clearly established that excessive spreading is found at either very high or very low flow rates, and that the flow rate must obviously be established by compromise at some intermediate value. The principles enunciated by Wilson are applicable to all forms of chromatography even though the paper preceded the development of most present-day techniques.

The Nobel Prize winning paper[4] of Martin and Synge in 1941 followed closely behind that of Wilson. These authors developed the widely popular plate theory of chromatography. Although the plate theory is entirely inadequate for the current burdens of theoretical use, it was the first theory to describe the development or evolution of a zone profile under the influence of nonequilibrium and in the presence of a linear isotherm. It was noted that after a sufficiently long development time the zone profile would become Gaussian in shape. Using intuitive arguments, Martin and Synge deduced the rule that the plate height H is proportional to flow velocity and the square of the particle diameter (it is important to note that the plate theory, by itself, does not yield any such specific conclusions about the variation of H). These authors repeated Wilson's argument about the need for an intermediate flow velocity, citing the adverse effects of diffusion from plate to plate at low velocities.

In the decade following the work of Wilson's, outstanding advances were made in the mathematical description of chromatography. DeVault[5] and Weiss[6] in 1943 improved on the equilibrium theory originated in Wilson's classic paper. The next great step, however, was the break from the restrictions and limitations of the equilibrium theory. Several authors (Walter,[7] Sillen,[8] etc.) contributed to this advancement, but H. C. Thomas made the most general and outstanding contribution.[9] His first treatment of nonequilibrium appeared in 1944. A paper published in 1948, however, was probably the most significant of the decade. Starting with the equation of mass conservation (neglecting the longitudinal diffusion term), Thomas introduced a Langmuir rate mechanism (which reduces to linear kinetics at low solute concentration). The equations were solved to yield the effluent concentration as a function of time. Since this involved complex mathematical expressions, Thomas went on to obtain a simplified asymptotic solution for the cases in which the flow rate is slow enough to prevent large departures from equilibrium. (This close-to-equilibrium condition is an integral part of current chromatographic practice and theory, as we shall see in Chapter 3.) He thereby obtained formulas from which the adsorption and desorption rates could be obtained from the experimental solute concentration curves of chromatography.

A development in some ways parallel to the one just described was reported by Boyd, Myers, and Adamson in 1947.[10] These authors, advancing beyond the simple theory presented by Beaton and Furnas[11] in 1941, described ion exchange kinetics in terms of diffusion through a liquid film. (The same authors had previously outlined the rate controlling steps pertaining to ion exchange.[12]) One of the most interesting applications of their work was the attempt, apparently the first ever made, to predict zone structure in terms of independent rate and equilibrium constants. Although the independent rate constants were five to ten times too small, it was shown that the extreme complexity of the chromatographic process was not an insurmountable barrier to concept and theory.

The second decade of chromatographic theory (1950–1960) involved developments on an extremely wide front. Part of this diversity resulted from the proliferation of chromatographic techniques, nearly each of which demanded theoretical attention at one time or another. The limited description given here pertains only to the main currents of linear chromatography and to those concepts which, although perhaps originating in a specialized area, apply to the whole broad spectrum of the field.

In 1952 Lapidus and Amundson developed an exacting mathematical theory which contained as parameters an unspecified mass transfer coefficient and a longitudinal diffusion coefficient.[13] This theory was later to become the foundation for the well-known treatment of van Deemter, Zuiderweg, and Klinkenberg.[14] In subsequent years other exacting and complicated theories were reported, but the most significant trend of the decade was toward the development of asymptotic theories. These theories revolved around the concept that an effective and practical chromatographic operation requires sufficient running time to obtain narrow and well separated zones. When the time is sufficiently long for this, however, it is long enough to justify an important approximation, that is, that the sorption-desorption kinetics are proceeding with only a slight departure from equilibrium. (Note that Thomas[9] first used this approximation to advantage in 1948.) This long-time assumption† has been used in many disguised but nearly equivalent forms. One of these is that a concentration pulse will gradually acquire a Gaussian profile, as first noted by Martin and Synge.[4] Another is that the time is large enough to allow each molecule to undergo a large number of individual sorption and desorption steps (Giddings and Eyring,[15] 1955). Whichever form the assumption has taken, it has freed the theoretician of a great deal of mathematical detail. It has therefore led to simpler and more directly useful equations when starting with rather simple physical assumptions. When the underlying assumptions are widened to account more realistically for the complex processes actually occurring, the corresponding theory is essentially impossible to develop without the long-time assumption. This assumption has thus led to theoretical treatments with a much broader scope as well as to simplified expressions for describing chromatographic zone spreading.

Both the Martin-Synge and the Thomas papers established rudimentary concepts related to the long-time assumption. Then, in 1954, Glueckauf derived a simple equation based indirectly on this assumption which allowed for solute diffusion through ion exchange beads and their surrounding liquid.[16] This was the first comprehensive equation to account for most of the major factors pertaining to a chromatographic technique. His concepts were valid, of course, beyond the range of ion exchange chromatography.

† In some systems a running time of one second or less is relatively long; hours may be required in others.

In 1955 Giddings and Eyring introduced probability (statistical) concepts into the description of molecular migration in chromatography.[16] The theory was extended to the complex two-site adsorption problem, and a simplified expression was again shown to be valid for one-site adsorption after a sufficient time. This approach was the forerunner of the random walk theory of chromatography (Giddings,[17] 1958), the simplest known chromatographic theory and the one used in this chapter to present the elementary ideas concerning the role of various kinetic and diffusion processes.

In 1956 an equivalent of the long-time approximation was used by van Deemter, Zuiderweg, and Klinkenberg to simplify the mathematical treatment of Lapidus and Amundson. This treatment resembled Glueckauf's in that the zone spreading was expressed as the height equivalent to a theoretical plate. Although the important concepts of this work originated as far back as Wilson's paper (e.g., the adverse effects of both high and low flow rates), and were in large part anticipated by Glueckauf's plate height theories,[16,18] the work served to popularize some important theoretical concepts, particularly in the field of gas chromatography. The subsequent applications of the so-called "van Deemter equation" were largely responsible for the rapid development of efficient gas chromatographic techniques.

The simplified theories described up to this point were still deficient in one way or another insofar as they did not express the most useful combination of exactness and simplicity. The expressions that were rigorously correct in the limit of long-time periods for the physical model assumed (Thomas and Giddings and Eyring) were fairly simple, but were not yet written in terms of Gaussian concentration profiles (or any such profile originating from effective diffusion processes). The expressions which were given in the Gaussian or plate height form (Gleuckauf, van Deemter et al.) were not, on the other hand, asymptotically correct. In 1957, prompted by a paper outside of chromatography,[19] this last easy step was accomplished for a simple reversible adsorption-desorption reaction.[20]

Following the successful combination of exactness (as a limit for fairly long but practical time periods) and simplicity (in the form of Gaussian, plate height or effective diffusion concepts), the theory of chromatography mushroomed as a useful and applicable tool. Golay in 1958 first applied this successful combination to a diffusion controlled mechanism—the sorption and desorption of solute within an open capillary column, the walls of which were assumed to possess a uniform coating of liquid stationary phase.[21] Subsequent experimental work showed that the gap between theory and experiment, while not entirely closed, was much narrower.

In 1959 Giddings started the development of a generalized nonequilibrium theory of chromatography for the express purpose of calculating the influence of any complex combination of sorption and desorption steps, whether controlled by diffusion or single-step reactions.[22] In subsequent papers this theory was applied to a wide range of chromatographic problems,[23] from gas-solid chromatography to the influence of the flow pattern in large preparative columns. This approach meant that chromatographic calculations were no longer limited to oversimplified models of the stationary phase, flow profile, and adsorption-desorption kinetics. The limit of theoretical accuracy was now to be determined by how closely the structural and adsorptive features of the column could be categorized and described. Thus the main barrier to further progress was to become a problem in physical and chemical description, and not a problem in chromatographic theory, which, finally, has outreached the state of knowledge concerning columns themselves. The generalized nonequilibrium theory has led to the narrowest gap yet found between theory and experiment—a 4–64% error in the independent prediction of zone spreading (plate height) in gas chromatography using lightly loaded (0.5%) glass beads.[24,†] The goal of obtaining independent and successful theoretical predictions on chromatographic performance, first attempted by Boyd and co-workers,[10] appeared to be approaching a state of fruition.

Nonkinetic Problems

With few exceptions, the foregoing summary has dealt with the theoretician's increasing ability to deal with the complex sorption-desorption kinetics or mass *transfer processes* of chromatography, whether these kinetics are controlled by diffusion processes or single-step reactions. The emphasis on kinetics stems from the fact that this is the most important and challenging aspect of chromatographic dynamics. Those familiar with theoretical concepts may recall that zone spreading originates from three sources. Sorption-desorption (or mass transfer) kinetics contribute the most of any effect to zone spreading in high speed runs, and sometimes over the entire practical operating range of flow velocities. Ordinary molecular diffusion in the flow direction contributes most at low or very low velocities. This effect is simple compared to kinetic effects, and its spreading contribution can be calculated immediately in terms of diffusion coefficients and a structural parameter, γ, with a value near unity (see Chapter 6). The theory of γ has been evolved by Knox and McLaren.[25] The nature of the diffusion effect was established at an early date in the evolution of chromatographic concepts.[13,14,16,26]

† Recent work in the author's laboratory indicates an error of about 10%.

The third source of zone spreading comes under the general title of flow pattern effects, and includes channeling (and other effects of packing inhomogenity) and the so-called "eddy diffusion" term. These effects can be very important for some columns as was realized in the first decade of chromatographic theory.[13,14] Although flow pattern effects are naturally very complicated in the granular materials used for column packing (and much theory remains to be done), the scope of the theory has so far been more restricted than that of the kinetic processes. In recent years, however, there has been an acceleration of research in this area, particularly in the development of eddy diffusion concepts,[27] and in the study of large-scale columns.[28–30] This work has shown that the effects of flow pattern are strongly interrelated with the kinetic effects discussed earlier. The reader is referred to Section 2.10 or to the literature for a discussion of the advances which have been made.

2.3 Role of the Theoretical Plate Concept

The theoretical plate model has maintained a widespread popularity in the description of chromatographic zone spreading. This is unfortunate in view of the nearly total failure of the theoretical plate concept in describing the physical and molecular events occurring in chromatography. The fact that the plate model has played an important role in the development of chromatographic theory does not alter the conclusion that it is obsolete for present-day use. Because of its extensive application, the relationship of this model to the chromatographic process is described in some detail below.

The theoretical plate model was introduced into chromatography because of its effectiveness in describing distillation procedures. We could argue inconclusively and at great length whether or not the comparison of chromatography and distillation warrants the transfer of the plate concept from the latter to the former method.† But the question of direct importance here has nothing to do with the comparison of distillation and chromatography; it concerns only the validity and usefulness of the theoretical plate model in its particular application to chromatographic methods.

The theoretical plate model has two functions in current usage. First, it serves to describe the chromatographic processes of flow, equilibration, etc. Its use for this purpose, as described below, is unacceptable. Second, it provides a parameter—the plate height—which can be used to characterize chromatographic zone spreading and thus resolution. The use of this parameter, as shown in Section 7.2, is acceptable and in no way implies that the model describes the actual column dynamics.

† For an excellent discussion see reference 31.

The theoretical plate model has been different things to different investigators, a fact which makes its evaluation somewhat difficult. Martin and Synge[4] used the plate model to account for the nonequilibrium processes in a column and ruled against considering longitudinal diffusion as part of the model (paradoxically, these authors later considered longitudinal diffusion as contributing directly to the plate height). In contrast to this, Glueckauf[16,32] considered the plate theory applicable only when there is essentially complete equilibrium of solutes between phases. The theory was expected to be valid when zone spreading was dominated by longitudinal diffusion and channeling effects. The plate theory was later extended to include nonequilibrium phenomena. Mayer and Tompkins[33] visualized the plate concept in terms of an intermittant flow process rather than the continuous flow of Martin and Synge.

Because of these and other conflicting interpretations, the theoretical plate model will be considered within the framework of the original Martin and Synge treatment. These authors had an excellent grasp of theoretical concepts and their plate treatment, insofar as fundamentals are concerned, has not been significantly bettered.†

Following Martin and Synge the plate height may be defined "as the thickness of the layer such that the solution issuing from it is in equilibrium with the mean concentration of solute in the nonmobile phase throughout the layer." An essential feature of the subsequent mathematical treatment is the assumption of complete equilibrium (within a plate) of solute between stationary and mobile phases.‡ The mathematical development is not consistent with and does not include the effects of longitudinal diffusion.§

The effect of a linked series of plates as defined above would be exactly the same as a series of mixing cells or vessels each of length equal to the plate height. Thus if a theoretical plate were replaced by a cell in which total mixing occurred, the mixing would lead to a complete cell equilibrium just as postulated for the hypothetical plate. The mixing may be regarded

† The mathematical execution has been refined.[34]

‡ This assumption is a necessary prerequisite to the expression $\delta v/h(A_L + \alpha A_S)$ for the fraction of solute lost from a plate with the passage of a small mobile phase volume δv (the notation is that used by Martin and Synge). If we assume that the mean stationary phase concentration to mean mobile phase concentration within a plate is α' rather than the equilibrium partition coefficient α, the correct expression would by $\delta v/h[(\alpha A_L/\alpha') + \alpha A_S]$. Thus the plate-wide equilibrium of solutes between phases, at least in the mean, is an integral part of the mathematical development and hence of the final result of the Martin and Synge development.

§ Martin and Synge assume that a fixed fraction (above) of the cell's solute is lost to the next cell with the passage of a given volume increment. If longitudinal diffusion were occurring, this fraction would depend on the concentration gradient and thus would not be fixed.

simply as a means for providing the required equilibrium. Mixing cell models have been used for the study of fixed-bed processes in chemical engineering.[35–38]

The most obvious conflict of the plate model with the actual column processes lies in the assumption of plate-wide equilibrium. In actual fact equilibrium is not attained anywhere in the column (with the unimportant exception of a single point at the peak maximum). Not only does the assumed equilibrium fail to exist, but the degree of nonequilibrium is critical in determining zone spreading (see Chapter 3). Martin and Synge were well aware of the prominent role of nonequilibrium in chromatography and certainly realized the nature of their approximate treatment. The fact that later investigators have interpreted the plate model in a literal fashion, sometimes incurring serious error, is not the fault of the original authors.

Another assumption of the theoretical plate model in conflict with reality is the discrete and discontinuous nature of the plates. Since a uniform equilibrium concentration is presumable found in each plate, the concentration profile consists of a series of steps, contrary to theoretical and experimental evidence. Some authors have associated plate discontinuities with the existence of discrete adsorbent or support particles. This association is highly questionable in view of the fact that experimental plate heights vary from less than one up to several thousand particle diameters. Except perhaps in the turbulent flow regime (which rarely occurs in chromatography) where a direct mixing occurs, there is no close relationship between the processes occurring over the length of a single particle and within a hypothetical plate.

A third defect of the plate model is its failure to allow for ubiquitous longitudinal diffusion effects, as discussed earlier.

Some of the inconsistencies of the theoretical plate model can be avoided by choosing another definition of the plate height. A more satisfactory definition of this quantity equates it to the length of a cell whose mean concentration (not stationary phase concentration) is in equilibrium with its own effluent.[39] The first two of the above inconsistencies can be virtually eliminated through this redefinition, but the failure of the plate model in giving a really adequate description of chromatographic processes still persists.

The theoretical plate model is not designed to account for the basic effects of particle size, molecular structure, sorption phenomenon, temperature, molecular diffusion, flow pattern, and other quantities influencing zone spreading. Since nothing is assumed in plate theory about these parameters, nothing can be gained from it concerning their role. Thus the plate model fails in the most important test of all—the very practical matter of predicting zone dispersion as a function of the numerous variables open to manipulation by the investigator. In order to be of any value the plate model must be combined with concepts and theories related to the fundamental column parameters. Under these circumstances, however, the theoretical plate model adds nothing to the combination; the additional concepts are sufficient by themselves to describe zone spreading.

But what, specifically, does the plate theory provide? The prediction of plate theory concerning zone development is as follows. As the zone (started entirely in the first plate) is washed through the first several plates, a highly discontinuous concentration profile is obtained with solute distributed in plates following the Poisson distribution. At an intermediate stage of development (say, 30–50 plates) much of the abrupt discontinuity is gone because neighboring plates have fairly similar concentrations. The distribution is still Poisson. At a later stage of development (usually when the zone center has migrated through 100 plates or more) the concentration profile is reasonably smooth and continuous and, although still Poisson, can be approximated by a Gaussian or normal error curve. The standard deviation σ of the Gaussian (a direct measure of zone spreading equal roughly to the quarter-width of the zone) is found to be $\sqrt{HL}$ where H is the plate height and L the distance migrated by the center of the zone. The one usefully correct prediction of the plate theory is that after the initial stages of development the zone width, proportional to σ, increases with the square root of the zone migration distance L. This prediction does not provide a very compelling argument for the use of plate theory since nearly all physical processes leading to zone spreading, including random processes and diffusion phenomena, are governed by the same laws. Furthermore, the numerical value of the plate height must be obtained independently by experimental or theoretical means.

The initial stages of zone development are beyond the scope of both the plate model and most other theoretical treatments. There is no evidence that the actual concentration distribution is ever of the Poisson type. In fact when the sorption phenomenon is governed by simple kinetics, it is known that the concentration profile is given in terms of Bessel functions and is thus more complex than Poisson (see Section 2.13). It is fortunate that the practical needs of chromatography are served best when the effective number of plates is large, and various simplified expressions revolving around the Gaussian profile are entirely adequate.

Although the theoretical plate model is of little value in describing the basic chromatographic processes, the plate height is a useful and widely accepted parameter for the characterization of zone spreading and resolution (see Section 7.2). In practice the plate height is used to describe all zone spreading phenomenon, including both nonequilibrium and longitudinal diffusion effects. As noted earlier, the evolution of a zone profile occurs in the same manner irrespective of the underlying mechanism. Hence the use of a single parameter, the plate height, is adequate for the overall description of the zone. For a uniform column free from concentration and velocity gradients the plate height may be defined as

$$H = \sigma^2/L \tag{2.3-1}$$

In nonuniform columns the zone spreading factors vary from point to point, and it is necessary to think of the varying plate height in terms of its local value

$$H = d\sigma^2/dL \tag{2.3-2}$$

that is, the plate height is the increment in the variance σ^2 per unit length of migration. The latter definition is the more universal of the two since it applies to both uniform and nonuniform columns.

In elution chromatography the zone profile evolves at the column end with time as a variable. The concentration-time profile has a standard deviation of τ, analogous to σ except having the dimensions of time. The two can be related by considering the fact that a given part of the zone with length σ can be swept off the column in a time equal to σ/(zone velocity) $= \sigma/Rv$. This time—the appearance time for one σ—is clearly equal to τ. Thus $\tau = \sigma/Rv$ or $\sigma = Rv\tau$. For a uniform column in which $H = \sigma^2/L$, we see that $H = (Rv\tau)^2/L = L\tau^2(Rv/L)^2$. Length over zone velocity—L/Rv—is simply the elution time t. Thus for elution from a uniform column the plate height can be obtained as

$$H = L\tau^2/t^2 \tag{2.3-2a}$$

Nonuniform columns will be discussed further in Section 2.14.

Although it is possible to characterize zone spreading by means of the plate height, we can also describe it in terms of an effective longitudinal diffusion coefficient. The use of either concept leads to the same concentration profile after sufficient development of the zone. Since the mathematics of diffusion has been developed very extensively, it is simplest to calculate zone profiles in terms of diffusion theory or the equivalent probability theory (see Section 2.14). One can then avoid the long and tedious mathematical development of the zone profile as originally given by Martin and Synge and repeated in most standard texts. The concentration profiles originating in frontal analysis, plug flow, δ-function injection (where the zone originates as a negligibly thin band), etc., are readily available as well-established solutions of the diffusion equation. This matter will be pursued further in Section 2.14.

The equivalence of the plate and diffusional treatments of chromatography has been demonstrated by several authors.[32,39] We shall follow the general approach of Glueckauf[32] in showing this identity. (Material involving mathematical or detailed theoretical considerations, which the general reader may wish to pass over, have been set off from the general discussion by brackets.)

Consider a unit cross-sectional area of a column which is oriented along the z axis and within which the average mobile phase fluid velocity is v. The column is divided into theoretical plates of length H as shown in Figure 2.3-1. The total amount of solute per unit volume of column

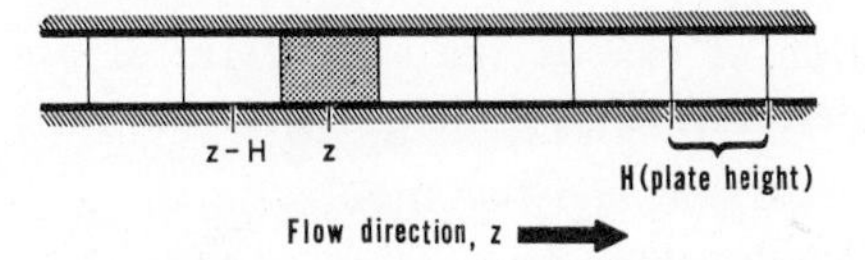

Figure 2.3-1 Division of column into theoretical plates.

(including both sorbed and migrating solute) is the overall concentration c. The amount in the mobile phase free to migrate with the velocity v is Rc. We will focus attention on the concentration change occurring in the shaded plate of Figure 2.3-1 in the very small time interval δt. The amount of solute passing across the unit area of plate boundary and into the plate is the distance moved by the mobile phase—$v\ \delta t$—times the concentration of freely migrating solute in the previous plate—Rc_{z-H}; we thus have $Rv\ \delta t\ c_{z-H}$. Similarly, the amount lost into the following plate in the time δt is $Rv\ \delta t\ c_z$. The net gain is

$$\text{net gain} = Rv\ \delta t(c_{z-H} - c_z) \tag{2.3-3}$$

The net gain of solute is also given as the concentration increase in the plate—$c_{t+\delta t} - c_t$—times the plate volume (per unit area)—H; or

$$\text{net gain} = H(c_{t+\delta t} - c_t) \tag{2.3-4}$$

If the plate height is small compared to the dimensions of the zone, the quantity c_{z-H} can be expanded in a Taylor's series about z

$$c_{z-H} = c_z - \left(\frac{\partial c}{\partial z}\right)_z H + \left(\frac{\partial^2 c}{\partial z^2}\right)_z \frac{H^2}{2} - \dots \tag{2.3-5}$$

with all further terms negligible. The quantity $c_{t+\delta t}$ can be similarly expanded to give

$$c_{t+\delta t} = c_t + \left(\frac{\partial c}{\partial t}\right)_t \delta t + \dots \tag{2.3-6}$$

where further terms are negligible because δt is very small and approaches zero in the limit (this is in contrast to the previous expansion where H does not approach zero). If these expansions are substituted back into equations (2.3-3) and (2.3-4), we obtain the equation

$$\left(\frac{\partial c}{\partial t}\right)_t = -Rv\left(\frac{\partial c}{\partial z}\right)_z + \frac{RvH}{2}\left(\frac{\partial^2 c}{\partial z^2}\right)_z \tag{2.3-7}$$

Since this equation is not limited to any specific distance z and time t, the subscripts may be ignored and the general equation written as

$$\frac{\partial c}{\partial t} = -Rv\frac{\partial c}{\partial z} + \frac{RvH}{2}\frac{\partial^2 c}{\partial z^2} \tag{2.3-8}$$

This is an equation of exactly the same form as used to describe the diffusion of a material which is undergoing flow or drift at the same time; that is,

$$\frac{\partial c}{\partial t} = -\mathscr{V}\frac{\partial c}{\partial z} + D\frac{\partial^2 c}{\partial z^2} \tag{2.3-9}$$

where $\mathscr{V}$ is the flow or drift velocity and D the effective diffusion coefficient. By directly comparing the last two equations, the mean flow velocity of the zone is $\mathscr{V} = Rv$ and the effective diffusion coefficient for the zone is $D = RvH/2$. Since a diffusion coefficient of value D will lead to a diffuse zone with standard deviation $\sigma = \sqrt{2Dt}$, the standard deviation is $\sigma = \sqrt{HL}$ as can be seen by substituting for D. This expression was used earlier without proof to relate σ and H.

2.4 Origin in Random Processes

The downstream migration of an individual solute molecule is highly erratic. Its exact path is determined by the interplay of random processes of various kinds. Included among these are ordinary molecular diffusion, the flow pattern effect called eddy diffusion, and sorption-desorption kinetics. The dispersion of a chromatographic zone is a direct consequence of these processes and their influence on the random nature of each molecule's migration through the column. Indeed, the following sections show that the extent of zone spreading can be estimated in terms of the statistical parameters connected with the migration process. The present section, however, will be concerned only with the qualitative relationship between zone spreading and the underlying random processes.

A random process is one in which individual movements occur unpredictably. The effects of this are best illustrated by ordinary molecular diffusion (not necessarily related to chromatography), either that occurring in the gas or in the liquid phase. In a gaseous system an individual molecule is subjected to numerous collisions with other molecules, and each collision causes a random change in the direction of motion of the given molecule. This influence will lead the given molecule to wander back and forth in a state of continual chaotic motion (for larger particles this is called Brownian motion). If a large number of molecules of a given type were to be started in exactly the same location, each would undergo chaotic motion with some ending up on one side and some on the other side of the beginning point. In this way an initially sharp concentration pulse would gradually become more and more dispersed. It is found that a plot of molecular concentration versus distance is Gaussian; that is, the concentration "pulse" or "peak" exhibits a Gaussian profile after the random processes get well underway.†

Diffusion in liquids is quite similar. The major difference is that the diffusing molecules spend most of their time locked into position by the crowding of neighboring molecules. Occasionally a given molecule will escape its cage and move to an adjacent position.[41] The direction of this movement is, again, unpredictable, and a sufficient number of such movements will yield the erratic path characteristic of all diffusion and related processes.

In chromatography a similar group of random processes lead to zone dispersion.[15,17] In contrast to pure diffusion, however, where only chaotic motion exists, chromatography involves the superposition of random or chaotic molecular motion onto the uniform drifting of the zone caused by the movement of the mobile phase. The solute molecules become dispersed around the center of the zone rather than around the beginning point. The zone's center, although moving, should be regarded as the point of reference about which spreading originates, and all random movements should be categorized according to the relative change in distance from this origin rather than from some stationary point.

A brief description of the three major sources of zone dispersion (mentioned at the beginning of this section) is given below. In the subsequent sections a detailed analysis will more clearly indicate the scope and importance of these effects.

(*1*) *Ordinary Molecular Diffusion.* Along with all the other molecular processes typical of chromatography, molecular diffusion is continually occurring in the flow or longitudinal direction. This diffusion not only occurs in gas and liquid phases, but is present to some degree at interfaces where molecules are often adsorbed though still capable of migration along the surface.

† A general quantitative treatment of this type of problem is given by S. Chandrasekhar.[40]

(*2*) *Sorption-Desorption Kinetics.* Whether partition or adsorption chromatography is being considered, the sorption and desorption of individual molecules occur after unpredictable intervals. Each time a molecule is sorbed, it immediately begins to fall behind the zone center as the latter continues its downstream motion. When the molecule is desorbed and becomes a part of the percolating fluid, it moves faster than the zone center (since the fluid velocity is always greater than the zone velocity). The sorption-desorption process thus leads the average molecule to execute a random back and forth motion with respect to the zone center. This motion resembles the chaotic back and forth motion of a molecule undergoing gaseous diffusion as just described. The result is the same: some molecules will purely by chance migrate ahead of the average and others will lag behind. This zone spreading process, like diffusion, will ordinarily lead to a Guassian concentration profile.

(*3*) *Flow Phenomenon.* The flow streamlines in a bed of granular particles veer tortuously back and forth in an effort to find relatively unobstructed pathways for fluid flow. A given fluid element following such a streampath will have its downstream velocity (as well as direction) fluctuate over wide limits because of the variable rate of fluid penetration through different interstices at different locations. A solute molecule being carried in such a streampath will, then, have its velocity erratically changed from high to low values. The velocity changes are random because the structure of the granulated material which controls the flow pattern is random. Hence some molecules, by following relatively open pathways, may acquire a greater than average downstream velocity, and others may become enmeshed in restricted channels and lag behind. The net result, of course, is zone dispersion. This effect is usually called eddy diffusion in the chromatographic literature.† The phenomenon is modified in practice because solute molecules can acquire a variable downstream velocity by diffusing laterally into adjacent streampaths as well as by following the same one throughout. These effects, related to the coupling theory of eddy diffusion, will be discussed in more detail in Section 2.10.

In the following sections a simple random walk model will be used to describe chromatographic migration. Later we will employ the more powerful nonequilibrium concepts to describe the sorption-desorption processes, and little reference will be made to the underlying random processes. This does not mean that they have ceased to exist, but simply that the laws describing them will be the macroscopic-statistical laws dealing with large numbers rather than with individual molecules. The comparison of the two approaches is best illustrated by the kinetics of

† The name originates in certain formal analogies between this process and one which involves true eddies.[42]

simple reactions. A chemical reaction occurs as a result of individual, random molecular events, any one of which occurs completely unpredictably. The statistics of large numbers, however, makes the course of the overall reaction predictable and the rate laws simple despite the random nature of the underlying process. The nonequilibrium theory will employ the macroscopic rate laws in much the same fashion as they are employed in chemical kinetics. With adsorption-desorption reactions in chromatography, for instance, reference will be made to the overall concentration changes at a given point due to these reactions rather than to the individual and random molecular process of attachment and detachment.

The quantitative treatment of random processes in chromatography can be made along one of two approaches. First, the rigorous probability theory, developed initially by Giddings and Eyring,[15] gives the precise influence of simple adsorption-desorption processes on the zone profile. Second, the random walk model,[17,43] used in the following sections, yields a simplified, approximate theory of zone dispersion in chromatography. The rigorous treatment, except where applied to slow kinetic processes as in Section 2.13, has been largely replaced by the greater convenience and versatility of the generalized nonequilibrium theory. The random walk model, however, still provides the simplest and most direct calculation of the effects of dynamic molecular processes on zone structure, and gives considerable insight concerning the origin of zone spreading. The following treatment is more comprehensive than any yet published.

2.5 The Random Walk Model

The so-called random walk is the simplest random process by means of which molecular displacement can occur.† In the *one-dimensional random walk* the object under consideration is displaced either forward or backward in a step of fixed length. Whether a given step is forward or backward is determined entirely by chance. If there is an exactly equal chance ($\frac{1}{2}$) for moving in either direction, the random walk is *symmetric*.

If a large number of molecules were started at a given origin, the operation of a many-step random walk would obviously lead to their dispersion. The random back and forth motion would lead each molecule to a different final position. The random walk is clearly similar to the random molecular processes discussed in Section 2.4, but it is simpler. The erratic back and forth motion is not executed in steps of equal length. The steps are, in fact, widely variable in length. Because of this, the

† The theory of random walk processes can be found in numerous books on probability, e.g., reference 44.

previous processes should actually be termed *random flight* processes.[45] This can be approximated as a random walk, however, if the length of the fixed step is taken as the average (root mean square) length of the actual displacement.

The advantage of the random walk model lies in the simplicity of its laws. If a large number of molecules (or other objects) start together on a random walk of many steps, the standard deviation (or quarter-width†) σ of the resulting Gaussian concentration profile is simply

$$\sigma = l\sqrt{n} \tag{2.5-1}$$

where l is the fixed step length and n the number of steps taken. This fundamental law will not be proved here, but several of its aspects will be explained. Equation (2.5-1) will soon be applied to obtaining the σ value of a chromatographic zone where the random steps are those described in Section 2.4. And from the σ value of this equation and the distance L of migration, a plate height value can be calculated directly from equation (2.3-1). Hence, the most important quantitative aspects of zone spreading are tied up in the simplicity of equation (2.5-1).

The first significant result of equation (2.5-1) is that the zone quarter-width σ is directly proportional to the step length l. This can be explained as follows. The displacement of a given molecule from the origin is made because more steps happen to be taken in one direction than the other. The actual distance of displacement (still looking at a single molecule) is the net number of steps in the one direction times the step length. If the length of step is doubled but each step is taken just as before, the final displacement from the origin will then be twice as large. If each object ends up twice as far from the origin, the final concentration pulse will be twice as wide and σ twice as great.

The next conclusion to be drawn from equation (2.5-1) is that σ increases only with the square root of the number n of steps. At first it might be argued that twice as many steps mean twice as much net displacement or spreading from the origin, and therefore σ must increase in direct proportion to n. The $\sqrt{n}$ dependence results, however, from a cancellation effect. Suppose a molecule happens to end up at a distance x from the origin after n steps. After another n steps the molecule may double its distance from the origin, or, since its subsequent movement is governed entirely by chance it may move back toward the origin. It may even pass through the origin and yield a final displacement in the other direction. On the average, due to a certain amount of return toward the origin,

† The bulk (about 96%) of a zone lies within a distance of 4σ; thus σ may be thought of in a loose way as the zone "quarter-width." This terminology is consistent with the common triangle approximation in which the width at base line is 4σ.

twice as many steps cannot possibly lead to twice the zone width. As it turns out, a much slower increase in σ is found as shown by its dependence on $\sqrt{n}$ rather than n. In practice n will be proportional to time or to the distance migrated. Hence we immediately conclude that the zone width increases only with the square root of these parameters.

It is often convenient to write equation (2.5-1) in terms of the *variance*, defined as σ^2. By squaring each side of the equation, we obtain

$$\sigma^2 = l^2 n \tag{2.5-2}$$

Suppose that two random walks occur one right after the other. In the first random walk each molecule takes n_1 random steps. The second random walk, of n_2 steps, is started with the molecules already dispersed around the origin from the first walk. The two random walks in succession give exactly the same displacement as a long random walk of $n = n_1 + n_2$ steps. Thus for the two random walks

$$\sigma^2 = l^2(n_1 + n_2) = l^2 n_1 + l^2 n_2 \tag{2.5-3}$$

The term $l^2 n_1$, as seen by comparison with equation (2.5-2), is the variance σ_1^2 due to the first random walk occurring alone. The term $l^2 n_2$ would be the variance if the second random walk were to occur by itself. Thus for the two random walks in succession

$$\sigma^2 = \sigma_1^2 + \sigma_2^2 \tag{2.5-4}$$

Now suppose that the two random walks were occurring simultaneously (as chromatographic processes do). The final displacement of any given molecule is determined by the total movement in one direction minus that executed in the other direction. It is unimportant whether or not the steps occur in a different order; as long as each and every step is repeated at one time or another, each molecule will end up at the same point and the concentration profile will be the same. Thus it doesn't matter whether the two random walks occur separately or whether the steps are intermixed with one another; the resultant σ and σ^2 will be the same for either, and equation (2.5-4) will be valid. The same arguments can be used to show that this basic summation rule holds as well for any number of simultaneous random processes; that is,

$$\sigma^2 = \sigma_1^2 + \sigma_2^2 + \sigma_3^2 + \dots \tag{2.5-5}$$

or

$$\sigma = \sqrt{\sigma^2 + \sigma_2^2 + \sigma_3^2 + \dots} \tag{2.5-6}$$

The last two equations are quite general: they hold for processes (such as the random flight) which are more complicated than the simple random walk. They apply as well whether each of the component random walks occurs in steps of the same length or of different length. The equations will be valid as long as the component random walks are independent—as long as one random walk does not interfere with the steps taken in the others.

As a final observation it should be noted that the width of a concentration pulse or chromatographic zone is not the sum of the widths caused by the component spreading processes. This can be seen in equation (2.5-4); if the component quarter-widths are $\sigma_1 = 1$ and $\sigma_2 = 1$, then the resulting quarter-width is obtained as $\sigma = \sqrt{2}$ rather than $\sigma = 2$. As before, this is a result of a cancellation effect. One random walk may accentuate the effects of another by moving individual molecules further from the origin, but it also may cancel the effects of another by moving them back toward the origin. The final result of this interplay is given mathematically in the last two equations.

Although the component σ values do not add together to give the resultant σ, the previous equations show that component σ^2 values are additive (provided the random walks occur independently). Since the plate height H of a uniform column is proportional to σ^2 ($H = \sigma^2/L$), the plate height is the additive sum of the contribution of each component mechanism. Because of their additive nature, the variance σ^2 and the plate height H are more convenient to work with than the standard deviation σ.

The application of the random walk model to chromatography proceeds, then, on the basis of relatively simple equations. A combination of equation (2.5-2), which gives σ^2 for each individual random process, and equation (2.5-5), which shows how the σ^2 of component processes adds together to form the final chromatographic zone, is sufficient to formulate the quantitative theory. There is, however, one additional equation which can be used to simplify the treatment of ordinary molecular diffusion as one of the sources of zone spreading. It is, of course, possible to treat molecular diffusion as a random walk process, and its magnitude can be obtained using equation (2.5-2) in terms of the step length and number. These parameters are rarely measured, however, and if σ^2 is to be related to a meaningful parameter, it must be related instead to the diffusion coefficient D. Einstein's equation does this as follows (see Chapter 6):[46]

$$\sigma^2 = 2Dt_D \tag{2.5-7}$$

where t_D is the time period over which diffusion occurs. This equation can replace (2.5-1) for the calculation of σ^2 values arising from molecular diffusion. Equation (2.5-5), showing the additivity of component σ^2 values, is unchanged.

Equation (2.5-7) is often used to estimate the "average" time which a molecule needs to diffuse a distance d from its starting point. If a group of molecules (started at a common origin) diffuse into a Gaussian profile in time t_D, a small fraction of these will get as far as the "outer reaches" of the profile—a distance 2σ or more from the center in either direction. The average (of the root mean square type in this case) displacement will be less than this, that is, about σ. Thus d may be set equal to σ. The time required on the average to generate such a displacement is then obtained from equation (2.5-7) as

$$t_D = d^2/2D \tag{2.5-8}$$

Basis of Chromatographic Separation

The random walk model provides a simple explanation of why separation occurs in spite of the ceaseless spreading of individual zones. As two neighboring components are carried along a column, the gap between them due to their different affinity for the stationary phase increases in proportion to the distance moved. The width of each zone, 4σ, increases with the square root of the number of steps, $\sqrt{n}$, and thus with $\sqrt{L}$. This holds true for multiple random processes as well since each component σ^2 is proportional to L. (The constant of proportionality in the expression σ^2 = constant $\times$ L is, of course, the plate height H. Consequently H is a measure of the generation of variance σ^2 per unit length of column.) Thus the gap Δz between component zone centers, increasing with the first power of L, eventually outdistances the reach 4σ of the spreading influences which increase only with $\sqrt{L}$. This is shown in Figure 2.5-1. Resolution becomes satisfactory only when the Δz line crosses the 4σ line. For those

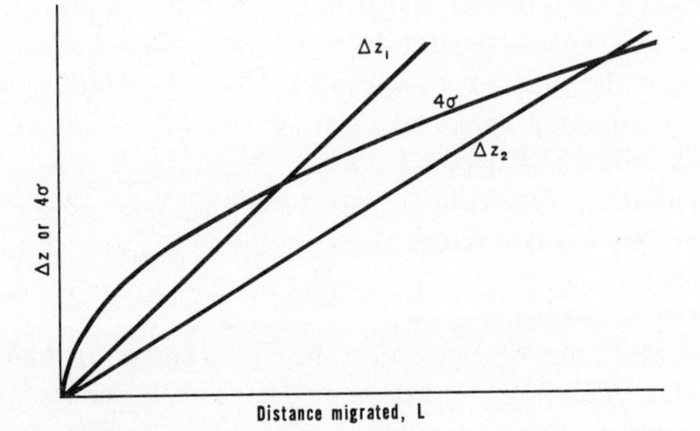

Figure 2.5-1 The increase of the separation gap Δz and the increase of zone width, 4σ as a function of the distance migrated. The Δz and 4σ lines must intersect before separation is achieved. Easily separated components, Δz_1, are separated in a shorter length than those showing a less selective retardation.

pairs of components which separate easily the Δz curve rises rapidly and intersects the 4σ curve after only a short migration. Other pairs which are more equally attracted to the stationary phase have a slow increase of Δz, and the separation (intersection) is not achieved until a considerable column length L has been used. In theory a pair of components with any degree of separation whatsoever, no matter how slowly the Δz line rises, will eventually have an intersection of Δz with 4σ. Thus increases in column length would seem to provide a solution to any separation problem. This conclusion is valid within certain practical limits. If the required column length is excessive (say, several hundred feet) it might be impossible

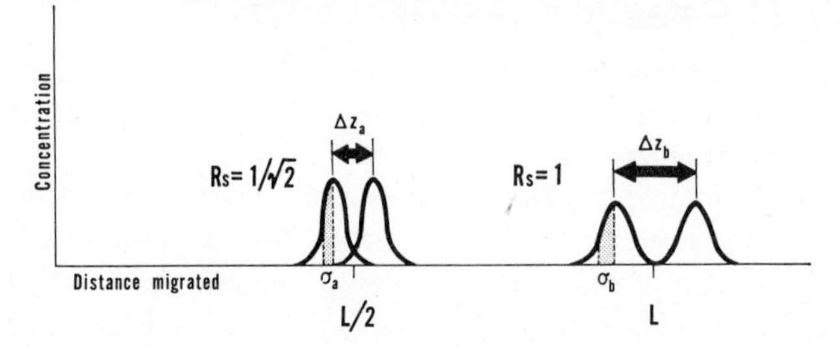

Figure 2.5-2 Concentration profiles of a given component pair at different points along the migration path. Separation is better for the most advanced pair since $\Delta z_b/\Delta z_a = 2$ while σ_b/σ_a has increased only to $\sqrt{2}$.

to force mobile fluid through such a long tube at the selected rate; it might take an unreasonably long time to complete the separation, and, when achieved, the zones might be so dilute as to be undetectable. Thus length increases are helpful as long as certain guidelines applicable to the particular system are followed. Regardless of the limitations, the foregoing discussion shows how a separation is achieved despite zone spreading, and shows that the square root dependence of zone width on n and L is essential to chromatographic methods. A full discussion of resolution concepts will follow in Chapter 7.

The quantitative discussion of separation often revolves around the ratio of Δz to 4σ, a quantity called the resolution[47,48]

$$Rs = \Delta z/4\sigma \tag{2.5-9}$$

An average value must be taken for σ since each component has a slightly different value. This ratio equals unity at the intersection point, thus indicating a satisfactory separation at $Rs = 1$. In uniform columns (to which this discussion pertains) Rs increases with the square root of length, thus always reaching unity if a sufficiently long L can be used. In Figure 2.5-2 a pair of component peaks is shown at two different locations, one twice as far along the column as the other. At the advanced location the resolution is unity. The same pair at the halfway point has a resolution of $1/\sqrt{2}$ and, while the peaks overlap considerably, the two peaks can be easily identified. Thus for some purposes (identification) resolution need not always be exactly unity; it is best not to set a rigid standard for this parameter although a value somewhere in the vicinity of unity will nearly always be required.

The ratio in equation (2.5-9) is composed of an equilibrium or thermodynamic parameter Δz (depending as it does on the difference in equilibrium distributions between mobile and stationary phases) and a dynamic zone spreading parameter 4σ. The equilibrium aspects of chromatography will be treated in detail in a companion volume by R. A. Keller; here, and particularly in the following sections, we will be concerned primarily with the factors which influence σ.

2.6 Longitudinal Molecular Diffusion

Our immediate objective is to obtain the component σ^2 values for the random molecular processes of chromatography. From this the plate height contributions can be directly obtained. We will first consider the contribution to the plate height made by molecular diffusion in the direction of flow.

The spreading of a zone due to molecular diffusion in the mobile phase can be calculated from equation (2.5-7). The diffusion time t_D is the time spent by solute molecules in the mobile phase while migrating through the distance L. While in the mobile phase the molecules are traveling with the average mobile phase velocity v, and the time thus spent in reaching L is $t_m = L/v$.† Thus the calculated σ^2 is $2D_m t_m = 2D_m L/v$, where D_m is the diffusion coefficient in the mobile fluid. In packed columns the tortuous and constricted path needed by molecules to skirt the granules reduces the actual distance diffused, and an obstructive factor γ, slightly less than unity, (about 0.6) is applicable (Section 6.4). Thus $\sigma^2 = 2\gamma D_m L/v$. The plate height contribution $H = \sigma^2/L$ is therefore

$$H = \frac{2\gamma D_m}{v} \tag{2.6-1}$$

† The time t_m, the column *dead time*, is the average time required for the passage of inert, nonsorbing molecules to position L. This is not the passage time of active molecules, since the latter spend much of their time affixed to the stationary phase, but it is that part of the passage time actually spent in motion. The total passage time to point L varies from one species to another (thus giving chromatographic separation) because of a varying time spent immobilized, but the average time spent in motion in reaching any point is the same for all species of molecules.

Under some circumstances longitudinal diffusion in the stationary phase also contributes a significant plate height. The value of σ^2 (corresponding to that above) is $2D_s t_s$ where D_s is the diffusion coefficient of solute in the stationary phase and t_s is the time spent in the stationary phase. The ratio of t_s/t_m is simply $(1 - R)/R$ (see Section 1.4), where R is the fraction of solute in the mobile phase, and thus $\sigma^2 = 2D_s t_m(1 - R)/R$. Since $t_m = L/v$ and $H = \sigma^2/L$, the plate height becomes

$$H = \frac{2\gamma_s D_s}{v} \frac{1 - R}{R} \tag{2.6-2}$$

This expression also contains an obstructive factor γ_s, which accounts for the fact that diffusion in the stationary phase cannot take place along a direct unobstructed route. The nature of γ and γ_s will be discussed in more detail in Section 6.4. In those cases where partition and adsorption both retard the solute, D_s must be considered as an average diffusion coefficient for all molecules held stationary.

The relative importance of the plate height contributions made by equations (2.6-1) and (2.6-2) depends on the particular system under consideration. For many systems, including virtually all gas chromatographic columns, the first of the two forms is the major contribution. Both equations, however, have the same inverse dependence on flow velocity, and thus group together as one term B/v when flow rate variations are being considered.

2.7 Adsorption-Desorption Kinetics

Intermittent capture and release of solute molecules by the stationary phase are controlled by two basically different mechanisms or some combination thereof. In regard to the mechanism discussed in this section, an abrupt molecular process is the critical step leading to sorption or desorption. This process is typified by molecular desorption from surfaces where molecules can detach, and then do so suddenly, if and only if they possess sufficient (activation) energy to cause the necessary rearrangement or rupture of chemical or physical bonding. Quite different in effect are the diffusion-controlled sorption-desorption kinetics (Section 2.8) where a change occurs only gradually as molecules diffuse in and out of localized regions.

Complex adsorption-desorption kinetics (involving sites with different adsorption energies, molecules which adsorb segment by segment, etc.) have been treated successfully only by using the generalized nonequilibrium theory of chromatography (Chapters 3 and 4). Simple kinetic mechanisms, which show the nature and importance of the factors concerned, can be formulated, however, as a random walk. In the simplest case solute molecules adsorb and desorb in accordance with the law of *first-order kinetics*: the rate of the process is proportional to the total number of molecules able to react. For each such process there is a *rate constant* or *transition rate*, k, which is the constant of proportionality in the rate law and denotes the fraction of the available molecules reacting in one second (Section 6.5). If the desorption rate constant k_d is, for example, 0.01, then $\frac{1}{100}$ of the solute adsorbed on the surface at time t will be desorbed within a second. Under such circumstances it would take approximately 100 seconds, which equals $1/k_d$, to run the desorption nearly to completion. The time $t_d = 1/k_d$ is the *mean desorption time* of the adsorbed molecules (this will be shown more precisely in Chapter 6). Similarly, $t_a = 1/k_a$ is the average time required for free molecules to adsorb. The following treatment is simpler if we deal only with the mean times t_a and t_d, but it should be remembered that these are related to the corresponding rate constants which, in turn, can be formulated in terms of basic properties such as temperature, surface area, bonding energy (see Chapter 6).

We will now consider a chromatographic zone which is migrating along the column. During this migration adsorption and desorption steps occur frequently. Each occurrence is a random event controlled by the erratic transfer of activating energy in and out of the molecular bonds or ties that are to be altered. A desorption process is equivalent to a random step forward because it releases the molecule to the mobile phase where it can migrate ahead freely. An adsorption process is a step backward because it leads to a period of immobility in which the molecule loses ground with respect to the center of the zone. The total number of random steps, as the zone migrates the distance L along the column, is the number of forward steps plus the number of backward steps executed during this migration. Since a desorption step must follow each adsorption step, the total number of steps is just twice the number of adsorptions occurring in the migration time.

As an average, a molecule in the mobile phase will remain in that phase a time t_a before its stay is terminated by adsorption. During this time it is moving along with the general flow at the mean flow velocity v. The distance covered before adsorption is thus vt_a. Migrating in segments of mean length vt_a, a total of L/vt_a segments are required to reach L. Since an adsorption terminates each such segment, the number of adsorptions is also L/vt_a. The total number of adsorptions and desorptions, equal to the number of random walk steps, is twice this, or

$$n = 2L/vt_a \tag{2.7-1}$$

At first it would appear that the length of step in our random walk is just the segment length discussed above. But as explained in Section 2-4, a step length must be a measure of how far a molecule moves relative to the zone center. The length of the segments discussed above is, in contrast, measured along the fixed column axis. While the desorbed molecule is moving a distance vt_a in the time t_a, the zone center is also moving. The length of step of the free molecule relative to the zone as a whole (or zone center) is thus vt_a minus the advance made by the zone in the same period. The zone itself migrates only a fraction R as fast as the free molecule; thus its advance in the time t_a is Rvt_a as compared to vt_a. Consequently, the relative displacement or step length is $vt_a - Rvt_a$, or

$$l = (1 - R)vt_a \tag{2.7-2}$$

Since from equation (2.5-2), $\sigma^2 = l^2 n$, we have

$$\sigma^2 = (1 - R)^2 v^2 t_a^2 \times 2L/vt_a = 2(1 - R)^2 vt_a L \tag{2.7-3}$$

and the plate height, equal to σ^2/L, becomes

$$H = 2(1 - R)^2 vt_a \tag{2.7-4}$$

Although this equation is complete as it stands, it is more convenient to express H in terms of the mean desorption time t_d. Now the ratio t_a/t_d is the time spent in the mobile phase divided by the time spent attached to the surface. Since the average molecule spends a fraction R of its time in free migration and a fraction $1 - R$ attached, this ratio is simply

$$\frac{t_a}{t_d} = \frac{R}{1 - R} \tag{2.7-5}$$

as shown in equation (1.4-10). Substituting t_a from this expression into equation (2.7-4), it is found that H is

$$H = 2R(1 - R)vt_d \tag{2.7-6}$$

This equation is an extremely important one. Although derived from the approximate random walk theory, it turns out to be rigorously correct. It will be shown in Chapter 3 that the equation applies to heterogeneous surfaces containing wide variations in properties from site to site as well as to perfectly uniform surfaces. It can also be used as a starting point for the random walk treatment of diffusion-controlled kinetics (Section 2.8).

Equation (2.7-6) shows several interesting characteristics whose essential features are repeated over and over again in the study of kinetic (or nonequilibrium) effects in chromatography. First, the plate height contribution due to such effects is generally proportional to the flow velocity

v. Second, H is also proportional to the time, t_d or t_a, necessary to make a transition between the mobile and stationary phases (we will thus find that all possible effort should be made in practice to increase the rate of transition). Third, H is proportional to $R(1 - R)$, a quantity which acquires a maximum at $R = \frac{1}{2}$, and is relatively small near the extremes $R = 0$ and $R = 1$.

2.8 Diffusion-Controlled Kinetics

Generally, diffusion-controlled sorption and desorption rates may originate either in the stationary phase or in the mobile phase. The comparative importance of the two types depends on relative diffusivities, the R value, and other column characteristics. This will become more evident later.

Diffusion in the Stationary Phase

We will first discuss those cases in which stationary phase diffusion is important. This includes, of course, only the partition systems (gas-liquid or liquid-liquid) since only in these cases does a bulk stationary phase (as compared to a surface) exist. (Two-dimensional diffusion may occur on a surface, but the process does not usually control adsorption and desorption rates.) The process of capture and release by the stationary liquid in partitioning systems qualitatively resembles the capture-release process on adsorptive surfaces. In each case a certain average time is required to sorb the molecule and then to release (desorb) it again. Thus the simple random walk picture is like that just presented for adsorption-desorption kinetics. We are led to the same final expression, equation (2.7-6), which gives the plate height H in terms of R, v, and the mean desorption time t_d. The major difference between the release by diffusion and the desorption from surfaces is that the rate of the former is determined by the magnitude of the diffusion coefficient and the rate of the latter is determined by the rate constant k_d. If diffusional desorption is brought about by molecules diffusing from the interior of a pool or droplet of stationary liquid of depth d, then the mean desorption time t_d is approximately the average diffusion time t_D, obtained from equation (2.5-8), required to diffuse the distance d. Thus $t_d = d^2/2D_s$, where D_s is the diffusion coefficient for solute in the stationary phase. When this is substituted into equation (2.7-6), the plate height contribution is found to be

$$H = 2R(1 - R)d^2v/D_s \tag{2.8-1}$$

In Section 2.7 we saw that H is reduced by increasing the sorption-desorption rate (decreasing t_d). In this case an increased rate is most easily effected by reducing d to its absolute minimum. This means that the stationary liquid should be somehow dispersed in very small units. It is also desirable to choose liquids within which the solute diffusion coefficient D_s is large.

Equation (2.8-1) is of the correct general form to describe diffusion-controlled kinetics in the stationary liquid. The more exact nonequilibrium treatment, developed in Chapters 3 and 4, shows that this equation should be written as

$$H = qR(1 - R)d^2v/D_s \tag{2.8-2}$$

where the *configuration factor* q replaces the integer 2 of equation (2.8-1). The value of q depends on the precise shape of the pool of partitioning liquid.[49] A uniform liquid film is characterized by $q = \frac{2}{3}$. Diffusion in rod-shaped (paper chromatography) and sphere-shaped (ion exchange) bodies is governed by $q = \frac{1}{2}$ and $q = \frac{2}{15}$, respectively. This matter will be developed more fully in Chapter 4.

Diffusion in the Mobile Phase

We have so far assumed that zone spreading is due to three independent sources: longitudinal diffusion, sorption-desorption kinetics, and flow pattern effects. The actual situation is somewhat more complex than this, as the next few sections will show. The process which we are about to discuss is, in a sense, a hybrid of sorption-desorption kinetics and flow pattern effects. The zone spreading itself originates in the velocity inequalities of the flow pattern, but the extent of spreading is governed largely by diffusion between fast and slow streampaths, a process analogous to sorption-desorption kinetics. We will make no further attempt to resolve the question of nomenclature here, but will proceed to a discussion of the nature of the phenomenon.

Diffusion in the mobile phase is more complicated than in the stationary phase. Only recently has it been possible to account quantitatively for some of the major components of zone spreading caused by these diffusion processes.[50,51] The quantitative theory is still not in satisfactory shape although approximate results can be obtained.[50] Part of the difficulty can be traced to the fact that the diffusion is occurring in a very complex network of interconnected channels and void spaces. In addition, the flow velocity varies over wide limits moving across a channel or from one channel to another. The fact that flow velocity varies in the direction of flow (as well as perpendicular to it) leads to the rather complex coupling phenomenon which relates mobile phase diffusion to eddy diffusion.[52] Accordingly, the discussions of eddy diffusion and coupling in Sections 2.9 and 2.10 are very closely related to the material presented here.

In spite of the serious complications mentioned above, a few clear facts emerge which allow us to group the various diffusion processes into distinct categories and to discuss the approximate theory relative thereto.

The random walk model tells us that zone spreading occurs whenever high and low velocities exist side by side with molecules exchanging between. (It is through such velocity inequalities that random steps occur back and forth.) As we will see, the complex network of channels in column packing materials leads to five different types of velocity inequalities which satisfy the above criterion. In the preceding discussion of rate-limiting diffusion processes in the stationary phase we treated only one broad category of velocity inequality—that existing between the stationary phase and that wide spectrum of velocities present in the mobile fluid. The five types of velocity inequalities considered here, however, exist between points in the mobile phase. In reality the total velocity spectrum, including the immobile contribution of the stationary phase, affects the diffusion processes of the mobile phase. In the following treatment the contribution of the stationary phase will not be considered. By excluding this, we are committing ourselves to inert solute components, that is, those washed directly through the column with no sorption. Chapter 3 will show that many of the plate height terms depend on the degree of sorption, or R value, but our plate height terms will show no such dependence. Our justifications for this omission are (1) simplicity, (2) the R dependence of the most significant mobile phase terms is small or altogether negligible, and (3) a shift in degree but not in kind is caused by sorption processes. It is not the role of an approximate theory of the random walk type to examine subtle variations. For the purposes of this theory sorption may be considered as a process that temporarily interrupts the random walk being executed in the mobile phase. With desorption the random walk proceeds again in the normal way. The plate height contribution of the mobile phase is thereby little affected.† The interruption, as well as its duration, does have a direct bearing on that random walk which is controlled by stationary phase diffusion or sorption-desorption. This part of the effect was discussed earlier.

Velocity inequalities which lead to zone spreading originate within the mobile phase in the following ways:[50]

† Different degrees of sorptive retention will remove the respective solute components from the random walk for varying amounts of time. This will change the number of steps n that can be taken in a fixed period. This by itself does not affect the plate height for the following reason. The number of random steps n is reduced by adsorption to exactly the same extent that the velocity of the zone and thus the distance L it travels is reduced (quantitatively, both are proportional to R). Since the variance σ^2 is proportional to n, the ratio σ^2/L is unaffected by changes in R. This ratio is simply the plate height H.

1. In each interstitial flow channel a high velocity exists in the center and a low velocity near the walls. As in all the cases discussed below, diffusion through the mobile fluid leads to the exchange of molecules between these regions. The zone spreading due to this process will be termed the *transchannel* contribution, a term derived from the fact that the interchange occurs across single channels only.

2. In some forms of partition chromatography (notably gas-liquid chromatography) the mobile phase occupies part of the space within the porous solid support particles. This part of the mobile phase is essentially stagnant, but is surrounded by mobile fluid in a state of motion. The interchange occurring through the stagnant mobile phase gives rise to a *transparticle* effect.

3. The direct observation of packed granular materials (Chapter 5) shows that small, tightly packed regions (several particle diameters across) are joined together by a rather loosely filled space composed of large "channels" in which unusually high flow velocities may occur. A significant velocity differential occurs between the large, open channels and the smaller surrounding ones, and leads to a *short-range interchannel* effect.

4. The foregoing pattern of large channels alternating with small ones is erratic in detail, no single undulation of the pattern being exactly repeated throughout a given cross section of the column. Variations exist in the average velocity within each undulation or *repeating flow unit*, and the diffusion back and forth among these units leads to a *long-range interchannel* effect.

5. Velocity differences are often found to exist between the outer regions and the center of the column, or between one outer region and another (the former may result from the influence of the wall on packing density, while the latter can originate in the bending of a column). The transfer of solute molecules between velocity extremes must occur through diffusion on a column-wide scale. The term *transcolumn* is used to describe this effect.

Figure 2.8-1 shows schematically the distance which molecules must diffuse to exchange between velocity extremes. The actual exchange process is occurring in a three-dimensional system (even in paper chromatography the paper thickness, equal to 100 or so fiber diameters, is sufficient to be regarded as a third dimension). While exchange may occur by penetration through the particles if they have an open porous structure, it will ordinarily occur by skirting in front of, behind, above, and below the particles which are obstacles in the direct path of exchange.

So as to gain an order-of-magnitude appreciation of these effects, we employ the random walk treatment of Section 2.5. We will assume that a certain "escape" or "exchange" time t_e is needed to transfer a molecule from one velocity extreme to another. Since the distance between velocity extremes (Figure 2.8-1) depends on the particular category among the five effects listed above, we will assume for the general case that a molecule must diffuse a distance $\omega_\alpha d_p$ to reach one extreme from another. The multiplying term ω_α, as indicated by the figure, acquires values from well

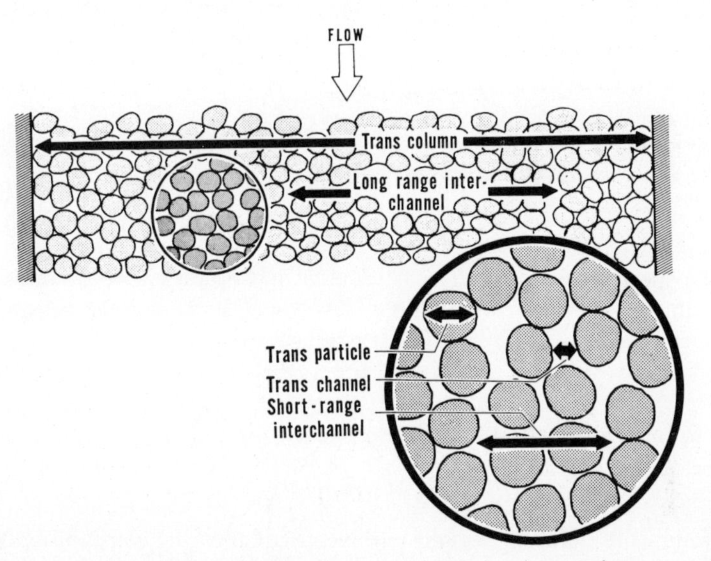

Figure 2.8-1 The location and distance covered by the various exchange processes between velocity extremes in the mobile phase.

below to well above unity. The average time t_e needed to diffuse the distance $\omega_\alpha d_p$ is obtained from equation (2.5-8) as

$$t_e = \omega_\alpha^2 d_p^2 / 2D_m \tag{2.8-3}$$

where D_m is the diffusion coefficient of solute in the mobile phase.

The length of step l in this general case is the distance gained or lost with respect to the mean because of the temporary residence in one of the velocity extremes. This distance is $\Delta v t_e$, where Δv is the difference between the extreme and the mean velocity. In general Δv will be some fraction of the mean velocity v, that is, $\Delta v = \omega_\beta v$, where ω_β, like ω_α, depends on the particular effect categorized above. Thus the step length is

$$l = \omega_\beta v t_e \tag{2.8-4}$$

This quantity may also be thought of as the simple product of the distance $S = vt_e$ which a molecule is carried without severe velocity changes, and the velocity ratio $\omega_\beta = \Delta v/v$ which gives the fraction of that distance actually gained with respect to the average. Thus

$$l = \omega_\beta S \tag{2.8-5}$$

an equation which is in the correct form to apply directly to eddy diffusion (Section 2.9).

The number of random steps is simply the number of paths of length $S = vt_e$ comprising the total migration distance L; that is,

$$n = L/S = L/vt_e \tag{2.8-6}$$

On combining these expressions with $\sigma^2 = l^2 n$ and $H = \sigma^2/L$, we obtain in the general case (using the S symbol)

$$H = \omega_\beta^2 S \tag{2.8-7}$$

In the specific case of interest here, after substituting $S = vt_e$ and using equation (2.8-3) for t_e, we obtain

$$H = \frac{\omega_\alpha^2 \omega_\beta^2}{2} \frac{d_p^2 v}{D_m} \tag{2.8-8}$$

which, for a given effect i, can be written

$$H = \omega_i d_p^2 v / D_m \tag{2.8-9}$$

where ω_i is a highly important parameter equal to the constant $\omega_\alpha^2 \omega_\beta^2/2$ as evaluated for category i. The sum of such terms $\Sigma\omega_i$ should yield the experimentally observed value ω. We will now proceed to estimate the magnitude of ω_i for the five categories listed above. The precise value of each ω_i may have an R dependence, but this will not be of concern here.

1. Each interstitial flow channel has a diameter roughly equal to $d_p/3$. The velocity extremes, however, occur at the center and the side, thus requiring a diffusion distance of $d_p/6$ for molecular exchange. Thus $\omega_\alpha = \frac{1}{6}$. The ω_β term is approximately unity since the extreme velocities are approximately zero (at the channel wall) and $2v$ (in the center). Thus ω_i for the transchannel effect, or ω_1 as we shall term it, is of approximate magnitude

$$\omega_1 \sim 0.01 \tag{2.8-10}$$

2. Diffusion within a support particle must occur between the particle center and the particle boundary in order to effect an exchange between stagnant and moving fluid. This distance is $d_p/2$ and thus $\omega_\alpha = \frac{1}{2}$. While in the stagnant interior, a solute molecule is falling behind the average gas displacement by a velocity $\Delta v = v$. Thus $\omega_\beta = 1$. These values for

ω_α and ω_β show that transparticle diffusion leads to a term of the following order of magnitude

$$\omega_2 \sim 0.1 \tag{2.8-11}$$

3. The short-range interchannel effect requires diffusion through a distance of roughly $1.25d_p$ to exchange molecules between the tightly packed regions and the large channels.[50] Thus $\omega_\alpha = 1.25$. Present evidence indicates that the velocity difference between these extremes is rather high, being greater than the mean velocity itself. Then ω_β is slightly less than unity, perhaps 0.8. The ω_i of equation (2.8-9) is therefore

$$\omega_3 \sim 0.5 \tag{2.8-12}$$

4. The long-range interchannel effect is very difficult to estimate, and the values given here must be treated with all possible caution. Assuming that an exchange occurs between regions that are three or four flow units apart with each flow unit being $2.5d_p$ in diameter, the magnitude of ω_α is about 10. The velocity differences between mean and extreme will not be as great as before, perhaps one-fifth of the mean velocity. Thus $\omega_\beta \sim 0.2$. The combined effect of these leads to the order of magnitude

$$\omega_4 \sim 2 \tag{2.8-13}$$

5. It is difficult to generalize concerning the transcolumn effect because so many variables are changed from case to case. For straight columns there is some indication that the major transcolumn inhomogeneity is due to the larger support particles settling near the outside and the smaller ones near the center.[53,54] The greater permeability of the large particles induces an increased flow rate in the outer regions. We can term this a *size-separation* effect. (In addition to this there is a *wall effect*,[55] probably of lesser magnitude,[53] caused by the narrow gap between the wall and adjacent particles.) In a typical case we might expect the velocity near the outside of the column to be 10% greater than average and that near the center to be 10% or so less (see Chapter 5). Thus $\omega_\beta \sim 0.1$. The diffusion distance between velocity extremes, in this case the wall and center of the tube, is the column radius r_c. This distance, however, is equal to $md_p/2$, where m is the number of particle diameters in one column diameter. Thus $\omega_\alpha = m/2$. The overall contribution of this effect is given, then, by

$$\omega_5 \cong \frac{1}{2}\left(\frac{\Delta v}{v}\right)^2\left(\frac{m}{2}\right)^2 \sim 0.001m^2 \tag{2.8-14}$$

When m exceeds 25 or so, the value of ω_5 becomes comparable to ω_3 and ω_4. Further increases are undesirable, a fact which will ordinarily make a narrow column preferable to a wide one. The importance of this topic to large preparative-scale columns is obvious.

The bending or coiling of a chromatographic column (common in gas chromatography) also leads to a transcolumn effect.[56] A molecule near the inside bend is progressing more rapidly than its outside counterpart because (1) its path is shorter, and (2) its velocity is greater (this results from a larger pressure gradient which is a consequence of the total column pressure drop being distributed over a shorter path). The ω_β factor is obtained by considering that a molecule's progress on an inside path is the ratio of path lengths greater than that along the center path (where the mean velocity prevails); this ratio applies once because of (1) and once because of (2). Thus the relative progress along inside and center paths is

$$(R_0 + r_c)^2/R_0^2 \tag{2.8-15}$$

where r_c is the tube radius and R_0 is the coil radius or radius of curvature of the bend. This ratio is approximately $1 + 2r_c/R_0$ when $R_0 \gg r_c$. Thus $2r_c/R_0$ is the fractional increase in velocity between mean and extreme, a term equal to ω_β.

The distance between velocity extremes is, in this case, the full distance across the tube—$2r_c$ rather than r_c as before. This distance is md_p, thus giving $\omega_\alpha = m$. With ω_α and ω_β so determined, we have

$$\omega_5 = 2m^2r_c^2/R_0^2 \tag{2.8-16}$$

The two preceding equations for ω_5 are somewhat misleading because, written as they are in terms of ω, equation (2.8-9), they express the plate height in terms of the particle diameter. Transcolumn effects, however, should be written in terms of the column diameter (or radius). This can be corrected by substituting these ω_5 values back into equation (2.8-9) and using $m = 2r_c/d_p$ to convert from column to particle diameter. We obtain

$$H = 0.004r_c^2v/D_m \tag{2.8-17}$$

for the transcolumn variation in permeability, replacing equation (2.8-14), and

$$H = 8r_c^4v/R_0^2D_m \tag{2.8-18}$$

for the bending or coiling of columns, replacing equation (2.8-16). More exact forms of these equations will be derived in Chapter 3; although the first equation is a good approximation, the factor 8 in the second should be replaced by $\frac{7}{12}\gamma$.

On the whole the foregoing ω_i values agree well with information obtained from the more exact nonequilibrium treatment and from experimental data. The dependence of the ω_i terms on R, ignored in this order of magnitude theory, is not sufficient to alter our general conclusions at either high or low R values. The largest R dependence is shown by ω_1, a term which by analogy to a capillary column may increase by about 10 from high to low R values. Since the value given for ω_1, 10^{-2}, is nearer the maximum than the minimum value, this term is always negligible anyway. As we go to the more significant ω_i parameters, the R dependence decreases. It is probably very slight for ω_4 and may be considered zero for the two calculations of ω_5. The experimental conclusion[50,51,57] seems to be that $\omega = \Sigma\omega_i$ increases by a factor of only about 2 as we go from complete inertness, $R = 1$, to complete retention, $R = 0$.

2.9 Flow Pattern and Eddy Diffusion

The nature of flow through a chromatographic medium is of fundamental importance in determining the structure and width of component zones. Flow by itself will always lead to zone spreading, as illustrated in Section 2.8, because solute molecules caught up in the fast streampaths will be displaced further downstream than those caught in the slow streampaths. That a point-to-point difference exists in flow velocity is a fundamental property of fluid flow under the most general conditions. We will reserve for Chapter 5 a more detailed discussion of the flow characteristics of chromatography, but will briefly outline some of these characteristics here. Then we will discuss the eddy diffusion phenomenon and show its close relationship to the flow and diffusion-controlled processes.

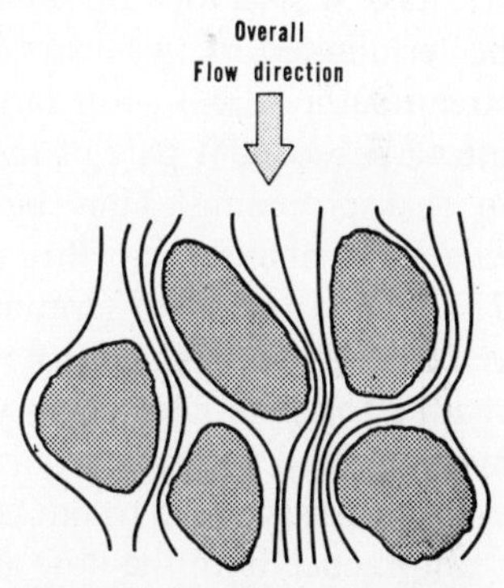

Figure 2.9-1 Schematic diagram showing the flow pattern (array of streampaths) in a small region of column.

Figure 2.9-1 shows schematically how the mobile phase flows around the granules or fibers in a very small region of the column. The solid lines indicate the flow or streampaths—those paths followed by small elements of fluid or by any object set adrift in the fluid (providing the object is not carried away from the streampath by diffusion or buoyancy). The pattern of flow resembles that in a river, the bed of which is beset with fixed obstacles. Since the flow avoids these obstacles along paths of least resistance, the velocity fluctuates over wide limits. The tight narrow channels have a lesser average velocity than the open unrestricted ones. The fluid velocity approaches zero at the surface of each particle because of viscous drag. Thus the velocity varies from channel to channel because of varying flow resistance, and from the outside to the center of each channel because of the viscous restraint of the containing walls.

The flow in a chromatographic column is generally laminar in nature, that is, without the turbulence characteristic of very high flow rates (Section 5.4). Hence the flow pattern (Figure 2.9-1) is independent of the flow rate (the velocity but not the direction of flow at a given point, nor the relative velocity at any two points, changes with the flow rate). The flow pattern is also independent of the nature of the mobile fluid, temperature, viscosity, etc. In short, the flow pattern of slow, laminar flow is essentially independent of mobile phase properties, and depends only on the structure of the porous support material and the incorporated flow space. Thus we speak of the flow pattern, and anything else entirely dependent on it, as a *structural property*.

The phenomenon of eddy diffusion is caused by the point-to-point variations in flow velocity. A molecule in a fast streampath, for instance, will take a step forward with respect to the zone. This high velocity can be terminated in two ways.[58] First, if the molecule follows its original streampath, it will soon (after one or so particle diameters) be carried into a subsequent part of the channel network with a velocity unrelated to that preceding. Thus by virtue of the flow alone, a molecule will erratically change velocities after each particle diameter or thereabouts. The second means of terminating abnormally high (or low) downstream velocities is through diffusion of the molecule laterally into a new flow channel or to a new velocity extreme. By virtue of the lateral diffusion mechanism, a molecule will randomly change velocities in just such a time as is necessary for it to diffuse to a new velocity regime.

In practice both the flow and lateral diffusion mechanisms are working simultaneously to exchange molecules between flow paths of unequal velocities. Under certain circumstances either of the mechanisms may dominate the other.

Most discussions of eddy diffusion have revolved around the flow mechanism of exchange only.[59,17] The theory describing this mechanism by itself will be called the "classical theory of eddy diffusion."[60] Since the flow pattern (and nothing else) is involved in this mechanism, the eddy diffusion term obtained in this theory will be a structural property or *structural parameter*, as defined above. The random walk approach to the classical theory of eddy diffusion will be given below, and will be followed by the "coupling theory of eddy diffusion" which allows for both flow and diffusive exchange.

Classical Theory of Eddy Diffusion

As emphasized previously, eddy diffusion is a consequence of the same velocity inequalities which give rise to the plate height term, equation (2.8-7), for diffusion-controlled kinetics or mass transfer. The sole difference resides in the mechanism of terminating a given velocity bias. The general equations developed in Section 2.8 are applicable, provided that the persistence-of-velocity span S is properly evaluated. (The reader may wish to refer to the discussion leading to equations (2.8-5) and (2.8-7).) In addition, a separate eddy diffusion term may be identified with each of the five categories of velocity inequalities—transchannel, transparticle, short-range interchannel, long-range interchannel, and transcolumn (Section 2.8). For each of these the appropriate S value leads to the plate height contribution given by $H = \omega_\beta{}^2 S$, equation (2.8-7).

The evaluation of the S quantities is complicated by the complex structure of column packing material. It should be possible to obtain approximate values. But the attempt to do this must be regarded as furthering our understanding of the relative importance of the different terms, and not as endeavoring to affix precise numerical values thereto. The science of flow in porous materials is not yet far enough advanced to accomplish the latter objective.

The first step in evaluating S is to write this quantity as

$$S = \omega_\lambda d_p \tag{2.9-1}$$

where ω_λ is a structural parameter, near unity in most cases. The argument for this proportionality is simply that a given change in particle size which leaves the relative "structure" untouched will change the length of each channel, or the distance between given flow constrictions, by the same ratio. It can be shown quite rigorously that an increase in one (particle diameter) is matched by the increase in another (channel length) under these ideal circumstances. In practice the relative structural factors will change some as d_p varies, but this change is probably small. Thus ω_λ may be regarded as nearly constant for each category in the velocity inequality classification.

If equation (2.9-1) is substituted into the general plate height expression, equation (2.8-7), the resulting plate height becomes

$$H = \omega_\beta{}^2 \omega_\lambda d_p \tag{2.9-2}$$

For a given velocity inequality i this may be written as

$$H = 2\lambda_i d_p \tag{2.9-3}$$

where

$$\lambda_i = \omega_\beta{}^2 \omega_\lambda / 2 \tag{2.9-4}$$

and ω_β as well as ω_λ are evaluated with respect to the particular category i. Recall from Section 2.8 that ω_β is defined as the ratio $\Delta v/v$, the fractional departure of a velocity extreme from its mean value. With this information we can attempt the approximate evaluation of λ_i terms.

1. It was estimated in Section 2.8 that $\omega_\beta \sim 1$ for the transchannel effect. The distance S is probably about d_p, thereby giving $\omega_\lambda \sim 1$. (The velocity inequality between the center and wall regions of a channel may be envisioned as lasting along the length of a single particle after which the flow will split up into several subsequent channels.) These values lead to a λ value, from equation (2.9-4), of

$$\lambda_1 \sim 0.5 \tag{2.9-5}$$

which, of course, gives a plate height contribution of $H \sim d_p$.

2. The calculation of λ for the transparticle effect requires a very careful consideration of the flow and velocity exchange phenomenon. A molecule entrained in the mobile fluid which occupies the pore system of a support particle is virtually motionless (remember that classical eddy diffusion is based on the molecular displacement resulting from fluid flow, and does not allow for the superimposed motion caused by diffusion). This is a consequence of the fact that the mean flow velocity through a network of interconnected pores increases with the square of the average pore diameter. Thus the small pores within a support particle (assuming that they are interconnected), with a diameter sometimes as small as 100 times less than that of the channels between support particles, will restrain fluid motion to a velocity v' of some 10^4 times less than average v. Hence a molecule entrained in this nearly quiescent fluid will require a long period of time t_e to escape from the particle and thus effect an exchange of velocities. This time will be approximately $t_e = d_p/v'$—the time needed to carry a molecule the distance d_p to the downstream particle boundary at a mean velocity v'. Once the molecule is spilled out into the interparticle flow space, it will be carried downstream for quite a distance until it is again in intraparticle flow after an approximate time t_e. (The time spent in interparticle space will roughly equal the time t_e spent in intraparticle space, provided that there is about the same amount of mobile fluid in each type of space. This assumption is particularly valid in gas chromatographic systems using diatomaceous earth support particles.)

The persistence-of-velocity span S is not a fixed quantity in this case as can be approximated for most other cases. It is approximately d_p for molecules entrained in intraparticle fluid. It is a considerably larger distance (about 10^4 particle diameters) for molecules in interparticle space because the occupancy time t_e, during which flow displacement is occurring, is quite large. An average of the two distances will be assumed here for S. This average is simply the mean displacement occurring in the exchange time t_e at a mean velocity v (this velocity is the average of both interparticle and intraparticle fluid velocities) or $S = vt_e$. Since $t_e = d_p/v'$, $S = (v/v')d_p$. As a consequence of equation (2.9-1), $\omega_\lambda = v/v'$, a quantity of order 10^4. Since ω_β was shown in Section 2.8 to be near unity, the λ value, from equation (2.9-4), may be as large as

$$\lambda_2 \sim 10^4 \tag{2.9-6}$$

This means that the plate height term for this phenomenon may be 10^4 or so particle diameters. This large value is completely unreasonable if it is assumed that it contributes directly to the observed plate height. This anomaly will be explained shortly by means of the coupling theory of eddy diffusion.

3. The short-range interchannel effect is associated with $\omega_\beta \sim 0.8$. The persistence of path parameter ω_λ is probably slightly greater than unity, since unrestricted and restricted flow channels seem to persist for something more than a single particle diameter. Thus $\omega_\lambda \sim 1.5$. The λ value consequently becomes

$$\lambda_3 \sim 0.5 \tag{2.9-7}$$

4. The long-range interchannel effect associates itself with $\omega_\beta \sim 0.2$. Since this case involves velocity fluctuations over regions of 5–10 particle diameters, velocities are likely to persist for at least 5 particle diameters, $\omega_\lambda = 5$. Thus

$$\lambda_4 \sim 0.1 \tag{2.9-8}$$

leading to a plate height of one-fifth d_p.

5. The λ value for transcolumn effects must be estimated by more roundabout methods than before because of the unusual way in which exchange occurs solely by a flow mechanism. Interchange between high and low velocity extremes must in this case involve a transfer of molecules laterally over a large distance. Such lateral transfer is not usually required of the flow process, since in the normal case all velocity fluctuations can be reached one below another. Some investigators would probably question the existence of such lateral transport by flow alone, but there clearly exists a mechanism for this transfer. An argument against a flow mechanism for lateral transport might be based on the fact that any penetration of mobile fluid sideways will require a counterpenetration by an equal amount of fluid. This results because the column walls do not allow a net lateral motion. It is thus necessary to find a driving force for lateral penetration. Such a driving force will arise as a consequence of normal flow through a random packing. Some groups of particles will be found with a screw-like channel configuration. Since only a half turn is needed for full exchange, and lesser turns will accumulate in effect, this mechanism of interpenetration does not require an elaborate structure which would be ruled out on the basis of chance.

It is possible to give this mechanism a quantitative foundation. For order-of-magnitude purposes an exchange across one particle diameter may occur while flowing past every 20 or so particles.† An exchange across m' particle diameters would then occur only after a downstream flow displacement of length

$$S' = 20m'^2d_p \tag{2.9-9}$$

Taking equation (2.9-9) for the downstream displacement needed for velocity exchange, the transcolumn effects can be estimated. The center-outside inequality, for instance, requires interchange across $m/2$ particle diameters (m is the number of particle diameters in a column diameter); thus $m/2$ can replace m'. The velocity-persistence length S can then replace S'. Thus $S = 5m^2d_p$, an equation which leads to $\omega_\lambda = 5m^2$. A value of 0.1 was previously established for ω_β, so that $\lambda = \omega_\beta{}^2\omega_\lambda/2$ becomes

$$\lambda_5 \sim 0.02m^2 \tag{2.9-10}$$

This contribution is significant for columns with more than 10 particles in a tube diameter. This includes most of the chromatographic columns used in practice.

The tube-bending effect can be calculated on similar grounds. Here $m = m'$, S is $20m^2d_p$, and ω_λ is $20m^2$. Since $\omega_\beta = 2r_c/R_0$, the λ value becomes

$$\lambda_5 = 40r_c{}^2m^2/R_0{}^2 \tag{2.9-11}$$

a term once again of significant magnitude ($\lambda > 1$) in most coiled columns with 10 or more particle diameters per tube diameter.

2.10 The Coupling Theory of Eddy Diffusion

The classical theory of eddy diffusion was based on the assumption that solute molecules are locked into fixed streampaths, and are thus forced to take displacement steps of approximate length d_p along with the surrounding fluid. But in fact solute molecules are not locked in with the surrounding fluid, and they often diffuse quite rapidly to regions of the mobile phase where there is an entirely different downstream velocity. The step length of the random walk, roughly proportional to the distance

† (This is an agreement with measurements made on lateral diffusivity in rapidly flowing systems; see reference 61.

S traveled with a fixed or nearly fixed velocity, may be drastically reduced as the molecule cuts short its residence within a given flow path. This process is illustrated in Figures 2.10-1 and 2.10-2. In Figure 2.10-1 the normal streampaths (solid lines) are continually interrupted by lateral diffusion (dashed lines) which carries the molecule into new flow channels. As a consequence the molecule may travel only a short distance within each channel (and so have a short step length), and its path may be highly

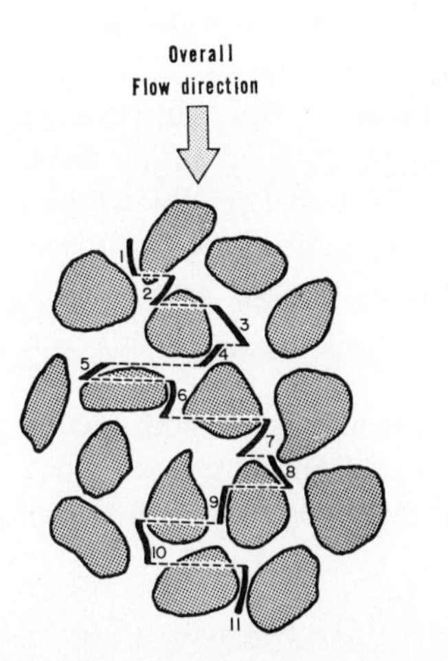

Figure 2.10-1 Schematic illustration of typical molecular pathway through column packing. The streampaths, interrupted by lateral diffusion, consist of 11 steps.

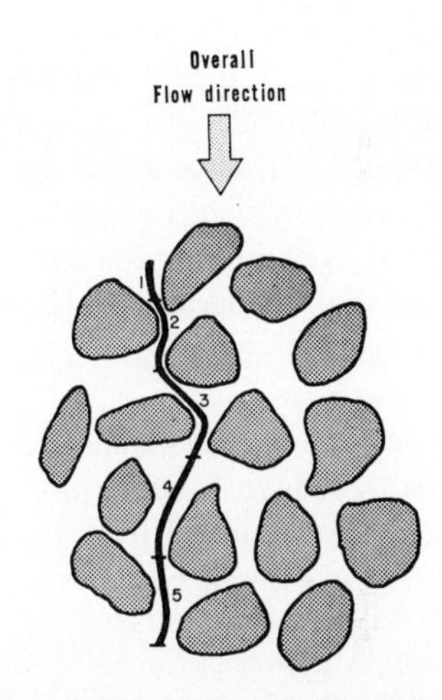

Figure 2.10-2 Schematic illustration of typical molecular pathway through column packing without diffusion. Note that fewer, but longer steps are required.

segmented (thus containing many random walk steps). In Figure 2.10-2, when diffusion is absent, fewer steps—only 5—are required, but their length is proportionately greater. Since the plate height increases with the square of the step length and only with the first power of step numbers, the plate height is least for the highly segmented path.

While diffusion may cut short the steps of a flow-dominated random walk, the opposite effect also exists in which normal flow variations will prematurely terminate the steps of a diffusion controlled random walk. The interaction between flow and diffusive exchange mechanisms serves in each case to reduce the velocity persistance span S and thus the step length which is proportional to S according to equation (2.8-5). The plate height, also proportional to S according to equation (2.8-7), is thereby reduced (fortunately) below either of the limits predicted by the individual theories.[60] The assumption once accepted that H was the sum of the two limiting expressions, $H = H_f$ (flow mechanism) + H_D (diffusive mechanism), is untenable in view of the underlying mechanism.

Quantitative Formulation

We shall now look at the quantitative aspects of the combined flow-diffusive exchange. To simplify things, let us assume at first that only one type of velocity inequality exists (e.g., the short-range interchannel velocity bias). The downstream distance S which a molecule is carried at one of the velocity extremes without interruption was shown to be the critical parameter for this spreading process. This quantity may be written as L/n, since n such spans make up the total migration distance L. The number n of exchanges between velocity extremes, equal to the number of steps, is the sum of the number of exchanges due to the flow mechanism n_f and the number due to the diffusive mechanism n_D. Thus

$$S = L/(n_f + n_D) \tag{2.10-1}$$

This equation shows quantitatively how the activity of both mechanisms tends to reduce S below the value resulting from either one alone. The plate height resulting from the combined mechanisms, shown by equation (2.8-7) to be $\omega_\beta^2 S$, becomes

$$H = \omega_\beta^2 L/(n_f + n_D) \tag{2.10-2}$$

a value also less than either limit because n_f and n_D are added in the denominator.

A more useful plate height expression may be obtained by seeking out explicit expressions for n_f and n_D. The number of steps n_D was shown by equation (2.8-6) to equal L/S_D, or L/vt_e, where S_D here means the persistence length pertaining only to the diffusive mechanism vt_e. With t_e as given by equation (2.8-3) we have

$$n_D = 2LD_m/v\omega_\alpha^2 d_p^2 \tag{2.10-3}$$

Similarly, n_f is given by L/S_f. The S (actually S_f) value from equation (2.9-1) gives

$$n_f = L/\omega_\gamma d_p \tag{2.10-4}$$

When these quantities are substituted into equation (2.10-2), and the latter rearranged, we have

$$H = \frac{1}{1/2\lambda_i d_p + D_m/\omega_i v d_p^2} \tag{2.10-5}$$

where, as in the previous sections, ω_i (for velocity inequality i) has been substituted for $\omega_\alpha^2\omega_\beta^2/2$ and λ_i for $\omega_\lambda\omega_\beta^2/2$. Considering the fact that diffusive exchange leads to the plate height expression (Section 2.8)

$$H_D = \omega_i d_p^2 v / D_m \tag{2.10-6}$$

and flow exchange leads to (Section 2.9)

$$H_f = 2\lambda_i d_p \tag{2.10-7}$$

the foregoing H equation is equivalent to

$$H = \frac{1}{1/H_f + 1/H_D} \tag{2.10-8}$$

An equation of this type applies to all categories of velocity inequality, thus yielding a summed expression,[62] ΣH, for the final plate height term:

$$H = \sum_i \frac{1}{1/2\lambda_i d_p + D_m/\omega_i v d_p^2} \tag{2.10-9}$$

The foregoing equations are at wide variance with the classical form $H = H_f + H_D$, and have properties which contrast significantly with those of the latter. The value of H is less than either H_f or H_D, and as shown by Figure 2.10-3 approaches the smaller of the two terms from below as the difference between them increases (the physical basis of this is that the fastest mechanism of exchange, which leads to the smallest plate height, is the controlling factor). Thus at high velocities H approaches the constant value $H_f = 2\lambda d_p$. At low velocities H approaches $H_D = \omega d_p^2 v / D_m$, and is thus proportional to velocity. Figure 2.10-3 shows the dependence of H on velocity and the nature of the dominant exchange mechanism in each region.

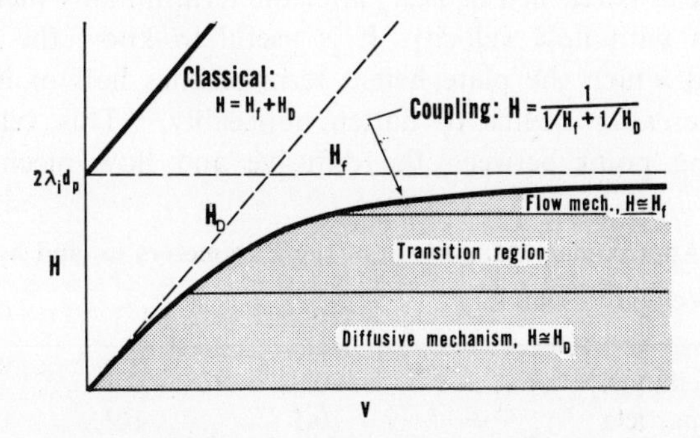

Figure 2.10-3 Contrast between coupling and classical expressions for plate height as a function of velocity. The coupling H is always less than H_f and H_D, while the classical H is always greater than either of these. Similar curves would be found if the sum of all velocity-inequality effects were considered, with the exception that the transistion region would be considerably extended (see Figure 2.10-4).

The coupling theory of eddy diffusion violates the summation rule in which $H = H_1 + H_2 + \dots$. As noted in Section 2.5, this rule is qualified by the condition that one random walk must not interfere with the steps of other random walks. Such interference does not generally occur, but it does occur in the eddy diffusion phenomenon, and is responsible for the unusual form of equation (2.10-8). A random step in eddy diffusion consists of a velocity fluctuation, which can be terminated by one of two mechanisms. If the step is terminated by one mechanism, it drastically changes the length of step which would have otherwise been taken in accordance with the other mechanism. Among random walk type processes, eddy diffusion is unusual (but not unique) because the termination of a step (or displacement) is a key random event as much as the displacement itself.

Although additivity does not govern the relationship between H_f and H_D, it does apply approximately to the various categories of velocity inequality (i.e., as in the sum in equation (2.10-9)). This is a consequence of each velocity and the displacement it leads to, being a sum of the component velocities (the components must, in fact, be evaluated in such a way that this remains true).

Magnitude of the Coupling Terms

The numerical significance of the component terms of the coupling equation, (2.10-9), can be assessed in terms of the approximate ω_i and λ_i values of Sections 2.8 and 2.9. For convenience these values are summarized in Table 2.10-1. Recall that each term, $1/(1/H_f + 1/H_D)$, in the summed coupling equation depends on flow velocity as shown in Figure 2.10-3; that is, a plateau is reached at $2\lambda_i d_p$ after the term initially increases in a linear fashion with flow velocity. It is useful to know the *transistion velocity* $v_{1/2}$ at which the plate height term reaches half of its limiting value and thereafter begins to flatten noticeably.[63] This velocity is a rough dividing point between the diffusive and flow mechanisms of

Table 2.10-1 Approximate Magnitude of the Parameters ω_i and λ_i

Type of Velocity Inequality	ω_i	λ_i	$\nu_{1/2} = 2\lambda_i/\omega_i$
1. Transchannel	0.01	0.5	10^2
2. Transparticle	0.1	10^4	2×10^5
3. Short-range interchannel	0.5	0.5	2
4. Long-range interchannel	2	0.1	0.1
5*a*. Transcolumn	0.02–10	0.4–200	40
5*b*. Transcolumn (coiled)	$0–10^2$	$0–10^3$	40

exchange at lower and higher velocities, respectively. This velocity is reached when $H_f = H_D$. Taking the ratio of H_D and H_f, equations (2.10-6) and (2.10-7), we find

$$\frac{H_D}{H_f} = \frac{\omega_i}{2\lambda_i} \frac{d_p v}{D_m} \tag{2.10-10}$$

and $v_{1/2}$ is the value of v when this ratio is unity

$$v_{1/2} = \frac{2\lambda_i}{\omega_i} \frac{D_m}{d_p} \tag{2.10-11}$$

The ratio $2\lambda_i/\omega_i$ is given in Table 2.10-1 for each type of velocity disparity, thus permitting the direct calculation of $v_{1/2}$ in terms of D_m and d_p.

A more fundamental approach to this problem begins with the nature of the dimensionless quantity $d_p v/D_m$ appearing in equation (2.10-10). The value of this group is a criterion of whether an average solute molecule, after being displaced a distance d_p, was moved primarily by diffusion or by flow, $d_p v/D_m < 1$ signifying the former and $d_p v/D_m > 1$ signifying the latter.† If two flow systems, say, a liquid and a gas, have the same values for $d_p v/D_m$, then the relative role played by diffusion and flow is the same in each even if the actual flow velocity differs, as it may, by ten-thousand-fold. Since $d_p v/D_m$ as a group tells more about molecular displacements in a column than any of the terms individually, this group merits a separate symbol

$$\nu = d_p v/D_m \tag{2.10-12}$$

and will be called the *reduced velocity*.[64] The importance of the variable ν to the study of gas and liquid chromatography will gradually be made clear.

With the new variable ν the ratio in equation (2.10-10) may be written as $(\omega_i/2\lambda_i)\nu$. This ratio becomes one when the value of ν is

$$\nu_{1/2} = 2\lambda_i/\omega_i \tag{2.10-13}$$

Thus the *reduced transistion velocity* $\nu_{1/2}$ is a dimensionless number dependent only on the structural factors λ_i and ω_i and independent of diffusivity and mean particle diameter. It is thus the same for liquids as for gases, and is the same in gas chromatography, for instance, for all

† The mean time required to diffuse a distance d_p is approximately $d_p^2/2D_m$. A similar flow displacement required a time of d_p/v. The ratio of the two times is $d_p v/2D_m$ or to good approximation $d_p v/D_m$. When this quantity is less than one, the diffusion time is less than the flow time—indicating that displacement by diffusion is the more rapid of the two. The opposite conclusion is reached for a ratio greater than one.

possible carrier gases. Thus if $\nu_{1/2}$ is determined for any one mobile phase, its value will be fixed once and for all in that column for any other possible mobile phase.

A better understanding of the use of ν in comparing chromatographic systems to one another might be found in the following examples. A typical gas chromatographic system in which D_m (or D_g) is 0.25 cm^2/sec and d_p is 0.025 cm (about 60 mesh) has a velocity of 10 cm/sec when $\nu = 1$. Thus ν is roughly a measure of the number of multiples of $v = 10$ cm/sec acquired by the flow velocity. The $\nu_{1/2}$ values shown in Table 2.10-1, then, show the multiples of $v = 10$ cm/sec at which the plate height velocity curve reaches half its final height. For a short-range interchannel, for example, this velocity would be 20 cm/sec or thereabouts. This transistion velocity will be somewhat larger for He carrier and small particles than for N_2 and large particles.

A typical liquid chromatographic system in which D_m is 1.5×10^{-5} and d_p is 0.015 cm (about 100 mesh) will have a velocity of only 10^{-3} cm/sec when $\nu = 1$. As far as relative diffusion and flow displacement are concerned, this velocity corresponds exactly with the 10 cm/sec in gas chromatography. Working in the small multiples of 10^{-3} cm/sec, the transistion velocity for a short-range interchannel is only 2×10^{-3} cm/sec, and the other velocities are reduced correspondingly. This contrast forms the principal difference between the dynamics of gas and liquid chromatographic systems. (See Section 7.5.)

It is useful to consider the role of ν in the full plate height equation, (2.10-9), which expresses the combined results of flow pattern effects. If each side is divided by d_p and ν substituted for $d_p v/D_m$, we obtain

$$h = \sum_i \frac{1}{1/2\lambda_i + 1/\omega_i \nu} \tag{2.10-14}$$

where $h = H/d_p$ is the *reduced plate height*.[64] The reason for introducing h is that the plate height is often compared in value to d_p, its minimum possible value is limited by d_p, and, most importantly, it yields the single universal equation above which, with only the structural parameters λ_i and ω_i, is approximately applicable to the flow effects in nearly all chromatographic systems.[64,65]

The wide applicability of equation (2.10-14) makes a detailed plot of h versus ν well worthwhile. This plot, Figure 2.10-4, is simply a plate height-flow velocity plot in the reduced variables h and ν. Like the equation, it is essentially applicable to all systems. The range of ν values is such as to cover the optimum operating conditions of chromatography (i.e., in the neighborhood of the minimum plate height). The individual curves are constructed with the parameters indicated in Table 2.10-1. The column is assumed straight, or at least with negligible bending, so that a **5*b*** curve does not appear. Curve **5*a*** is plotted for a column with a diameter of $12d_p$ ($m = 12$), giving, as shown in the previous sections, $\omega \sim 0.14$ and $\lambda \sim 2.8$. This is the one curve which varies widely from column to column, first because the values of ω and λ vary with m^2, and second because these parameters will be a sensitive function of the method of packing the column

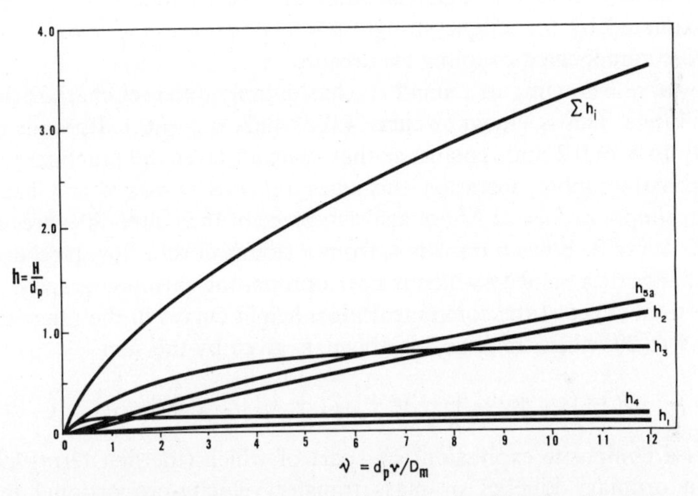

Figure 2.10-4 A plate height vs. flow velocity plot, using the reduced coordinates *h* and *ν*, for various categories of velocity inequalities in the mobile phase.

and the uniformity of the support particles. The other curves, representing effects **1–4,** should be reasonably constant for different column packings.† Effect **2,** however, will be absent in those packings which do not have porous particles which entrain some of the mobile phase. The study and classification of all these effects are recent undertakings hence the actual numerical magnitude of the parameter and the resulting curves must be regarded as approximate.

Also shown in Figure 2.10-4 is the sum (upper line) of the component curves. This indicates the order of magnitude of the total flow velocity effect. It is in rough argeement with experimental values, thus giving confidence that the above theory is based on the correct approach. The detailed comparison with experimental results will be postponed for the volumes on gas chromatography and liquid chromatography. It should

† Recent work by Sternberg and Poulson[66] has shown that the reduced plate height is considerably lessened in columns with $m < 5$. In these cases the column walls may terminate some nonequilibrium effects, as postulated by Giddings and Robison.[60]

be noted that the upper curve in Figure 2.10-4 is not the expected experimental plot because the contributions of longitudinal diffusion and sorption-desorption kinetics have not yet been added in. This will be done in Section 2.11.

The individual curves of Figure 2.10-4 demonstrate the importance of the $\nu_{1/2}$ value. Where $\nu_{1/2}$ is large (curves **1**, **2**, and **5*a***), the reduced plate height continues to increase with velocity over the entire range of the plot. Unless excessively high velocities are used, these curves can be approximated by the simple straight-line expression, $\omega_i \nu$, instead of by the more complicated coupling expression.

The curve belonging to a small $\nu_{1/2}$ has entirely different characteristics in this range. This is shown by curve **4** for which $\nu_{1/2} = 0.1$. Its value rises quickly to $h = 0.2$ and remains at that plateau. Over the practical range of chromatographic operation this term remains constant and can be written simply as $h = 2\lambda_4$, once again in place of the coupling expression.

Only curve **3** shows a transition from a diffusion to a flow mechanism of exchange at a velocity which is near optimal for chromatography.

The total effect of the component plate height curves in the range from $\nu = 0.5$ to 20, when simplified as above, is given by the sum

$$h = (\omega_1 + \omega_2 + \omega_5)\nu + 2\lambda_4 + 1/(1/2\lambda_3 + 1/\omega_3\nu) \quad (2.10\text{-}15)$$

This is a composite expression[64,67]—part of which (the first term) looks like an ordinary kinetics or mass transfer velocity-proportional term, part of which (second term) looks like a classical eddy diffusion term, and part of which (last term) retains its coupling characteristics. The velocity limits $\nu = 0.5$–20 should be kept clearly in mind in using this equation.

Detailed Considerations

As a final discussion on the coupling theory, it should be emphasized that Figure 2.10-1 and 2.10-2 are schematic and do not illustrate the full complexity of molecular migration in chromatography. First, although Figure 2.10-1 shows separate segments for the molecule's flow path and diffusion path, these two movements will actually occur simultaneously. The loss of a molecule from a channel will occur gradually rather than discontinuously at a fixed point. In fact, a channel's boundaries are themselves ill resolved except at the solid support surface, and only a gradual change in average flow velocity will occur as the molecule diffuses through the mobile fluid from one channel to another. Because of the ill-defined boundaries, the 5 discontinuous steps illustrated in Figure 2.10-2 should be replaced by 5 continuous velocity changes. Discrete steps as such do not occur, and the physical motion we are attempting to describe is one of gradual change in which the same *approximate* velocity is held throughout a given length approximated by the step length. Beyond this length the velocity is no longer correlated with its previous value. One definition of the effective step length, in fact, could be given as "that distance traveled before the correlation in downstream velocity is essentially lost."

A second simplification, found in Figure 2.10-1, concerns the straight, perfectly lateral diffusion paths (dashed lines). No diffusion actually occurs in a straight line. The path jerks and continually varies in direction just as with Brownian motion. The "up and down" component of diffusion, not shown, is a part of the longitudinal diffusion contribution to the plate height. This displacement component also serves as part of the exchange mechanism, along with the lateral component. Due to the limitations of a two-dimensional figure, the "in and out" component of diffusion is not shown either. The diffusion paths may actually be envisioned as skirting the particles, in front or in back, and not passing through them (unless the particles are permeable). A two-dimensional figure makes the flow channels appear to be more separate and distinct than they actually are; by going behind or in front of adjacent particles, each channel is connected continuously with every other channel at the same level. The term *flow channel* should henceforth be taken to signify a region of rather uniform flow intimately joined to other flow regions rather than a physically isolated flow passageway.

Finally, Figures 2.10-1 and 2.10-2 have been simplified by not showing the sorption-desorption process. Adsorption or absorption does little more than interrupt the unhindered migration path at intervals. After desorption the downstream motion continues without change, for the solute molecules have no memory which would make their further migration dependent on past attachments.

2.11 Summary of Applications to Chromatography

The random walk model has shown us the nature of the plate height terms which control zone spreading in chromatography. First, we found that longitudinal molecular diffusion gave a contribution inversely proportional to flow velocity. We shall abbreviate this term as B/v, where B is a constant related to the diffusivity of the solute (see equations (2.6-1) and (2.6-2)). Then we found that sorption-desorption kinetics contributed a term proportional to velocity. We shall write this term as $C_k v$ if originating in an adsorption-desorption process, and C_s if originating as a diffusion-controlled process in the stationary phase. The sum of all such terms will be written simply as Cv. Finally, we found coupling terms arising in the inequalities of flow velocity in the mobile phase. Over a limited range some of these terms were found to be constant (these will be written simply as A), some proportional to velocity (these will be written as $C_m v$ and grouped with the other Cv terms), and others were found to have the usual coupling transistion between the latter and the former. Expressed in the most straightforward possible way, the equation for the sum total of these effects can be written as

$$H = A + \frac{B}{v} + Cv + \sum \frac{1}{1/A + 1/C_m v} \tag{2.11-1}$$

which, as mentioned, applies only to a limited, although practical, range of flow rates. This equation is particularly useful in showing how H depends on flow velocity v. For numerical use the various constant terms must be obtained.

If we take a very broad range of flow velocities, some of the terms will not appear velocity-independent; that is, the initial A term and part of the C term (the C_m component) will be absorbed back into the coupling expression of equation (2.11-1). In those cases where the most general equation is needed, then, H should be written as

$$H = \frac{B}{v} + Cv + \sum \frac{1}{1/A + 1/C_m v} \tag{2.11-2}$$

It was shown in Section 2.10 that with reduced variables we could write a single equation for the mobile phase in which the constant terms (equivalent to A and C) become universal parameters roughly independent of particle size, nature of the mobile phase, temperature, etc. A reduction to universal parameters is not found for the stationary phase coefficients (C_s, C_k, and part of B), but casting these terms in reduced form makes possible an immediate numerical comparison of the stationary phase with the mobile phase, and it is thus possible to conveniently pinpoint (and hopefully remedy) the major sources of zone spreading in practical chromatography.[65] The reduced form of the equation has the important advantage of automatically translating some of the lessons learned on one chromatographic system to the practical operation of other systems. By comparison, the apparent simplicity of equations (2.11-1) and (2.11-2) is largely lost once the details of the system are considered because of the enormous variation (ten-thousandfold or more between the extremes of gas and liquid chromatography) of the B and C parameters.

The reduced variables to be used are the reduced plate height $h = H/d_p$ and the reduced velocity $\nu = d_p v/D_m$. The last or coupling term of equation (2.11-2) has already been cast in reduced form, equation (2.10-14).

The others can be done by a simple substitution of the above variables. The final result reads as follows

$$h = \frac{2\gamma}{\nu}(1 + \beta_s) + \Omega\nu + \sum \frac{1}{1/2\lambda_i + 1/\omega_i\nu} \tag{2.11-3}$$

where the λ_i's and ω_i's are universal structural parameters with the approximate values shown in Table 2.10-1, and γ is an obstructive factor for diffusion in the mobile phase, a parameter also having only minor variations from system to system. The terms β_s and Ω are dimensionless parameters but depend specifically on the rate and diffusion constants of the stationary phase.

The three terms of equation (2.11-3) correspond exactly with the three terms of equation (2.11-2). The term $2\gamma/\nu$ describes longitudinal molecular diffusion in the mobile phase and $2\gamma\beta_s/\nu$ describes that in the stationary phase. The parameter β_s can be shown to equal

$$\beta_s = (1 - R)\gamma_s D_s / R\gamma D_m \tag{2.11-4}$$

The two terms together are equivalent to the B/v term of equation (2.11-2).

The $\Omega\nu$ term is the kinetic or mass transfer term originating in the stationary phase. It is composed of a kinetic term $\Omega_k\nu$ and a stationary phase diffusion term $\Omega_s\nu$, corresponding to the $C_k v$ and $C_s v$ stemming from the Cv term of equation (2.11-2). Both the individual and composite Ω parameters can be written

$$\Omega = CD_m/d_p^2 \tag{2.11-5}$$

The terms appearing in the last expression in equation (2.11-3) have already been discussed in connection with coupling.

A well-designed chromatographic column has a plate height not larger than 1–5 particle diameters (h = 1–5). A significant part of this is always contributed by the mobile phase. This can be shown by taking the parts of equation (2.11-3) which apply only to the mobile phase

$$h = \frac{2\gamma}{\nu} + \sum \frac{1}{1/2\lambda_i + 1/\omega_i\nu} \tag{2.11-6}$$

The coupling part of this expression has already been plotted as a function of reduced velocity in Figure 2.10-4. If we add to this the $2\gamma/\nu$ term, with γ set at 0.6, the lower curve of Figure 2.11-1 results. This reduced plate height curve may change slightly from column to column, but for well-designed systems the variations will be relatively small. Thus as a starting point, considering the mobile phase alone, we are generally faced with an h value which even at its minimum is approximately 2.† The stationary

† When $m < 5$, h may be less than unity; that is, $H < d_p$. This effect has been observed by Giddings and Robison,[60] and by Sternberg and Poulson.[66]

phase terms must be added to this. These terms are generally more flexible and the choice thus more critical. A bad choice—a stationary phase with undesirable kinetic features—can easily cause h to rise beyond practical limits. A desirable choice will not increase h more than twofold or so beyond the ground value of the mobile phase. First, this choice should include a β_s term which is less than unity. Fortunately, this requirement is not usually demanding; in many systems, including nearly all of gas chromatography, this term is negligible. Second, the contribution

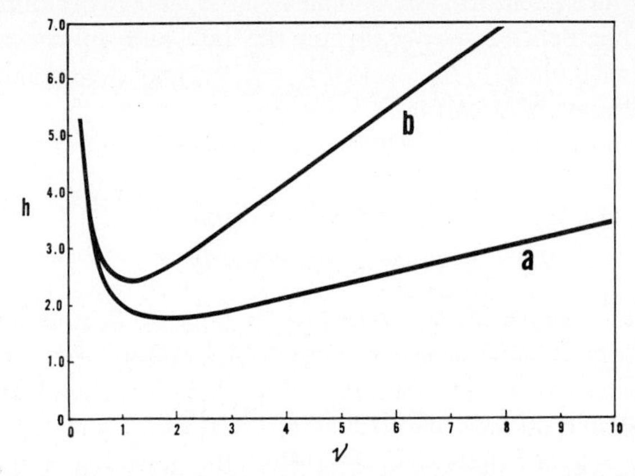

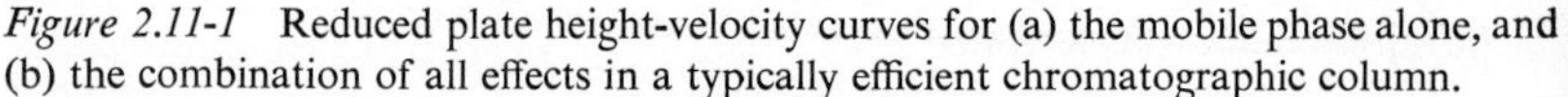
Figure 2.11-1 Reduced plate height-velocity curves for (a) the mobile phase alone, and (b) the combination of all effects in a typically efficient chromatographic column.

of $\Omega\nu$ at the point where the mobile phase curve reaches a minimum should not more than double the latter (the minimum will, of course, shift when $\Omega\nu$ is added, but this rule nonetheless gives a rough ceiling for the desirable magnitude of Ω). Since the particular mobile phase curve plotted here reaches a minimum of $h \sim 1.75$ at $\nu \sim 1.75$, Ω should as an approximation not be allowed to increase much beyond unity for an effective system. This simple criterion, applicable to most chromatographic systems, can be applied by using equation (2.11-5) for Ω and the expression derived earlier for the C terms of the stationary phase. It is, of course, necessary to estimate or measure the kinetic parameters that enter C, including the value of D_m, to apply this criterion. Such estimates can often be obtained from chromatographic results. Although little effort has been made in the past to transfer useful results from one chromatographic system to another, the method shows great promise.

The upper curve of Figure 2.11-1 shows a complete plate height curve including a stationary phase contribution. The value of Ω is assumed to be 0.5, while β_s is equated to zero. The minimum plate height under these circumstances is 2.4 particle diameters; $h \sim 2.4$. This occurs at $\nu \sim 1$. In the last section it was shown that the value $\nu = 1$ corresponded to $v \sim 10$ cm/sec in gas chromatography and $v \sim 10^{-3}$ cm/sec in liquid chromatography. These values should, then, approximately coincide with the highest separation efficiency of the respective systems. The enormous difference in the velocities of highest efficiency is mainly due to the large ratio (10^4–10^5) of the diffusion coefficient D_m in gases as compared to liquids. This same factor is responsible for the requirement that C terms in gas chromatography must be from 10^4 to 10^5 times smaller than in liquid chromatography if Ω is to be kept near unity. This very severe kinetic requirement limits the types of adsorbents or absorbents which can be used successfully with gas chromatography, whereas the requirements for liquid chromatography are very generous.

2.12 Criticism of the Random Walk Model

Many of the advantages and disadvantages of the random walk model should already be clear. Its chief disadvantage is inexactness. It is an approximate theory of discrete forward and backward steps, whereas most chromatographic processes involve a more subtle, continuous molecular motion.

The random walk model's major advantage is simplicity. No other approach has led to valid approximate equations with so little mathematical difficulty. In some cases where the processes are enormously complicated—in the coupling mechanism—the random-walk model might remain for years as the only theoretical approach to the problem. In other cases it serves as a check on the more abstract theories, insuring that the quantitative assumptions and detailed mathematical steps are correctly made. Its latter use has often served an important purpose in the author's laboratory.

By way of a general summary, the random walk model may be said to present a simple picture of chromatography with a large element of statistical truth. As an approximate statistical theory, however, its limits of validity must be carefully watched or else they will be occasionally overstepped.

Some rather specific remarks about the random walk model can be made on the basis of the equations derived in this chapter. From all that is now known about the chromatographic process, including the more rigorous theories, the random walk model may be said to have yielded expressions which have essentially the correct dependence on flow velocity, diffusivities, rate constants, particle diameter, stationary phase dimensions, and relative retention. The dependence is correct, that is, after a reasonably long time period in which the number of random steps is sufficient to form a near-Gaussian profile. But the expressions are generally in error by some numerical coefficient. Under the best of circumstances the error does not exceed 2 or 3 (in some cases—equation (2.7-6)—the expressions are fortunately exact). The reasons for the numerical inexactness were apparent from the beginning: the divergence of the actual processes from those of the model. The continuous nature of the actual velocity changes has already been mentioned. In some cases an arbitrary decision must be made on what to take as the length and number of steps. In addition, the model is based on a fixed number of steps for all participating molecules, while in reality, particularly in reference to sorption-desorption kinetics, a variable number of steps are taken. The length of step, also, to the extent this can be defined for some of the continuous processes, is notoriously variable.

Some attempts have been made to carry the random walk treatment beyond these difficulties. Jones, particularly, has extended this general method in a detailed way to gas chromatography.[68] The view taken in this book is that the random walk model should serve where it is to most advantage—in providing a simple qualitative and quantitative grasp of the principal effects—and that the more subtle and complex processes can be treated much more successfully in other ways, particularly by the generalized nonequilibrium theory.

2.13 Rigorous Stochastic Theory: Kinetic Effects

The stochastic, approach† was first developed by Giddings and Eyring[15] as a method for determining the effects on zone structure of the random sorption-desorption processes of chromatography. The theory has since been extended by Giddings,[69] Beynon, Clough, Crooks and Lester,[70] and McQuarrie.[71] The approach, based on probabilistic ideas, considers the random migration of a single solute molecule down a chromatographic column. Due to the random events there is a certain probability, which can be explicitly written, that after a time t the molecule will have arrived somewhere in the interval from z to $z + \delta z$ along the column. If this probability, derived for one solute molecule, is now multiplied by the total number of such molecules, the expression obtained will be the final number of molecules in the δz interval. This automatically gives a concentration profile for the zone.

The stochastic approach may be described as rigorous rather than comprehensive. It generally begins with a simple, precisely defined

† A "stochastic process" and a "random process" are essentially the same thing. Feller points out that the word "stochastic" is more often used, as here, when time is a variable.[44]

mechanism and proceeds to a rigorous expression for the concentration profile resulting from that mechanism. This expression can then be simplified into a limiting form valid for long time intervals (see Section 2.2). Due to the rigorous approach to the problem, the theory is difficult to apply to complex mechanisms of exchange. It has been applied to one-, two-, and multisite adsorption process[15,69,71] (thus allowing for a heterogeneous surface of adsorption), but not to diffusion-controlled processes. The generalized nonequilibrium theory, described in Chapter 3, is much more successful in the analysis of complex exchange processes. The latter approach, however, is valid only as a limit for long time periods. Although this limit is usually adequate for practical purposes, the stochastic approach, which escapes this limitation, is useful when the long time assumption cannot be made. Zone asymmetry in chromatography and multiple zones which originate with a slow reaction mechanism in paper chromatography are examples of phenomena which are not described by the limiting theories.[72] The stochastic approach can be used to study these cases provided the kinetics are, or can reasonably be assumed to be, governed by simple, tractable processes.

The stochastic approach is not, of course, the only rigorous approach to the kinetics of chromatography. Much earlier work was done using the material-conservation equations as a starting point. Thomas, particularly, developed this method in an elegant manner.[73] Although the two approaches yield much the same results, the stochastic approach has now been more extensively developed (having been applied to multisite adsorption) and has been more often applied to practical problems. For these reasons the stochastic approach will be discussed here.

The development given in this section will follow the treatment by Giddings and Eyring. The goal is to obtain the concentration profile (at elution) for a single kind of adsorption site, or, more generally, for any first-order exchange process which by some mechanism or another leads molecules from one velocity state to another.† The more general case may occur as a result of a chemical interconversion between two species each with a different rate of downstream migration in the chromatographic column.[72]

The treatment begins with the assumption[15] that "there is a definite, nonvarying constant, k_a, representing the probability per unit time that the molecule will adsorb on the surface." This first-order rate constant is accompanied by a first-order desorption constant k_d. The assumption that k_a and k_d are nonvarying constants—independent of time or of the molecules history—causes a departure from diffusion-controlled kinetics.

† This includes the effects of changes in ionic charge which leads to a mobility inequality in electrophoresis or electrochromatography.[19]

With the latter the chance for attachment to the surface, for example, is greater right after detachment because the molecule is still in the near vicinity of the surface (diffusion-controlled processes always occur in a way which is governed by relative proximity). Thus application of the simple stochastic theory to diffusion-controlled processes can be made only as an approximation. The theory is best adapted to adsorption-desorption–controlled kinetics.

While migrating to the end of a column of length L, solute molecules spend an average time $t_m = L/v$ in the mobile phase. In this time a given solute molecule will adsorb on the surface a fixed number of times. The chance that this number is r is given by the Poisson distribution expression[74]

$$\psi_r = \frac{(k_a t_m)^r}{r!} \exp(-k_a t_m) \tag{2.13-1}$$

Now if the molecule started its migration in the mobile phase, one desorption must accompany each adsorption in order for the molecule to end up in the mobile phase immediately preceding elution. Each molecule adsorbing r times must therefore desorb r times as well. The chance that the rth desorption will occur after spending a time between t_s and $t_s + dt_s$ on the surface is the chance that $r - 1$ desorptions occur in time t_s multiplied by the chance $k_d\, dt_s$ that the rth desorption occurs somewhere in the interval dt_s.[15] This probability is

$$dP_r = \frac{(k_d t_s)^{r-1}}{(r-1)!} \exp(-k_d t_s) \cdot k_d\, dt_s \tag{2.13-2}$$

The first part of the right-hand side is an expression equivalent to equation (2.13-1). The expression dP_r is a conditional probability which gives the distribution in time spent on the surface providing exactly r adsorption processes have occurred. To remove this restriction, dP_r must be multiplied by the chance ψ_r that r adsorptions occur and then summed over all possible r values; that is,

$$dP = \sum_{r=1}^{\infty} \psi_r\, dP_r \tag{2.13-3}$$

When ψ_r and dP_r from the two previous equations are substituted into this summation and rearranged, we obtain

$$dP = \exp[-(k_d t_s + k_a t_m)]\left(\frac{k_d k_a t_m}{t_s}\right)^{1/2} \sum_{r'=0}^{\infty} \frac{(X/2)^{2r'+1}}{r'!\,(r'+1)!} \cdot dt_s \tag{2.13-4}$$

where

$$X = (4k_a k_d t_m t_s)^{1/2} \quad (2.13\text{-}5)$$

The summation term is in a form equivalent to the Bessel function, $I_1(X)$, of imaginary argument.[75] Thus the probability distribution† of t_s, given by $P = dP/dt_s$, becomes

$$P = (X/2t_s)I_1(X) \exp(-k_d t_s - k_a t_m) \quad (2.13\text{-}6)$$

When the quantity X is large, the first term of the asymptotic expansion of $I_1(X)$ can be used. This gives

$$P = \frac{k_d k_a t_m}{2\sqrt{\pi}} \frac{\exp[-(\sqrt{k_d t_s} - \sqrt{k_a t_m})^2]}{t_s^{3/4}} \quad (2.13\text{-}7)$$

It should be noted that X is twice the geometrical mean of the expected (or average) number of adsorptions, $k_a t_m$, and the expected number of desorptions in time t_s, $k_d t_s$. Since adsorptions and desorptions are numerous in any useful chromatographic procedure, the condition that X be large will generally be fulfilled.

The summation in equation (2.13-3) concerns all cases where one or more adsorptions occur. There is a finite chance, however, that no adsorptions (or reactions) will occur in the designated migration. This condition is particularly true when slow reaction or interconversion processes are occurring. The chance that no adsorption occurs in $\exp(-k_a t_m)$. This leads to the following distribution function, to be added to the P already obtained.

$$P = \exp(-k_a t_m)\, \delta(t_s) \quad (2.13\text{-}8)$$

The Dirac delta function,[77] $\delta(t_s)$, indicates that the time t_s is precisely zero for that fraction of molecules going without adsorption.

A slightly different probability distribution from the above expressions is obtained if the molecule starts its migration in an adsorbed state. This may occur, for example, in paper chromatography where the zone (spot) is applied to the paper and allowed to adsorb before migration commences. If a molecule is initially adsorbed, and r additional adsorptions (calculated as before) occur during migration, $r + 1$ desorptions will be required before elution can occur. Proceeding much as before, the probability distribution equivalent to (2.13-6) becomes

$$P = k_d I_0(X) \exp(-k_d t_s - k_a t_m) \quad (2.13\text{-}9)$$

† In the terminology of modern mathematics P is known as a probability *density function* rather than a *distribution function*. The latter is reserved for the integral of P from $-\infty$ to a given time t_s. The word "distribution" has, however, become well entrenched as a description of P, and this commonus age will be essentially continued.[76,77]

The asymptotic expansion of $I_0(X)$ leads to an expression quite similar to (2.13-7)

$$P = \frac{k_d}{2\sqrt{\pi}(k_a k_d t_m)^{1/4}} \frac{\exp\left[-(\sqrt{k_d t_s} - \sqrt{k_a t_m})^2\right]}{t_s^{1/4}} \tag{2.13-10}$$

The foregoing expressions are probability distribution (or density) functions in the time t_s spent adsorbed. All other terms in the expressions for P are assumed constant, including the time t_m spent in the mobile fluid (this restriction will be discussed later). We are often interested in zone profiles, which are measured in terms of the total time t rather than t_s. However, since $t_s = t - t_m$, we conclude that the probability profiles are of the same dimensions in t as t_s, but displaced further along the axis by a time t_m. In gas chromatography t_s is simply the time beyond the "air peak," and, indeed, in any form of elution chromatography t_s is merely the time measured from the appearance of an inert or nonsorbing solute zone.

Since a probability density profile multiplied by the total number of solute molecules becomes a concentration profile, all of the previous expressions for P are (within a constant factor) solute concentration profiles.

Limiting Forms

The asymptotic expansion of the Bessel functions appearing in P is valid only when X, equation (2.13-5), is large. It will be recalled that X is twice a geometrical average of the expected number of adsorptions and desorptions. Roughly, then, X is the total number of adsorptions and desorptions occurring on a column. The magnitude of X can be large only if the elapsed time is large compared to the time required for adsorption or desorption. These conditions lead to the various limiting expressions for "long times," and, indeed, the two asymptotic forms are a type of limiting expression. But we want to write these expressions in a form which corresponds to the Gaussian profile which eventually emerges from all sorption-desorption processes. The general procedure followed here led to the first rigorous limiting form for Gaussian profiles.[20]

The asymptotic expressions in equations (2.13-7) and (2.13-10) both contain the variable t_s in the exponential term and in the denominator. The latter is a slowly varying form whose variation with t_s can be ignored compared to the exponential expression.† Thus to good approximation

† This will become apparent as we find that the rapid variation of the exponential term can, under long time conditions, change P significantly over intervals of t_s that are small compared to t_s itself.

the two asymptotic equations have the same dependence on t_s after a sufficient time has elapsed. This might have been expected, since the difference in the two was merely that of whether the molecule started as a detached or an adsorbed species; after a long time period this detail of the molecule's history becomes insignificant compared to the hundreds or thousands of subsequent adsorptions.

The quantity appearing in the exponent of the two asymptotic expressions is of immediate interest to us. This quantity is written as

$$f(t_s) = (\sqrt{k_d t_s} - \sqrt{k_a t_m})^2 \qquad (2.13\text{-}11)$$

The value of this function is zero when $t_s = k_a t_m / k_d$. Looking back at the asymptotic equations, it is clear that the P profile has a maximum very near to this point.† As it turns out, this value is the mean t_s value (after a long time interval) and can be written as $\bar{t}_s$.

A Taylor's expansion of $f(t_s)$ will now be made around $\bar{t}_s$. It is found that the first two terms of the expansion are zero and that the subsequent terms give

$$f(t_s) = \frac{k_d \, \Delta t_s^{\,2}}{4\bar{t}_s} - \frac{k_d \, \Delta t_s^{\,3}}{8\bar{t}_s^{\,2}} + \frac{5k_d \, \Delta t_s^{\,4}}{64\bar{t}_s^{\,3}} - \ldots \qquad (2.13\text{-}12)$$

where Δt_s is the departure of t_s from the mean $t_s - \bar{t}_s$. It will now be assumed that all terms but the first are negligible in the range of actual interest (this assumption will be discussed shortly). The asymptotic equations, which depend on t_s through the exponential $\exp[-f(t_s)]$, now acquire the form

$$P = \text{constant} \times \exp(-k_d \, \Delta t_s^{\,2} / 4\bar{t}_s) \qquad (2.13\text{-}13)$$

which is identical to the Gaussian profile

$$P = \text{constant} \times \exp(-\Delta t_s^{\,2} / 2\tau^2) \qquad (2.13\text{-}14)$$

in its dependence on the time Δt_s. By comparison of the two equations we find that the standard deviation in time units τ (equivalent to σ in distance units) is $\sqrt{2t_s/k_d}$. The plate height, equal to $H = L\tau^2/t^2$ (see Section 2.3), becomes

$$H = 2L\bar{t}_s / k_d t^2 \qquad (2.13\text{-}15)$$

The ratio $\bar{t}_s/t$ is the average fraction of time spent by molecules in the stationary phase, and is thus $1 - R$. The time t is simply the column length L divided by the mean velocity of the zone Rv. With these substitutions H becomes

$$H = 2R(1 - R)v/k_d \qquad (2.13\text{-}16)$$

† The $t_s^{1/4}$ and $t_s^{3/4}$ in the respective denominators shifts the maximum to slightly smaller values of t_s. An expression for this shift can be found in ref. 15.

This equation is identical to equation (2.7-6) as derived from the random walk model, provided that $1/k_d$ is replaced by the mean desorption time t_d.

We will now return to prove the assumption that all terms beyond the first in equation (2.13-12) are negligible in the usual range of interest. Suppose we wish to compare the magnitude of these terms at some point where Δt_s is a multiple α of τ; specifically, $\Delta t_s = \alpha\tau$. When τ is replaced by the value $\sqrt{2\bar{t}_s/k_d}$, then multiplied by α and substituted for Δt_s in equation (2.13-12), we get

$$f = \frac{\alpha^2}{2} - \frac{\alpha^3}{(8k_d\bar{t}_s)^{1/2}} + \frac{5\alpha^4}{16k_d\bar{t}_s} - \cdots \qquad (2.13\text{-}17)$$

The average number of desorptions $k_d\bar{t}_s$ is a large number under the conditions postulated here. Thus for an α value whose magnitude is in the vicinity of 1 or 2, the first term is much larger than the subsequent terms. Larger α values correspond to the dilute outer edges of the zone, and the dominance of the later terms is not usually very significant here. Thus for the bulk of the zone the assumption that f can be represented by the first term, only, is a good approximation.

Note that the second term of equations (2.13-12) and (2.13-17) contributes a negative term when $t_s > \bar{t}_s$ (Δt_s positive) and a positive term when $t_s < \bar{t}_s$. The former corresponds to the tail of the zone and the latter to its leading edge. Since the concentration profile varies as exp $(-f)$, the extra diminishing of f at the tail and addition to f at the front tilts or *skews* the concentration profile so that it is higher in the trailing part of the zone. To whatever extent the second term does contribute to the profile, then, it is responsible for the kind of zone asymmetry called *tailing*. The latter subject will be discussed later in this section.

Nonelution Chromatography

The consequences of the stochastic theory are occasionally important for zones which are stopped before they reach the end of the chromatographic bed. This problem has been discussed by Giddings.[69] Mathematically, we must formulate the distribution in t_m or t_s when their sum, the total time t, is fixed (t corresponds to the end of the experiment) rather than, as before, formulating a t_s (or t) distribution with t_m fixed. The former case is more difficult and will not be treated in detail here. The results, in equation form, are identical; equation (2.13-7) is the probability distribution for molecules which begin and end in a desorbed state. But for the nonelution case the actual range of variables, and sometimes the variable itself is different.

Since we wish to write down the equations for zone structure in the most general possible way, we will postulate an interconversion reaction between two species or *states of existence* A and B

$$A \underset{k_2}{\overset{k_1}{\rightleftharpoons}} B \tag{2.13-18}$$

In the cases we have discussed so far A and B would be the desorbed and adsorbed state, respectively. However, the theory is equally valid if A and B are chemical species (perhaps isomers) and it is desired to know the probability that the time spent in the A or B form lies in a particular time range. If A and B both undergo chromatographic migration, but at a different rate, the probability distribution in time becomes a distribution in distance (e.g., if it is known that a particular molecular entity spends more time than average in the faster moving form B, then it will also migrate further than average). The probability distribution theory is also valid in electrophoresis, where A and B may have different charges and thus different mobilities.

For the most general use it is convenient to write the probability equations in terms of x, where x is the fraction of time spent in the B form

$$x = t_B/t_A \tag{2.13-19}$$

To find the probability that x lies in a certain interval, say, x_0 to $x_0 + dx$, we simply take the appropriate probability density P as expressed below, using x_0 in place of x, and multiply this by the interval size dx; that is, $P(x_0)\,dx$.

The four density functions given below were obtained from a paper by Keller and Giddings[72] (the parameter x in the paper was incorrectly defined as the fraction of time spent in the A form; B should replace A as shown here). These correspond to molecules that begin as A and end as B, $P_1(x)$; molecules that begin and end as A, $P_2(x)$; molecules that start as B and end as A, $P_3(x)$; and molecules that begin and end as B, $P_4(x)$.

$$P_1(x) = a \exp\,[-a(1 - x) - bx] I_0 \sqrt{4abx(1 - x)} \tag{2.13-20}$$

$$P_2(x) = [ab(1 - x)/x]^{1/2} \exp\,[-a(1 - x) - bx] I_1 \sqrt{4abx(1 - x)} \tag{2.13-21}$$

$$P_3(x) = b \exp\,[-a(1 - x) - bx] I_0 \sqrt{4abx(1 - x)} \tag{2.13-22}$$

$$P_4(x) = [abx/(1 - x)]^{1/2} \exp\,[-a(1 - x) - bx] I_1 \sqrt{4abx(1 - x)} \tag{2.13-23}$$

The parameters a and b are k_1t and k_2t, respectively, where t is the total time being considered. There is a finite chance that A or B will remain unchanged throughout the experiment. The chance of the former, for which $x = 0$, is

$$P(x = 0) = \alpha \exp(-a) \tag{2.13-24}$$

the chance of the latter, leading to $x = 1$, is

$$P(x = 1) = \beta \exp(-b) \tag{2.13-25}$$

where α and β are the respective probabilities that the initial state is A and B. The total probability distribution may now be written as the sum of the two expressions just given plus the four previous expressions, weighted by the probability α or β that the initial state is the correct one

$$P(x) = \alpha[P_1(x) + P_2(x)] + \beta[P_3(x) + P_4(x)] \tag{2.13-26}$$

As mentioned earlier the results of the stochastic theory are most significant when the long-time limiting forms are not valid; say, for some slow reaction or adsorption process which may adversely affect zone structure. An example of this is found when two forms of a migrating species, each with a different migration rate, convert slowly back and forth into one another. Depending on the interconversion rate this species may finally show up as a single, double or occasionally a triple zone.[72] The zones may be nearly symmetric or have severe tailing or leading parts. One of the examples calculated by Keller and Giddings is shown in Figure 2.13-1. Two distinct concentration peaks are found, but the

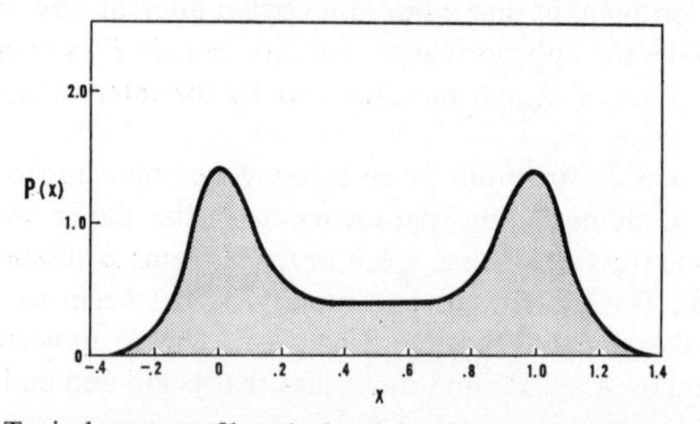

Figure 2.13-1 Typical zone profile calculated for the chromatographic migration of two interconverting species. The two zones tail toward one another, as seen by the increased concentration in the center.

probability profile (which may be interpreted as a concentration vs. distance profile) shows a considerable quantity of material distributed rather evenly between the peaks. Further examples, along with the resulting spot shapes in the case of paper chromatography, will be given in Part III.

Another example of zone distortion which arises in slow kinetics and is subject to the above theory is tailing. This phenomenon is a practical nuisance because it leaves traces of a component far to the rear where overlap may occur with other components. It is particularly unwelcome in gas chromatographic trace analysis where small peaks can be swamped by the large tail of preceding peaks.

Origin of Tailing in Kinetics

The appearance of tailing has usually been attributed to the presence of a nonlinear isotherm, a condition which arises, especially in adsorption chromatography, when the sample size is too large and the column thus *overloaded.* It is undoubtedly true that much tailing has originated in nonlinearity, but it is probably true that tailing nearly as often arises from kinetic effects where the kinetics are strictly linear. The suggestion that tailing originates in kinetics was first made by Giddings and Eyring[69] in 1955 to explain experimental data that apparently were not associated with a nonlinear isotherm. Keller and Giddings[72] later showed that numerous forms of asymmetry, including tailing, are associated with slow, linear kinetics. The stochastic theory was used for the calculation of numerical examples. The view that tailing is often a kinetic phenomenon has been strengthened by recent gas chromatographic work with ultra-sensitive detectors where the sample size is so small as nearly to preclude nonlinearity. It has also been confirmed by the ion exchange work of Sen Sarma et al.[78]

Earlier it was shown that a slight departure from a Gaussian elution profile, in the direction of tailing, was expected even with rapid exchange kinetics. Similar considerations show that zones which remain on the column may show a slight asymmetry in either direction. These effects are not ordinarily large however. Other sources of tailing are found in large volumes of dead space within the column or the detection unit (this is analogous to the kinetic origin of tailing) and in the release and absorption of heat accompanying the sorption-desorption process. The latter has been shown to produce an effect similar to that of a nonlinear isotherm,[30] an effect that is negligible at sufficiently low concentration. This problem has been treated in some detail by Scott[79] for gas chromatography. Also in gas chromatography the change in gas volume accompanying solute absorption causes tailing[80] (the effect is probably negligible in liquid chromatography).

The kinetic mechanism suggested by Giddings and Eyring is as follows.[81] It is proposed that there are two (or more) different kinds of adsorption sites. One of these is the *normal adsorption site*, leading to rapid exchange,

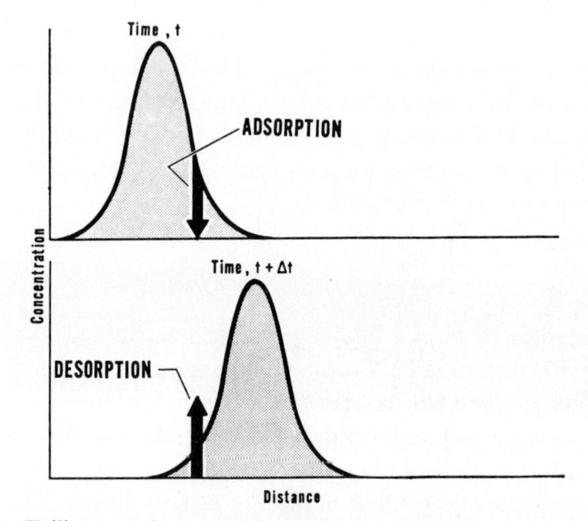

Figure 2.13-2 Tailing may be caused by molecules which absorb on a tail-producing site from somewhere within the zone (upper figure), and then desorb (lower figure) after the zone has moved on over the site of adsorption.

which is responsible for the main chromatographic effect. It is postulated that a second type of site exists, a *tail-producing site*, which is relatively scarce but has a slow desorption rate. Molecules are thus adsorbed only infrequently, but once adsorbed are held strongly and released only after the bulk of the zone has gone by. The release of solute into the

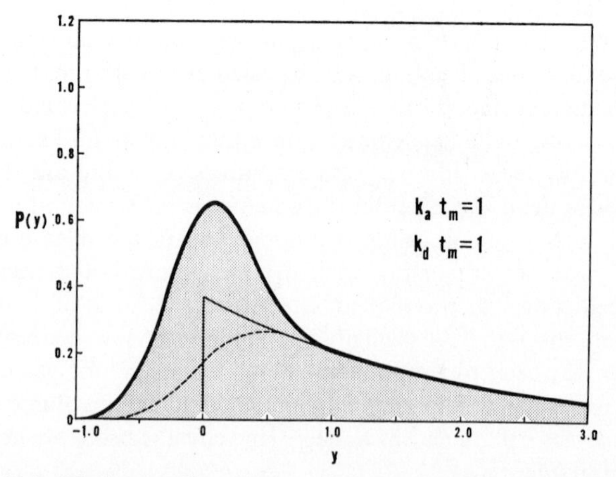

Figure 2.13-3 Elution profile with $k_a t_m = 1$ and $k_d t_m = 1$. In this and subsequent profiles the light solid line is a direct consequence of the tail-producing sites, the dashed line is the same except as it is smeared by column processes and the heavy solid line is the final concentration profile.

trailing portion of the zone is expected to build up the tail and thus cause an observable tailing effect.† This process is illustrated in Figure 2.13-2.

The theoretical approach to this kinetic mechanism of tailing can be made in several ways. We could simply use the two-site stochastic theory which has been developed by several authors.[69,71] A simpler approach is to use a limiting method to discern the effects of the rapid exchange sites and the rigorous one-site theory for the slow exchange sites. It will be recalled that the limiting forms, which by themselves lead to Gaussian profiles, describe a zone spreading process which is equivalent to a simple diffusion process. It is only necessary then to superimpose this diffusion process (and the concomitant retention) onto the zone profile calculated from one-site theory. This procedure has worked successfully in the description of multiple zone formation, discussed earlier, and has subsequently been used for the calculation of tailing profiles.[81] Some of these profiles are shown in Figures 2.13-3 to 2.13-5. The tail-producing site, following the one-site theory given earlier in this section, yields the discontinuous profile indicated by the light solid lines. In addition to this there is a narrow spike of material ($R = 0.1$) which did not adsorb on these sites. When an effective diffusion process equivalent to 1000 plates is superimposed onto the light solid line, the dashed line results. When this diffusion process is applied to the light solid line plus the narrow spike (not shown), the final concentration profile indicated by the heavy solid line and the shaded area results.

Figures 2.13-3 and 2.13-4 are elution profiles calculated in terms of the time relative to the dead time $y = t/t_m$. In Figure 2.13-3 the fact that $k_a t_m = 1$ means that there is an average of one adsorption to the tail-producing site per molecule (some molecules adsorb more than once and others, responsible for the narrow spike mentioned above, not at all). The fact that $k_d t_m = 1$ means that a molecule, once adsorbed, requires on the average one dead time t_m to desorb again. As seen by the figure, this desorption time is long enough to form a considerable tail on the profile. Figure 2.13-4 is different only by having a larger desorption rate, $k_d t_m = \sqrt{10}$, which means that molecules stick to the tail-producing sites for only $1/\sqrt{10}$ times the dead time. The length of the tail is reduced, but is still sufficient to be noticeable. The choice of $\sqrt{10}$ for the desorption parameter was made in order to render the mean desorption time equal to the standard deviation of the zone in time units τ. It was found that more rapid desorption than this did not cause a significant degree of tailing.

† It is remarkable how closely this mechanism corresponds to the one proposed independently by Sen Sarma, Anders, and Miller to explain tailing in ion exchange chromatography.

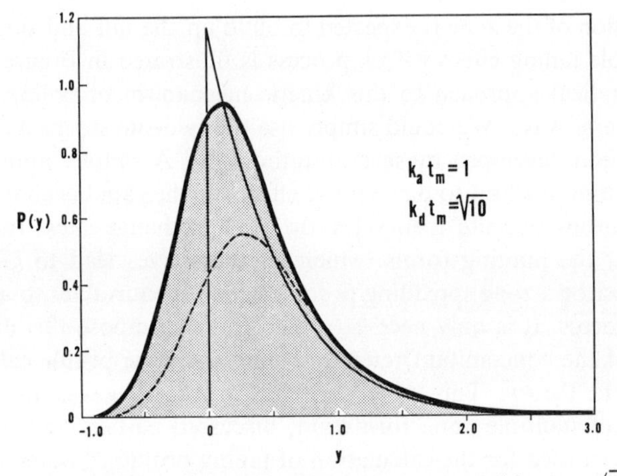

Figure 2.13-4 Elution profile with $k_a t_m = 1$ and $k_d t_m = \sqrt{10}$.

Figure 2.13-5 shows a nonelution profile analogous to the elution case just discussed. The parameters a, b, and x are those used for equations (2.13-19) to (2.13-26). The parameters are such that an average of one adsorption ($b = 1$) occurs to the tail-producing site during chromatographic development, and the mean time for desorption equals this same time. The figure indicates that a significant tail is formed. By adjusting the adsorption and desorption rates, a tail of almost any proportions could be generated.

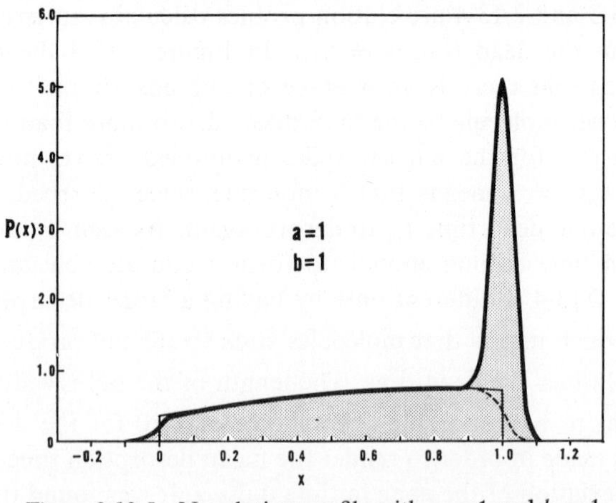

Figure 2.13-5 Nonelution profile with $a = 1$ and $b = 1$.

Additional Considerations on the Stochastic Theory

The stochastic theory has been treated in the literature in far greater detail than here. An attempt has been made to limit the present discussion to those aspects of the theory which are applicable to the present experimental side of chromatography. The future may bring several aspects not considered here into practical focus. We will briefly describe the kinds of problems which have been treated with the stochastic theory but which did not seem to merit a detailed discussion at this time.

One aspect of the stochastic theory which has been pursued from the beginning is the effect of a nonuniform surface with different kinds of adsorption sites. The mathematics rapidly becomes intractable, however, as we pass from the sheltered simplicity of one-site theory. Giddings and Eyring,[15] as well as Giddings,[69] formulated two-site theory in terms of complex double summations. Giddings developed a somewhat simplified integral expression from which a sample profile was computed. McQuarrie[71] has formulated both the two-site and the general n-site problem, but the results are too complex to be readily usable.

Another problem connected with practical operation is the finite width of the initial zone. General integral equations have been given for solute entering the column with an arbitrary concentration profile.[15] A number of specific cases have also been treated.[71] A related question, concerning the effect of a variable mobile phase time t_m, has also been discussed.[71]

One of the most promising developments of the stochastic theory is the expansion of the probability (concentration) profiles into a Gram-Charlier series.[71] The first term of such a series is a Gaussian profile, and the subsequent terms are related to the departures from this ideal profile. Since all chromatographic profiles approach Gaussians at long times, a series with such a leading term is particularly fitted to the study of chromatographic zones.

2.14 Zone Spreading in Nonuniform and Linked Columns and in Regions External to the Column

The previous sections deal with the contributions to zone spreading which originate with flow through uniform regions of column. If a column is nonuniform, due to changes along its length in diameter, sorptive strength, flow velocity, etc., the earlier plate height expressions will apply individually and only to those segments of the column which are uniform or nearly uniform. Such segments can always be found, for even with a continuous variation in column properties a sufficiently short segment will show relatively little change from end to end. Thus there is a collection of local plate height terms, $H_1, H_2, H_3, \ldots$ etc., each possibly different because it is obtained for a different segment. From the individual value H_i, we know that the zone spreading in segment i would be of magnitude $\sigma_i = \sqrt{H_i L_i}$, but this information is useless unless we know how this combines with spreading in the other segment to yield the final zone spreading parameters σ (nonelution) or τ (elution). This section will be primarily devoted to a study of the combination of the local plate height terms to form the end product, the experimentally determined *apparent plate height* $\hat{H}$. The study will account for changes along a column's length but not for changes with respect to time as found in programmed temperature gas chromatography (Part II) or gradient elution chromatography (Part III).

To clarify our goal, several examples of "nonuniform" columns of practical interest will be mentioned. The subsequent theory will be applicable to these except in special cases. First, columns are often linked in series, either to take advantage of several sorptive materials in the same run or to bring about changes in the column diameter. Each link in the series has, of course, a different plate height value. Second, in gas chromatography the decompression of gas produces an increased velocity as the column outlet is approached. Since the plate height is velocity dependent, the plate height varies from point to point. Finally, the "segments" used for introducing the sample and for sample detection may be considered as extensions of the column within which a particular degree of zone spreading is occurring.

The problem of combining individual components of zone spreading is approached in one of two ways, depending on the relationship of the segments. First, and most commonly, we may assume that the dispersion processes in each segment are independent of those in any other segment (later we will find that the *zone spreading* contributed by one segment can be influenced by the *zone velocity* in the final segments, but this is another problem). Second, we may assume some mutual interaction between the processes in adjacent segments. The latter can be illustrated in terms of the transcolumn velocity differential. It will be remembered from Section 2.8 that consistent and apparently reproducible differences in permeability lead to velocity variations over the column cross section, these velocities usually being highest near the outside of the column. Now if the exchange of solute molecules between the fast flowing outside regions and the retarded center is slow, we can envision a given group of molecules at the outside gaining an appreciable lead on their counterpart in the center. The gap between the two groups, a measure of zone spreading, may continue to enlarge through several segments (remember that a segment's boundaries are often rather arbitrary; they are defined in such a manner that each segment will have nearly uniform properties). Thus the zone spreading in the sequence of segments will be related, with a common origin in the existence (at the beginning) of one group of molecules near the center and the other near the outside. This mutual dependence would not exist if the average exchange time were a good deal less than the residence time in each segment because each exchange implies an independent start for the downstream displacement of molecules. Thus the independence of dispersion processes is found when all exchange mechanisms are rapid or when the segments are long. Independence may usually be assumed if each segment contains at least several theoretical plates, but there are some notable exceptions to this rule in large diameter preparative columns where the dependence between segments may be excessive due to the slow transcolumn exchange. Dependent processes will not be treated in a general way here because they are likely to be significant in only a few special cases such as preparative columns.

In those cases where uniformity can be assumed over fairly long intervals (corresponding to the first and most common of the two categories above), a relatively simple and yet general theory of zone spreading can be given. This theory is based on the treatment of the problem given by Giddings.[82]

It will be recalled from Section 2.3 that the plate height for elution chromatography in a uniform column is $H = L\tau^2/t^2$ where L is the column length, τ the standard deviation in time units, and t the total time. For a nonuniform column this expression is written as an apparent plate height, $\hat{H}$

$$\hat{H} = L\tau^2/t^2 \tag{2.14-1}$$

to determine $\hat{H}$ it is only necessary to discover how the final L, τ, and t values depend on the segment parts. The total column length, obviously, is a sum of the segment lengths

$$L = \sum L_i \tag{2.14-2}$$

and the total residence time of the zone is the sum of the component residence times

$$t = \sum t_i \tag{2.14-3}$$

But the linear addition rule applying to L and t does not apply to τ. The quantity τ is a statistical parameter much like its nonelution counterpart σ. It will be remembered that σ obeyed the addition rule $\sigma^2 = \sigma_1^2 + \sigma_2^2 + \sigma_3^2 + \ldots$, provided that the events leading to one σ_i value did not disturb the other σ_i values. This provision is valid for τ_i values in the present case because the events in each segment are essentially independent insofar as their effects on τ is concerned (this will be fully discussed below). Thus†

$$\tau^2 = \sum \tau_i^2 \tag{2.14-4}$$

Upon substituting the last three equations into (2.14-1) we obtain for $\hat{H}$

$$\hat{H} = (\sum L_i)(\sum \tau_i^2)/(\sum t_i)^2 \tag{2.14-5}$$

The local value of τ_i obtained from segment i can be related to the local

† If the segments are short so that relatively few random steps occur in each, the zone profile originating within any given segment may be non-Gaussian. However as enough such segments are added together the resulting profile, as shown by the central-limit theorem, will approach a Gaussian as a limit. (See ref. 77.)

plate height in this segment by $H_i = L_i\tau_i^2/t_i^2$. If we also use the relationship $t_i = L_i/\mathscr{V}_i$ we can write the foregoing equation in terms of the length L_i of the segments, the zone velocity $\mathscr{V}_i$, and, as we originally hoped, the local plate height values H_i

$$\hat{H} = (\sum L_i)(\sum H_i L_i/\mathscr{V}_i^2)/(\sum L_i/\mathscr{V}_i)^2 \tag{2.14-6}$$

If the variations in column properties occur continuously rather than stepwise, the summations may be replaced by the appropriate integrals

$$\hat{H} = L\int_0^L (H/\mathscr{V}^2)\,dz \bigg/ \left(\int_0^L dz/\mathscr{V}\right)^2 \tag{2.14-7}$$

This expression is particularly interesting when the local H is constant throughout the column. (This situation exists for the A, B/v, and $C_m v$ components of H in gas chromatography when the nonuniformity is caused by the column pressure drop.) It can easily be shown that the expression reduces to the form[82]

$$\hat{H} = H\overline{(1/\mathscr{V})^2} \;/\; \overline{(1/\mathscr{V})}^2 \tag{2.14-8}$$

Consequently $\hat{H}$ is greater than H by the mean square of $1/\mathscr{V}$ divided by the square of its mean. Since this quantity is always equal or greater than unity, $\hat{H} \geq H$. $\hat{H}$ is equal to H only when $\mathscr{V}$ is constant throughout, when the column is uniform. These equations demonstrate the fact that zone spreading in a nonuniform column is greater than might be expected by the local plate height values. A physical interpretation of this phenomenon has been given.†

† The following explanation is taken from reference 82: "In order to magnify the [nonuniformity] effects so that they are most noticeable, let us assume that the . . . velocity at the outlet is nearly infinitely greater than at the inlet. The column can then be divided roughly into two regions (not necessarily equal in length), the first of which involves normal flow rates and the last of which involves a very fast flow. The local plate height, H, is constant throughout. It is clear that nearly all of the residence time is spent in the first region where flow is not abnormally rapid. Since $H = L_i\tau_i^2/t_i^2$ is constant and thus the ration τ_i/t_i is constant for a given segment length, τ will be negligible wherever t is, namely in the final, high-flow region of the column. Thus the following conclusion can be drawn regarding the terms of the apparent plate height, $\hat{H} = L\tau^2/t^2$. The τ^2 and t^2 terms will be contributed entirely by the first region of the column. The length however will be contributed by both regions. Thus $\hat{H}$ will approximately equal $(L_1 + L_2)\tau_1^2/t_1^2 = L_1\tau_1^2/t_1^2 + L_2\tau_1^2/t_1^2$. The first term of the last expression, $L_1\tau_1^2/t_1^2$, is by itself the local plate height, H (assuming near uniformity in each region), and thus $\hat{H} = H + L_2\tau_1^2/t_1^2$. Thus $\hat{H} > H$. This inequality arises, therefore, because the rapid-flow regions of the column are contributing to the length (which gives a proportional increase in $\hat{H}$), but not to the terms in τ^2/t^2, which normally would diminish at a rate sufficient to compensate for the increase of $L(\tau^2/t^2$ is of such a character that additional contributions actually diminish its value since t increases more rapidly than τ). The above reasoning applies also when the [nonuniformity is not so exaggerated], but the arguments are not as clear cut nor the effects as great in magnitude."

The Addition Rule for τ^2 and σ^2

It is an interesting fact that the statistical addition rule is valid for τ, $\tau^2 = \Sigma\tau_i^2$, but is not, under nonuniform circumstances, directly valid for σ, $\sigma^2 \neq \Sigma\sigma_i^2$.[82–84] If a zone encounters a velocity increase or decrease on entering a new segment, it will immediately expand or compress to new dimensions (this width change is independent of the normal zone spreading processes). Thus a certain gain in σ^2 might be magnified or reduced on passing to the next segment, and thereafter changed again on entering each new segment until the column end is reached. The relative change in zone width at each transition is simply in the ratio of the new zone velocity to the old.[82–84] If a group of successive changes occur in passing from segments 1 to n, σ^2 or gain in σ^2 will change as follows:

$$\frac{\sigma^2\,(\text{final})}{\sigma^2\,(\text{original})} = \left(\frac{\mathscr{V}_2}{\mathscr{V}_1}\frac{\mathscr{V}_3}{\mathscr{V}_2}\frac{\mathscr{V}_4}{\mathscr{V}_3}\cdots\frac{\mathscr{V}_n}{\mathscr{V}_{n-1}} = \frac{\mathscr{V}_n}{\mathscr{V}_1}\right)^2 \tag{2.14-9}$$

where $\mathscr{V}$ is the mean zone velocity (not the mobile phase velocity which is v) in the indicated segment. Thus each contribution to the final σ^2 is amplified by the factor $(\mathscr{V}_n/\mathscr{V}_1)^2$. This contribution is independent of the zone spreading which occurs in the intervening segments: each subsequent contribution to σ^2 is also added on as $\sigma_i^2(\mathscr{V}_n/\mathscr{V}_i)^2$. Summing all such contributions, we have the following addition rule

$$\sigma^2 = \sum \sigma_i^2(\mathscr{V}_n/\mathscr{V}_i)^2 \tag{2.14-10}$$

An equation of this form was first used to evaluate gas compression effects in gas chromatography.[82–84] The presence of the velocity ratio complicates this expression, but such a ratio generally needs to be considered since nearly any nonuniformity in column properties, not just changes in fluid velocity, will influence $\mathscr{V}$. Changes in the sorptive power of the stationary phase will, for instance, alter the zone velocity without causing concommitant changes in the mobile phase velocity.

The addition rule $\tau^2 = \Sigma\tau_i^2$ is valid as it stands because a transistion between segments does not enlarge nor decrease τ or τ^2 values. If one molecule falls behind another by a certain time Δt due to the spreading processes in a particular segment, it will remain that much behind except as this interval is influenced by the independent dispersion processes in later segments. Changes in zone velocity will not effect this time lag. If, for instance, the zone passes into a new segment where its velocity is suddenly decreased by a factor of two, the zone will be compressed to half of its original width and the distance between the two molecules in question will also decrease by two. Due to the decrease in velocity, however, it will take just as long for the rear molecule to cover the distance between the two, and the time lag will thus remain the same.

A quantitative formulation of the above effects can be made as follows. Any value of σ, or even increment of σ, is related to the value (or increment) of τ by $\sigma = \mathscr{V}_\tau$. (This expression, shown in Section 2.3, simply expresses the fact that Δ distance = velocity $\times$ Δ time.) The σ^2 generated in segment i, σ_i^2, is therefore given by $\sigma_i^2 = \mathscr{V}_i^2 \cdot \tau_i^2$. The final σ^2 is likewise related to the final τ^2 by $\sigma^2 = \mathscr{V}_n^2\tau^2$. When these expressions are substituted into equation (2.14-10) we obtain

$$\mathscr{V}_n^2\tau^2 = \sum \mathscr{V}_i^2\tau_i^2(\mathscr{V}_n/\mathscr{V}_i)^2 \tag{2.14-11}$$

after canceling $\mathscr{V}_n^2$ and $\mathscr{V}_i^2$ terms, this expression becomes

$$\tau^2 = \sum \tau_i^2 \tag{2.14-12}$$

as assumed.

Whether one regards equation (2.14-10) as an exception to the usual form of the statistical addition rule $\sigma^2 = \Sigma\sigma_i^2$ is a matter of choice, making little practical difference. It is possible to think of the *real* contribution of segment i as being the "amplified" value, $\sigma_i^2(\mathscr{V}_n/\mathscr{V}_i)^2$. In this case the addition rule, involving a sum of such terms as in equation (2.14-10), may be considered valid.

Spreading in External Regions†

Zone spreading will occur in every part of the chromatographic system, from the beginning point to the point of detection. The original zone, as it is placed on or injected into the system, may, perhaps due to its finite size, already have a considerable width. Thus due to one disturbance or another each zone, at the very beginning of its migration, will have a certain profile of measurable width. In paper chromatography, for instance, the application of the solute spot prior to development leads to a small circle or ellipse containing solute. In gas chromatography the finite sample size coupled with the mixing which occurs in the injection port will lead to a considerable zone width before the column is entered. In all elution forms of chromatography some additional zone spreading will occur between the end of the column and the point of detection.

As a practical matter it is always advisable to reduce extra-column contributions to zone spreading as much as possible. Such contributions serve only to destroy resolution. Thus it is advisable to reduce to a minimum

† Since completion of this section a comprehensive manuscript[85] on extra-column effects has been made available to the author. See also ref. 86.

the dimensions of the original sample. The volume of the injection port preceding the column (if any exists) and the volume of the detector plus the connecting tubes leading to it should in all cases be reduced to an absolute minimum.

To a first approximation these effects can be assessed by evaluating their contributions τ_e^2 to the total variance in elution time, and adding these to the column contribution τ_c^2. The final value

$$\tau^2 = \tau_e^2 + \tau_c^2 \tag{2.14-13}$$

is, of course, larger than τ_c^2 alone. Gaussian zones result whenever the external source of spreading leads to Gaussian profiles, or whenever $\tau_e^2 \ll \tau_c^2$, whether the external contribution is Gaussian or not (the latter conclusion is a result of the central-limit theorem where small component displacements, Gaussian or otherwise, lead to a Gaussian end product subject to the usual addition law for τ^2).† When neither of these conditions holds—when extra-column spreading is severe and also badly distorted from the usual Gaussian—a special method must be used to compute the final profile. The addition of two or more arbitrary probability (concentration) distributions has long been practiced in mathematical statistics. The method described below is an extension of that proposed for elution chromatography by Giddings and Eyring[15] in 1955. McQuarrie has recently dealt with this subject in considerable detail.[71] See also Said[34] who approached the problem from the viewpoint of the theoretical plate model, and Johnson and Stross[87] who used a differential material balance equation. A powerful Laplace transform method has been described by Reilley, Hildebrand and Ashley.[88] (See also Sternberg.[85])

If the column were removed from a chromatographic system and the solute thus allowed to pass directly to the detector, a certain concentration profile would be observed. This may be written as a probability profile (which differs from the former only by having unit area) $P_e(t)$. Similarly, a profile $P_c(t)$ may be written for the column by itself (without accessory apparatus). This will usually be Gaussian, but may take on other forms as discussed in the last section. Together these two functions give a zone with the following profile

$$P(t) = \int_{-\infty}^{\infty} P_c(t - t')P_e(t')\,dt' \tag{2.14-14}$$

This equation is obtained by considering the contribution of each infinitesimal part of the P_e profile to the solute eluted in unit time at time t.

† Sternberg[85] shows that the summation rule for τ^2 is valid whether the zones are Gaussian or not, where τ is considered simply as the second moment of the profile around the center.

The equation is, of course, based on the assumption that the external and column spreading processes are independent.

Some examples of the application of the above equation will be useful here. Suppose that the sample is injected at time zero. A finite time t_e is required to pass through the external parts, but the spreading there is assumed negligible. A subsequent time of t_c is needed to get the center of the zone, which now forms a Gaussian, through the column. The two component profiles are thus

$$P_e(t) = \delta(t - t_e) \tag{2.14-15}$$

$$P_c(t) = \frac{1}{\sqrt{2\pi}\tau_c} \exp\left[\frac{(t - t_c)^2}{2\tau_c^2}\right] \tag{2.14-16}$$

where $\delta(t - t_e)$ is the Dirac δ-function. When these two expression are used in equation (2.14-14), the following Gaussian profile emerges

$$P(t) = \frac{1}{\sqrt{2\pi}\tau_c} \exp\left[-\frac{(t - t_e - t_c)^2}{2\tau_c^2}\right] \tag{2.14-17}$$

The zone width parameter τ_c is identical to the column contribution, but the time when the peak of the zone emerges, $t = t_e + t_c$, is a time t_e greater than the time taken by the column itself.

More commonly the external sources will contribute to zone spreading as well as zone retention. In some cases it is possible to write both probability functions, $P_c(t)$ and $P_e(t)$, as Gaussians. Thus writing $P_c(t)$ as in equation (2.14-16), and writing $P_e(t)$ in the same form but replacing τ_c by τ_e and t_c by t_e, we obtain the following integrated form from equation (2.14-14)

$$P(t) = \frac{1}{\sqrt{2\pi(\tau_e^2 + \tau_c^2)}} \exp\left[-\frac{(t - t_e - t_c)^2}{2(\tau_e^2 + \tau_c^2)}\right] \tag{2.14-18}$$

This is a Gaussian with a variance of $\tau_e^2 + \tau_c^2$ and retention time, as above, of $t_e + t_c$. This equation is a proof, for this special case, of the summation rule for variances, $\tau^2 = \Sigma\tau_i^2$.

If the volume of the injection port or detector is large so that a complete mixing of the contained sample occurs, the profile due to the port or detector will be exponential

$$P_e(t) = (\dot{V}/V_0) \exp(-\dot{V}t/V_0) \tag{2.14-19}$$

where V_0 is the port or detector volume and $\dot{V}$ the mobile phase flow rate in cm³/sec. The ratio $V_0/\dot{V}$ would simply be the time needed to empty the contained volume if replacement of the mobile fluid did not occur from upstream. The exponential distribution given above has been treated at some length by Johnson and Stross for gas chromatographic detectors.[88] Numerical plots of the final zone profile are shown, and it is clearly demonstrated that the width and the location of the maximum are distorted by detectors of large volume. The "complete mixing" hypothesis is, of course, idealization, but a rough idea of the damage done by large irregular volumes is well worth noting. Only if external volumes, at least those of considerable size, are well designed (without dead pockets) and thus lead to narrow peaks will the effects be tolerable. Some additional aspects of this problem have been treated elsewhere.[85,86,89]

Frontal Analysis

The law combining independent probability profiles, equation (2.14-14), also applies to the finite dimensions of the injected sample. In most cases the injected sample ideally approaches a δ-function and does not contribute appreciably to the zone width. Other initial profiles, however, are possible and, as pointed out by Reilley, Hildebrand, and Ashley,[88] may have useful characteristics. A full discussion of these is found in their paper. The most common initial profile not resembling a δ-function is that pertaining to frontal analysis (see Section 1.5). In this case the incoming concentration suddenly increases from zero to some constant level c_0, at time zero. The incoming profile is given mathematically by the following, where the effects of injector and detector volumes on $P_e(t)$ have been neglected.

$$P_e(t) = 0, \qquad t < 0 \tag{2.14-20}$$

$$P_e(t) = c_0, \qquad t > 0 \tag{2.14-21}$$

Since such a function is difficult to normalize to unit area we have converted $P_e(t)$ directly to a concentration profile by using c_0. When substituted into equation (2.14-14), the result is the observed concentration profile, $P(t) = c(t)$

$$c(t) = c_0 \int_0^\infty P_c(t - t')\, dt' \tag{2.14-22}$$

If the column contribution $P_c(t)$ is a Gaussian as in equation (2.14-16), the concentration profile becomes

$$c(t) = \frac{c_0}{\sqrt{2\pi}\tau_c} \int_0^\infty \exp\left[-\frac{(t - t' - t_c)^2}{2\tau_c^{\,2}}\right] dt' \tag{2.14-23}$$

Changing to the new variable $T = (t - t' - t_c)/\sqrt{2}\tau_c$ and rearranging, we find

$$c(t) = \frac{c_0}{2}\left[1 + \operatorname{erf}\left(\frac{t - t_c}{\sqrt{2}\tau_c}\right)\right] \tag{2.14-24}$$

where values of the error function (erf)

$$\operatorname{erf} x = (2/\sqrt{\pi})\int_0^x \exp(-T^2)\, dT \tag{2.14-25}$$

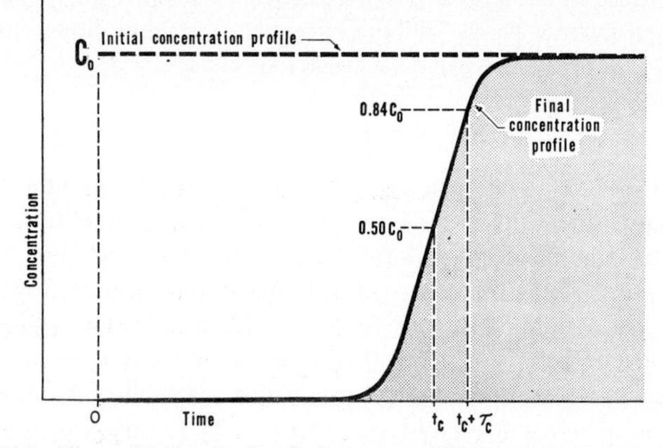

Figure 2.14-1 The solid line is the final zone profile for frontal analysis using an elution technique. The dashed line is the initial or injected concentration profile. The former differs from the latter, due to the intervening column, by having a less abrupt concentration rise and by having this rise appear a time t_c (the column residence time) later.

are tabulated in many statistics books.[90,91] The concentration profile $c(t)$ rises from zero to c_0 as t arrives at and then passes the column elution time t_c. It has, in fact, risen to half the final concentration c_0 when the elapsed time is precisely t_c (it should again be emphasized that external volume and retention, which would serve to increase the effective t_c, are assumed negligible). The rise of $c(t)$, as shown in Figure 2.14-1, is less abrupt than the sharp increase in the original concentration to c_0. This is a result of the zone spreading processes in the column as manifested in the parameter τ_c. The larger τ_c, the less abrupt is the increase; the time between t_c and $t_c + \tau_c$ always leads to an increase in concentration from $0.5c_0$ to $0.84c_0$. This result is demonstrated in Figure 2.14-1.

Finite Sample Width

The zone may arrive on the column with an already considerable width. This may be due to the finite volume it occupies or to additional diffusion processes that occur in the interval before column migration starts. The theoretical approach to this problem has already been outlined, but considerable interest has been shown for years in the maximum number of "theoretical plates" the zone can occupy without causing undue spreading. This special problem, then, will be briefly treated here. If the initial zone is of such a width as to contribute τ_e^2 to the variance (standard deviation squared) in elution time, and the column contributes τ_c^2, then the total is the sum $\tau^2 = \tau_e^2 + \tau_c^2$, as shown in equation (2.14-13). We obviously wish to make τ_e^2 as small as possible to reduce the overall zone width and so the ratio

$$\chi = \tau_e^2/2\tau_c^2 \tag{2.14-26}$$

should also be small. In terms of this ratio the sum composing τ^2 becomes

$$\tau^2 = \tau_c^2(1 + 2\chi) \tag{2.14-27}$$

and

$$\tau = \tau_c(1 + 2\chi)^{1/2} \cong \tau_c(1 + \chi) \tag{2.14-28}$$

the latter relationship holding approximately because χ is small. Thus the zone width, proportional to τ (usually considered as 4τ for Gaussian zones in time units), will be increased by the fraction χ beyond the minimum of τ_c. An upper limit of desirability, say 0.1 (corresponding to a zone 10% wider than its attainable minimum), will usually be imposed on χ. To express this limiting ratio in terms of observables, its τ_c^2 component will be written as

$$\tau_c^2 = t^2H/L \tag{2.14-29}$$

obtained directly from $H = L\tau^2/t^2$. Its τ_e^2 component is given by

$$\tau_e^2 = (\sigma_e/Rv)^2 = (N_eH/4)^2(t/L)^2 \tag{2.14-30}$$

obtained by writing the "width" of the zone, $4\sigma_e$, as the number N_e of plates initially occupied times their length, the plate height H, and by writing the zone velocity Rv as the column length L over the elution time t (we shall hereafter neglect external retention, unless specifically mentioned, and thus write the column retention time t_c simply as t). Using the last two expressions to form the ratio indicated in equation (2.14-26), we have

$$\chi = N_e^2/32N \tag{2.14-31}$$

where the number N of plates in the column has been substituted for L/H. We find then that

$$N_e/N = (32\chi/N)^{1/2} \tag{2.14-32}$$

This is the fraction of total plates in the column that the original sample may permissibly occupy, or alternatively, it is the volume occupied by the original sample divided by the total retention volume.† If χ were limited to 0.1, this fraction would be given by

$$N_e/N = 1.8/\sqrt{N} \tag{2.14-33}$$

If it were felt necessary to keep χ as small as 0.01, the constant 0.6 would replace 1.8. This corresponds roughly to the criterion evolved by van Deemter et al.[14]

We see from equation (2.14-33) that a column of 1000 plates can tolerate a sample spread initially over about 60 plates, and that a column of 10,000 plates will work acceptably with an initial sample occupying 180 plates. These are maximum values which ordinarily can and should be reduced to some degree.

The assumption made above that the zone width is 4 standard deviations, $4\tau_c$, is applicable to Gaussian profiles but not necessarily to other initial distributions of sample. If, for instance, the sample is introduced as a square plug, the plug width will be $\sqrt{12} \simeq 3.5$ standard deviations. The number 32 in equation (2.14-32) will become 24, a change which is not sufficient to affect significantly the above arguments.

Nonelution Chromatography

The concepts developed for elution chromatography can in most cases be applied to determining zone structure in nonelution chromatography. Thus the zone quarter-width σ may be taken as the velocity of the zone center $\mathscr{V}$ times τ, except in the rare case where the nonuniformity is so great that large fractional changes occur over the relatively short dimensions of the zone. Also τ must be calculated such that it pertains to the actual zone location; if the zone is only part way down the column, τ must be calculated as if the column ended and elution occurred at that point. This matter presents no obstacle to the calculations, but it does merit caution.

In discussing elution chromatography, the apparent (or experimental) plate height was defined as

$$\hat{H} = L\tau^2/t^2 \tag{2.14-34}$$

an expression which also equals the point-to-point (or local) plate height in uniform columns. It might seem logical to define an apparent plate height for nonelution chromatography as σ^2/L. This equals the above expression in uniform columns and also equals the point-to-point plate height in uniform columns. Pursuing the matter, however, we find that

† This ratio is also the permissible time interval for getting a sample completely into the fluid stream (injecting) divided by the retention time. The problem of finite injection times in gas chromatography has been treated in a manner similar to this by Guiochon.[92]

σ^2/L gives an expression not at all equivalent to equation (2.14-34) (see below derivation for equation (2.14-35)), and presents an entirely unrealistic picture of the degree of separation occurring in a nonelution process with serious column nonuniformites. Since zone spreading and plate height are primarily of interest as a measure of a column's ability to resolve components, σ^2/L is not a useful parameter here. This situation is essentially reversed with programmed temperature gas chromatography. This point will be more apparent in the discussions on optimum separations in gas and liquid chromatography. But here it is sufficient to deal solely with the quarter-width σ, and if a plate height expression is needed equation (2.14-34) can be used.

The expression σ^2/L may be written as $\mathscr{V}_n^2\tau^2/L$ where $\mathscr{V}_n$ is the zone velocity in the final (or nth) column segment. This turn may be written as $\bar{\mathscr{V}}^2\tau^2/L(\mathscr{V}_n/\bar{\mathscr{V}})^2$ where $\bar{\mathscr{V}}$ is the mean zone velocity (averaged over time). Since $\bar{\mathscr{V}}$ is simply L/t, this expression may be written as

$$\frac{\sigma^2}{L} = L\frac{\tau^2}{t^2}\frac{\mathscr{V}_n^{\ 2}}{\bar{\mathscr{V}}} = H\frac{\mathscr{V}_n^{\ 2}}{\bar{\mathscr{V}}} \tag{2.14-35}$$

a term which is $(\mathscr{V}_n/\bar{\mathscr{V}})^2$ larger than $\hat{H}$. For severe nonuniformities, then, σ^2/L may not even approach H. If the final zone velocity, $\mathscr{V}_n$, is very large, σ^2/L and thus the zone width may be very large. This does not necessarily detract from separability, since the high velocity may well displace one zone completely out of the final segment before another enters.

2.15 References

1. R. V. Blundell, S. T. Griffiths, and R. R. Wilson in *Gas Chromatography, 1960*, R. P. W. Scott, ed., Butterworths, Washington, 1960. p. 360.
2. F. Helfferich in *Advances in Chromatography*, Vol. **1**, J. C. Giddings and R. A. Keller, ed., Marcel Dekker, New York, 1965. Chap. 1.
3. J. N. Wilson, *J. Am. Chem. Soc.*, **62,** 1583 (1940).
4. A. J. P. Martin and R. L. M. Synge, *Biochem. J.*, **35,** 1358 (1941).
5. D. DeVault, *J. Am. Chem. Soc.*, **65,** 532 (1943).
6. J. Weiss, J. Chem. Soc., **1943,** 297.
7. J. E. Walter, *J. Chem. Phys.*, **13,** 332 (1945).
8. L. G. Sillen, *Arkiv. Kemi. Mineral. Geol.*, **A22,** No. 15 (1946); *Nature,* **166,** 722 (1950).
9. H. C. Thomas, *J. Am. Chem. Soc.* **66,** 1664 (1944); *Ann. N.Y. Acad. Sci.*, **49,** 161 (1948).
10. G. E. Boyd, L. S. Myers, Jr., and A. W. Adamson, *J. Am. Chem. Soc.*, **69,** 2849 (1947).
11. R. H. Beaton and C. C. Furnas, *Ind. Eng. Chem.*, **33,** 1501 (1941).

12. G. E. Boyd, A. W. Adamson, and L. S. Myers, Jr., *J. Am. Chem. Soc.*, **69,** 2836 (1947).
13. L. Lapidus and N. R. Amundson, *J. Phys. Chem.*, **56,** 984 (1952).
14. J. J. van Deemter, F. J. Zuiderweg, and A. Klinkenberg, *Chem. Eng. Sci.*, **5,** 271 (1956).
15. J. C. Giddings and H. Eyring, *J. Phys. Chem.*, **59,** 416 (1955).
16. E. Glueckauf in *Ion Exchange and its Applications*, Metcalfe and Cooper, Lmd., London, 1955. p. 34.
17. J. C. Giddings, *J. Chem. Ed.*, **35,** 588 (1958).
18. E. Glueckauf, *Trans. Faraday Soc.*, **51,** 34, 1540, (1955).
19. K. J. Mysels, *J. Chem. Phys.*, **24,** 371 (1956).
20. J. C. Giddings, *J. Chem. Phys.*, **26,** 1755 (1957).
21. M. J. E. Golay in *Gas Chromatography* 1958, D. H. Desty, ed., Academic Press, New York, 1958, p. 36.
22. J. C. Giddings, *J. Chem. Phys.*, **31,** 1462 (1959).
23. J. C. Giddings, *J. Chromatog.*, **3,** 443, 520 (1960); *Nature*, **188,** 847 (1960); *J. Chromatog.*, **5,** 46 (1961); *Anal. Chem.*, **33,** 962 (1961); **34,** 458, 722, 1186 (1962); **35,** 439 (1963); *J. Gas Chromatog.*, **1,** No. 4, 38 (1963); *J. Phys. Chem.*, **68,** 184 (1964).
24. J. C. Giddings, *Anal. Chem.*, **34,** 1026 (1962).
25. J. H. Knox and L. McLaren, *Anal. Chem.*, **36,** 1477 (1964).
26. B. Drake, *Anal. Chem. Acta.*, 3, 452 (1949).
27. J. C. Giddings, *Nature*, **184,** 357 (1959); *Anal. Chem.*, **34,** 885 (1962); **35,** 1338 (1963).
28. M. J. E. Golay in *Gas Chromatography*, H. J. Noebels, R. F. Wall and, N. Brenner, ed., Academic Press, New York, 1961, Chap II.
29. F. H. Huyten, W. Beersum, and G. W. A. Rijnders in *Gas Chromatography, 1960*, R. P. W. Scott, ed., Butterworths, Washington, 1958. p. 224.
30. J. C. Giddings, *J. Gas Chromatog.*, **1,** No. 4, 38 (1963).
31. D. Ambrose and B. A. Ambrose, *Gas Chromatography*, Van Nostrand, Princeton, 1962, p. 7–10.
32. E. Glueckauf, *Trans. Faraday Soc.*, **51,** 34 (1955).
33. S. W. Mayer and E. R. Tompkins, *J. Am. Chem. Soc.*, **69,** 2866 (1947).
34. A. S. Said, *Am. Inst. Chem. Engrs. J.*, **2,** 477 (1956).
35. H. Kramers and G. Alberda, *Chem. Eng. Sci.*, **2,** 173 (1953).
36. R. Aris and N. R. Amundson, *Am. Inst. Chem. Engrs. J.*, **3,** 280 (1957).
37. J. J. Carberry, *Am. Inst. Chem. Engrs. J.*, **4,** 13M (1958).
38. J. M. Prausnitz, *Am. Inst. Chem. Engrs. J.*, **4,** 14M (1958).
39. J. C. Giddings, *J. Chromatog.*, **2,** 44 (1959).
40. S. Chandrasekhar, *Revs. Mod. Phys.*, **15,** 1 (1943).
41. H. Eyring, K. J. Laidler, and S. Glasstone, *Theory of Rate Processes*, McGraw-Hill, New York, 1941.
42. A. Klinkenberg and F. Sjenitzer, *Chem. Eng. Sci.*, **5,** 258 (1956).
43. J. C. Giddings in *Chromatography*, E. Heftmann, ed., Reinhold, New York, 1961, Chap. 3.
44. W. Feller, *Probability Theory and its Applications*, Wiley, New York, 1950; Chap. 14.
45. S. Chandrasekhar, *Revs. Mod. Phys.*, **15,** 1 (1943).
46. A. Einstein, *Ann. der Physik*, **17,** 549 (1905).
47. T. Ellerington in *Gas Chromatography, 1958*, D. H. Desty, ed., Butterworths, London, 1958. See also page XI.

48. W. L. Jones and R. Kieselbach, *Anal. Chem.*, **30,** 1590 (1958).
49. J. C. Giddings, *Anal. Chem.*, **35,** 439 (1963); **33,** 962 (1961); *J. Phys. Chem.*, **68,** 184 (1964).
50. J. C. Giddings, *Anal. Chem.*, **34,** 1186 (1962).
51. R. H. Perrett and J. H. Purnell, *Anal. Chem.*, **35,** 430 (1963).
52. J. C. Giddings, *Anal. Chem.*, **35,** 1338 (1963); *Nature*, **184,** 357 (1959).
53. J. C. Giddings and E. N. Fuller, *J. Chromatog.*, **7,** 255 (1962).
54. J. C. Giddings and G. E. Jensen, *J. Gas Chromatog.*, **2,** 290 (1964).
55. M. J. E. Golay in *Gas Chromatography*, H. J. Noebels, R. F. Wall, and N. Brenner, ed., Academic Press, New York, (1961), Chap. 2.
56. J. C. Giddings, *J. Chromatog.*, **3,** 520 (1960); *J. Gas Chromatog.*, **1,** No. 4, 38 (1963).
57. S. Dal Nogare and J. Chiu, *Anal. Chem.*, **34,** 890 (1962).
58. J. C. Giddings, *Nature*, **184,** 357 (1959); **187,** 1023 (1960).
59. A. Klinkenberg and F. Sjenitzer, *Chem. Eng. Sci.*, **5,** 258 (1956).
60. J. C. Giddings and R. A. Robison, *Anal. Chem.*, **34,** 885 (1962).
61. A. L. Pozzi and R. J. Blackwell, 37th Meeting of the Society of Petroleum Engineers, Los Angeles, October 8–10, 1962.
62. J. C. Giddings, *Anal. Chem.*, **34,** 1186 (1962); *Anal. Chem.*, **35,** 1338 (1963).
63. J. C. Giddings, *Nature*, **184,** 357 (1959).
64. J. C. Giddings, *Anal. Chem.*, **35,** 1338 (1963).
65. J. C. Giddings, *J. Chromatog.*, **13,** 301 (1964).
66. J. C. Sternberg and R. E. Poulson, *Anal. Chem.*, **36,** 1492 (1964).
67. J. H. Knox and L. McLaren, *Anal. Chem.*, **35,** 449 (1963).
68. W. L. Jones, *Anal. Chem.*, **33,** 829 (1961).
69. J. C. Giddings, *J. Chem. Phys.*, **26,** 169 (1957).
70. J. H. Beynon, S. Clough, D. A. Crooks, and G. R. Lester, *Trans. Faraday Soc.*, **54,** 705 (1958).
71. D. A. McQuarrie, *J. Chem. Phys.*, **38,** 437 (1963).
72. R. A. Keller and J. C. Giddings, *J. Chromatog.*, **3,** 205 (1960).
73 H. C. Thomas, *Ann. N.Y. Acad. Sci.*, **49,** 161 (1948).
74. W. Feller, *Probability Theory and its Applications*, Wiley, New York, 1950, pp. 155 ff.
75. G. N. Watson "Theory of Bessel Functions," Cambridge Univ. Press, Cambridge, 1944.
76. W. Feller, *Probability Theory and its Applications*, Wiley, New York, 1950, p. 133.
77. C. Eisenhart and M. Zelen in *Handbook of Physics*, E. U. Condon and H. Odishaw, ed., McGraw-Hill, New York, 1958.
78. R. N. Sen Sarma, E. Anders, and J. M. Miller, *J. Phys. Chem.*, **63,** 559 (1959).
79. R. P. W. Scott, *Anal. Chem.*, **35,** 481 (1963).
80. C. H. Bosanquet and G. O. Morgan in *Vapor Phase Chromatography*, ed. D. H. Desty, Academic Press, New York, 1957, p. 35; also C. H. Bosanquet in *Gas Chromatography*, 1958, ed. D. H. Desty, Academic Press, New York, 1958, p. 107.
81. J. C. Giddings, *Anal. Chem.*, **35,** 1999 (1963).
82. J. C. Giddings, *Anal. Chem.*, **35,** 353 (1963).
83. G. H. Stewart, S. L. Seager, and J. C. Giddings, *Anal. Chem.*, **31,** 1738 (1959).
84. J. C. Giddings, S. L. Seager, L. R. Stucki, and G. H. Stewart, *Anal. Chem.*, **32,** 867 (1960).
85. J. C. Sternberg in *Advances in Chromatography*, Vol. **2,** J. C. Giddings and R. A. Keller, ed., Marcel Dekker, New York, 1966, Chap. 6.

86. G. Guiochon, *J. Gas Chromatog.*, **2,** 139 (1964).
87. H. W. Johnson, Jr., and F. H. Stross, *Anal. Chem.*, **31,** 357 (1959).
88. C. N. Reilley, G. P. Hildebrand, and J. W. Ashley, Jr., *Anal. Chem.*, **34,** 1198 (1962).
89. J. C. Giddings, *J. Gas Chromatography*, **1,** 1, 12 (1963).
90. H. C. Plummer, *Probability and Frequency*, Macmillan, London, 1940, p. 134 (definition) and pp. 270–274 (Table).
91. B. O. Peirce, *A Short Table of Integrals*, Ginn, Boston, 1929, pp. 116–120.
92. G. Guiochon, *Anal. Chem.*, **35,** 399 (1963).

Chapter Three

Nonequilibrium and the Simple Mass Transfer Terms

3.1 The Nature and Effect of Nonequilibrium

A chromatographic zone may be thought of as a moving concentration pulse. As the pulse moves through a given region of the column, the concentration within that region first increases to a maximum and then decreases back to zero. The initial increase in concentration, occurring at the leading edge of the zone, arises because the solute is carried in by the mobile fluid from the concentration-rich regions upstream. The increase in concentration within the mobile fluid does not instantaneously affect the solute concentration in the stationary phase. Instead, the sorption-desorption processes require a finite amount of time to bring the stationary phase concentration into equilibrium with the initial concentration jump. During this initial equilibrium period, however, additional concentration gains are made by the moving fluid, and still further equilibration must occur. As long as the concentration within the mobile fluid continues to increase (and it does so until the center of the zone reaches the given region), complete equilibrium remains just out of reach, with the stationary phase concentration lagging slightly behind the mobile phase concentration.[1,2]

These processes are exactly reversed after the center of the zone has passed through the region under consideration. Each new influx of the mobile fluid brings with it a more dilute solution from the concentration-starved regions of the tail of the zone. The stationary phase must divest itself of solute to reach equilibrium at this lower concentration. The finite time required for this equilibration, causes, once again, a concentration lag in the stationary phase as compared to the mobile phase. In this case, however, the stationary phase concentration is higher than that in the mobile phase because its lag is relative to a decreasing concentration. The stationary phase concentration is thus slightly higher than equilibrium, and the diluted mobile phase is beneath its equilibrium value.

A schematic diagram of equilibrium lag in stationary-phase concentration is shown at the top of Figure 3.1-1. The actual concentration profile for the stationary phase trails behind its equilibrium value. As noted earlier, these profiles show that there is a concentration deficiency (with respect to equilibrium) in the leading part of the zone and a concentration excess to the rear. The mobile phase concentration, shown at the bottom of Figure 3.1-1, is ahead of its equilibrium concentration.

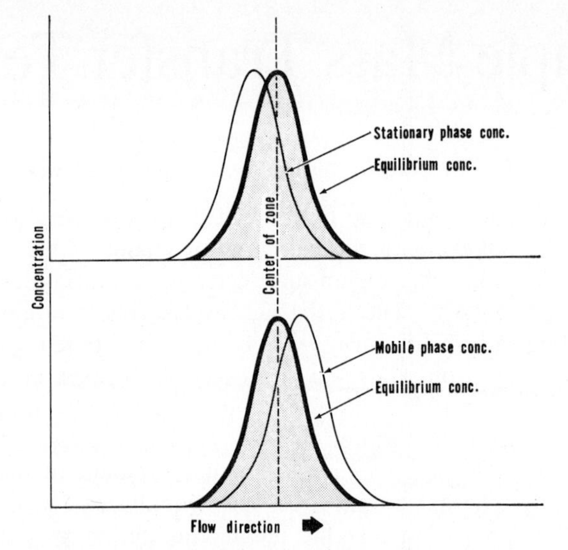

Figure 3.1-1 The upper diagram shows that the actual concentration profile for the stationary phase lags behind that which would exist at equilibrium (i.e., with an infinite exchange rate between phases). The lower diagram shows that the mobile phase concentration is displaced ahead of the equilibrium profile.

(The two equilibrium curves, shown by the heavy lines, superimpose on one another except perhaps for a scale factor, and represent the overall concentration profile which would be obtained experimentally if the zone were dissected.) The basic reason why the mobile phase concentration leads its equilibrium value is that the motion of the zone derives from the transport of solute in this phase. Hence each concentration value existing in this phase is constantly being transported ahead at a greater velocity than that acquired by the overall zone of solute. The diagram of mobile phase concentration illustrates the qualitative arguments presented earlier; the actual concentration exceeds equilibrium at the front of the zone and is deficient at the rear.

The degree of nonequilibrium, indicated by the gap between the related curves of Figure 3.1-1 will be slight if the rate of sorptive-desorptive exchange (or mass transfer) is large or if the zone migrates through slowly thus preventing rapid concentration changes. This is very important, for the degree of nonequilibrium largely determines the extent of zone spreading.

Nonequilibrium and Zone Spreading

It should be recalled from Chapter 1 that a zone's downstream migration is caused entirely by that portion of solute molecules in the moving fluid. The rate of migration, in fact, is directly proportional to the fraction of molecules in the mobile phase at a given point. The fraction of molecules in the mobile phase at equilibrium is, of course, R. Ahead of the zone center, as shown in the lower half of Figure 3.1-1, the fraction of molecules in the mobile phase is greater than the equilibrium value R. Hence the migration velocity of this part of the zone is greater than the normal velocity Rv. Behind the zone center the fraction of molecules in the mobile phase is below the equilibrium fraction R, and the zone velocity is thus less than the overall zone velocity Rv. Thus a situation exists in which the front of the zone is advancing more rapidly than the back. The zone is thus being widened by this velocity divergence found within its boundaries. The rate of widening increases with the degree of nonequilibrium, and this depends on the rate of sorption and desorption and on the velocity of migration. Chromatography has always been regarded as a procedure requiring a large area of contact between the two phases. This is explained simply by noting that sorption and desorption proceed more rapidly, and nonequilibrium is thus less severe, as the contact area increases. It has also been recognized from the time of Wilson's paper[3] that large flow velocities destroy chromatographic resolution. As seen above, this is a consequence of the increased nonequilibrium which accompanies high velocities.

The generalized nonequilibrium theory of chromatography[4-7] (Chapter 4) and its simplified precursors (this chapter) are based on precisely the same picture of nonequilibrium as just given. This qualitative view is given quantitative meaning by first calculating the degree of nonequilibrium in terms of the sorption-desorption rates and the concentration changes accompanying the moving gradients of the zone. Next, and still very much in accord with the qualitative picture, the velocity increase of the zone ahead of center, caused by the nonequilibrium, is written in terms of an excess forward flux (transport) of solute through a typical cross section. To the rear of center the forward flux is calculated to be deficient in amount. The theory shows that the excess (or deficiency) of solute flux is proportional to the negative concentration gradient. Hence solute is being transported away from the zone center in exact accordance with Fick's first law of diffusion (Chapter 6). The process is thus analogous to diffusion, and an effective diffusion coefficient can be used to describe the rate of zone spreading. The effective diffusion coefficient, of course, is related to the plate height H, as indicated in Section 2.3, and thus H can be calculated directly.

The nonequilibrium approach is based on entirely different considerations than the random walk approach. The single molecular events of the random walk description (and the stochastic theory) have been replaced by the gross processes of mass transfer in the nonequilibrium theory. The difference in approach was discussed thoroughly in Section 2.4.

Although the nonequilibrium approach is more powerful and precise than the random walk theory, it is used only for the calculation of the C (or rate) terms in the plate height expression (see Section 2.11). These terms are all related to the transfer of solute between regions having different velocities, and so they are often called the *mass transfer terms* (or, of course, nonequilibrium terms). Although these terms constitute only part of the total plate height expression, it is by far the most significant part from the point of view of practical chromatographic operation and from the standpoint of theoretical interest.

In analogy with most theories in the physical sciences, the generalized nonequilibrium theory was built on a foundation established earlier by many workers in several fields. The direct development of this theory can be traced back[2,4,8,9] to the mathematical treatment of another diffusion-reaction-flow system—flames. The efforts of Hirschfelder and co-workers are particularly important here.[10] The generalized nonequilibrium theory is also similar in some ways to the early treatment of steady countercurrent electrophoresis by Westhaver.[11]

Numerous theories have been developed by other means which rigorously account for some of the less complicated flow and kinetic features of chromatography and related systems. The most notable case is that of parabolic flow and simultaneous diffusion in an open tube as developed by Taylor.[12] This has been extended to more complex geometries by Aris.[13] Golay,[14] using an approach based on communication engineering, treated the open-tube case in which a uniform retentive layer was assumed to exist on the wall. The presence of retention made this, by contrast to the cases treated by Taylor and Aris, a true theory of chromatography. The Golay case was extended by Khan[15] to include an interfacial resistance term originating between the mobile and stationary phases.

Along somewhat different lines Mysels and Scholten,[16,17] as well as Bak and Kauman[18] and other workers, have treated single-step kinetic phenomena by different approaches. These authors have been particularly interested in migration within an electric field, and they have extended the theory to include rapidly alternating current (this is equivalent to considering an alternating back and forth flow in chromatography). In addition to these various theories the stochastic treatment of Section 2.13 has been developed by a number of authors. All of the foregoing theories have been or can be written in terms of effective diffusion or plate height.

This brief summary indicates that the underlying basis of chromatography is closely related to that of many other subjects. The transfer of information between fields, however, has been all too slow.

3.2 Derivation of Plate Height for Simple One-Site Adsorption Kinetics

In Chapter 4 we shall be compelled to demonstrate the broad scope and general utility of the nonequilibrium approach. Initial acquaintance with nonequilibrium theory, however, is best made with a few specific cases in mind, preferably those containing all possible elements of simplicity. The ones chosen here and in Section 3.3 will be of this type. The critical assumptions of the theory will not be belabored in these sections so as to avoid tedious discussions. The main object of these sections is to acquaint the general reader with the approach and the physical insight connected with this powerful theoretical tool. In Chapter 4 we will treat the problem more extensively and rigorously, and in Section 4.8 the practical applications of the theory will be summarized.

The simplest kinetic idealization of adsorption chromatography is that in which adsorption occurs on a uniform surface containing equivalent sites. Such a surface, illustrated in Figure 3.2-1, will be called a 1-site surface to distinguish it from the more complex (and more realistic) multisite surfaces to be considered later.

The simplest idealization of partition chromatography is that in which the stationary phase exists as a thin uniform film. This will be treated in the next section.

Nonequilibrium Parameters

The nonequilibrium theory is based on solute concentration and changes in concentration which result from the flow and kinetic processes of chromatography.[4,6] Thus it is well to recall the basic definition of the concentration term c. This quantity is equal to the amount of solute per unit volume of column packing, and is thus the concentration unit most accessible to experimental measurement. (This should not be confused with the concentration term c', introduced later, which designates the amount of material per unit volume of a particular phase rather than per unit of overall volume.) The concentration c may be broken into two contribution parts, c_m and c_s, corresponding to the amount of solute per unit volume of column packing which resides in the mobile and the stationary phases, respectively. When full equilibrium is attained, these terms will be designated by $c_m{}^*$ and $c_s{}^*$. It was indicated earlier that solute in a chromatographic zone was slightly out of equilibrium due to the moving concentration gradients and the inability of equilibration to keep up with this. Our main concern is with the extent of departure from

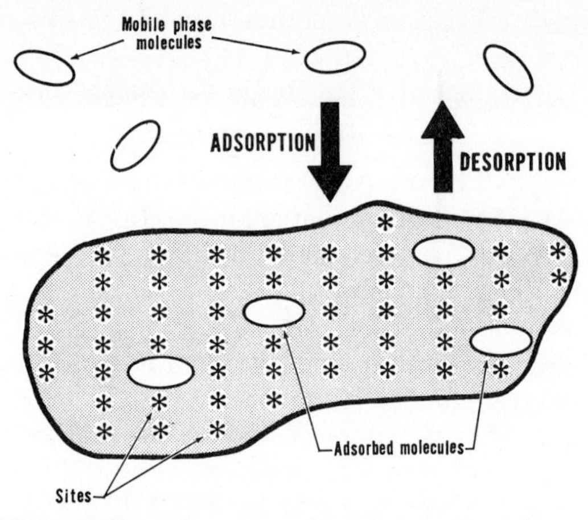

Figure 3.2-1 Schematic illustration of a patch of surface containing equivalent adsorption sites. A dynamic exchange between the free and adsorbed state occurs with the apparent first order rate constants k_a and k_d.

equilibrium since this departure was found to be directly responsible for zone spreading. Thus we introduce the *equilibrium departure terms* ϵ_m and ϵ_s by means of the following equations[4,8,9]

$$c_m = c_m{}^*(1 + \epsilon_m) \tag{3.2-1}$$

$$c_s = c_s{}^*(1 + \epsilon_s) \tag{3.2-2}$$

Rearranging, we have

$$\epsilon_m = \frac{c_m - c_m{}^*}{c_m{}^*} = \frac{\Delta c_m}{c_m{}^*} \tag{3.2-3}$$

$$\epsilon_s = \frac{c_s - c_s{}^*}{c_s{}^*} = \frac{\Delta c_s}{c_s{}^*} \tag{3.2-4}$$

Thus we see that each ϵ term is the fractional departure from equilibrium. For instance, if ϵ_m is 0.01, this indicates that the departure of c_m from equilibrium, $\Delta c_m = c_m - c_m{}^*$, is one one-hundredth of the equilibrium value itself; that is, c_m is 1% larger than its equilibrium value $c_m{}^*$. Negative values may be similarly interpreted.

The nonequilibrium theory will be framed primarily in terms of the ϵ and c^* values. These directly reflect the status of the concentration components and the former indicates at a glance the state of nonequilibrium within these components.

The nature of nonequilibrium is such that the two departure terms ϵ_m and ϵ_s are not independent of one another. It is clear that excess solute in one phase, over and above equilibrium, is accompanied by a deficit in the other phase; that is, there is a balance of nonequilibrium. The quantitative argument runs as follows. First, the sum of the two component concentrations equals the overall concentration, $c_m + c_s = c$. If both component concentrations were allowed to reach equilibrium, we would have $c_m{}^* + c_s{}^* = c$. Thus $c_m{}^* + c_s{}^* = c_m + c_s$. The use of equations (3.2-1) and (3.2-2) shows that $c_m + c_s$ equals $c_m{}^* + c_s{}^* + (c_m{}^*\epsilon_m + c_s{}^*\epsilon_s)$. The term in parenthesis must therefore equal zero

$$c_m{}^*\epsilon_m + c_s{}^*\epsilon_s = 0 \qquad (3.2\text{-}5)$$

this is the "balance-of-nonequilibrium" expression. Either ϵ term may be written in terms of the other

$$\epsilon_s = -c_m{}^*\epsilon_m/c_s{}^* = -R\epsilon_m/(1 - R) \qquad (3.2\text{-}6)$$

where the last expression results from writing $c_m{}^*/c_s{}^* = R/(1 - R)$.

One additional aspect of the equilibrium departure terms is that these terms are generally small, $\epsilon \ll 1$. Any efficient chromatographic system must allow a very rapid exchange of solute between the mobile and stationary phase. Except for very unusual circumstances (extremely high flow rates, etc.), the exchange is sufficiently rapid to maintain near-equilibrium conditions. This assertion is used as one of the cornerstones of nonequilibrium theory. It permits simplifications to be made where otherwise extreme difficulty would be encountered. The *near-equilibrium assumption* (to be examined in detail in Section 4.7) places the nonequilibrium theory among those asymptotic theories which directly or indirectly use the *long-time assumption*. The significance of this latter assumption, and the alternatives sometimes necessary, have been discussed in Sections 2.2 and 2.13.

In the following theoretical development we shall refer to *mass transfer processes* and *mass transfer rates*. These terms will refer only to that transfer of solute in and between phases which occurs as a response to the nonequilibrium, and not to such phenomena as diffusional transport up and down the column.

Theoretical Approach

The specialized nonequilibrium theory given here, as well as the later generalized form, proceeds by a series of steps, each designed to bring a particular concept into quantitative focus. The step-by-step approach for 1-site kinetics is as follows.

Step One: Relation of Mass Transfer Kinetics to Nonequilibrium. The rate of mass transfer between phases (through adsorption and desorption) increases as equilibrium is unbalanced more and more. We seek here to derive the equation relating the mass transfer rate to the equilibrium departure or ϵ terms. Since mass transfer is also related to zone migration (considered in step two), we will subsequently establish a relationship between ϵ values and zone migration using the mass transfer rate as the connecting link.

The rate of mass transfer for the mobile phase may be expressed in terms of the mass transfer term s_m, where

$$s_m = \left(\frac{dc_m}{dt}\right)_{\text{mass transfer}} \tag{3.2-7}$$

s_m is the rate of increase in c_m due to mass transfer processes only. The expression above is seen to be simply the net reaction rate—the rate of the reaction (desorption) producing mobile phase substance minus the rate of loss (adsorption) from the mobile phase. Thus

$$s_m = k_d c_s - k_a c_m \tag{3.2-8}$$

where k_d and k_a are the apparent first-order rate constants of reaction. When the concentrations are at equilibrium, the mass transfer rate s_m goes to zero

$$0 = k_d c_s^* - k_a c_m^* \tag{3.2-9}$$

Hence the existence of s_m as a finite term depends on the departure from equilibrium as expressed in the ϵ terms. If we substitute, then, $c_m^*(1 + \epsilon_m)$ for c_m and $c_s^*(1 + \epsilon_s)$ for c_s, we have

$$\begin{aligned} s_m &= k_d c_s^*(1 + \epsilon_s) - k_a c_m^*(1 + \epsilon_m) \\ &= \underbrace{k_d c_s^* - k_a c_m^*}_{\text{zero}} + k_d c_s^* \epsilon_s - k_a c_m^* \epsilon_m \end{aligned} \tag{3.2-10}$$

The expression which we have indicated to be zero is so because of equation (3.2-9). We are left with

$$s_m = k_d c_s^* \epsilon_s - k_a c_m^* \epsilon_m \tag{3.2-11}$$

Since we are concentrating here mainly on the mobile phase (and thus we have s_m rather than s_s), it simplifies the derivation to have just one equilibrium departure term, that for the mobile phase, ϵ_m. We may eliminate ϵ_s from this expression by using equation (3.2-6). This gives

$$s_m = -c_m^*(k_d + k_a)\epsilon_m \tag{3.2-12}$$

This is the desired expression relating the mass transfer rate s_m to the equilibrium departure term ϵ_m.

Step Two: Mass Transfer Rate and Flow. The mass transfer rate s_m is also related to flow. The reason for this, as will be seen below, is that flow supplies new solute to a region and that this solute must be redistributed between the phases. The rate of redistribution (or mass transfer) is obviously determined by the rate at which new solute is brought into (or taken away from) the region by flow.

The equation expressing the above fact is the *equation of mass conservation.* The concept of mass conservation has been the starting point for numerous theories of chromatography as well as for the theory of other hydrodynamic-kinetic problems.[19] The essence of the idea is very simple.[19] We merely relate the change in the amount of material (solute) within a given region to all of the sources and sinks which may add or subtract this material. The rigorous solution of the resulting differential equation (or equations) is not generally so simple, however. Without judicious approximations, the mathematics is impossibly complicated for realistic chromatographic systems. The approximation which removes the equilibrium theory from this utter complexity is, mainly, the approximation that the system is operating close to equilibrium. This will be demonstrated below.

We now write the equation of mass conservation for our system in its simplest form

$$\frac{\partial c_m}{\partial t} = \left(\frac{dc_m}{dt}\right)_{\text{mass transfer}} + \left(\frac{dc_m}{dt}\right)_{\text{flow}} \tag{3.2-13}$$

the rate of increase of mobile phase concentration c_m in a given region is equal to the increase caused by mass transfer plus the increase caused by the flow in and out of the region. These processes are illustrated in Figure 3.2-2. The first of these two expressions is simply the mass transfer rate s_m. The second expression, representing the gain of solute by flow, is given by

$$\left(\frac{dc_m}{dt}\right)_{\text{flow}} = -v\,\frac{\partial c_m}{\partial z} \tag{3.2-14}$$

where $\partial c_m/\partial z$ is the concentration gradient (the partial differential form simply indicates that there is more than one variable and that during the differentiation of one all others must be held constant). The basis of this equation is given fully in the footnote.† By way of an intuitive understanding, it is clear, first, that the rate of increase in c_m due to flow must be in direct proportion to the velocity of flow v. Second, the increase rate depends on whether flow is entering the region from a richer source of solute (as in the leading part of a zone) or from a solute-starved neighborhood. The measure of the richness of the concentration source a fixed

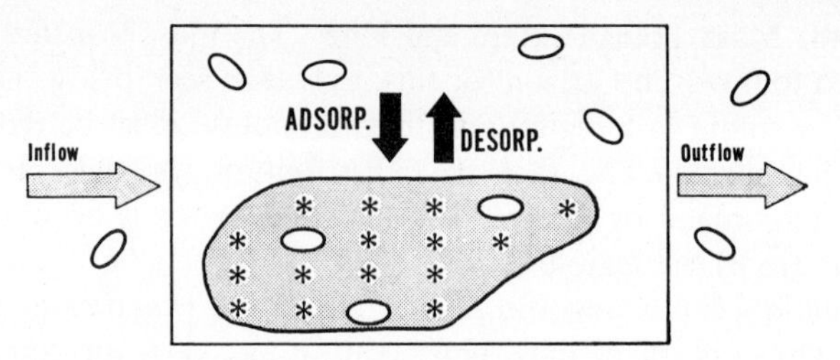

Figure 3.2-2 Changes of solute mass in the mobile phase of a small region of column. The rate of increase of mobile-phase solute is governed by (1) the amount desorbed minus the amount adsorbed in unit time and (2) the amount carried into the region minus the amount carried out by flow through the region in a given time. These two sources for mobile-phase solute correspond to the two terms in equation (3.2-13).

distance upstream is the concentration gradient $\partial c_m/\partial z$, and hence the flow term is proportional to this also. The minus sign originates because the concentration gradient is negative in the leading part of the zone where the flow increase term, above, is positive.

It should be noted that the equation for mass conservation, equation (3.2-13), does not contain a term for diffusion. In the presence of existing

† Consider a small slab of material within the column, faced normal to flow, with unit cross-sectional area and thickness dz. The volume of the slab, because of the unit area of the faces, will also be dz. In a given time dt the amount of solute flowing in through the upstream face is the concentration at the face (located at point z), $(c_m)_z$, times the distance, $v\,dt$, swept out by the flow in time dt. In this same time the outflow from the downstream face, located at $z + dz$, is by analogy to the above $(c_m)_{z+dz}\,v\,dt$. This expression, by a Taylor's expansion, equals $v\,dt[(c_m)_z + (\partial c_m/\partial z)\,dz]$. The net gain of solute in the slab, inflow minus outflow, is thus

$$\text{gain} = -v(\partial c_m/\partial z)\,dz\,dt \tag{3.2-15}$$

The gain rate, or gain per unit of time (gain/dt) is

$$\text{gain rate} = -v(\partial c_m/\partial z)\,dz \tag{3.2-16}$$

The gain rate in the slab over the slab volume, dz, is the time rate of increase of concentration within the slab

$$(dc_m/dt)_{\text{flow}} = -v\,\partial c_m/\partial z \tag{3.2-17}$$

This is the desired expression.

concentration gradients, a certain amount of solute will diffuse into the region under consideration. This solute source is generally negligible, however. While this will be explained in detail in Section 4.7, the following physical reasoning can be given. During a chromatographic run a zone is transported the full length of the column by flow. Longitudinal diffusion, which contributes to zone spreading (Chapter 2) transports solute, at most, over one zone width during the run. Thus the magnitude of diffusional transport is far less than that of flow transport, and may consequently be ignored.

Use of the foregoing facts and equation (3.2-13) gives a more explicit form of the mass conservation equation

$$\frac{\partial c_m}{\partial t} = s_m - v\frac{\partial c_m}{\partial z} \tag{3.2-18}$$

On solving for s_m, we have

$$s_m = \frac{\partial c_m}{\partial t} + v\frac{\partial c_m}{\partial z} \tag{3.2-19}$$

At this point it becomes necessary to apply the near-equilibrium approximation. Since c_m and $c_m{}^*$ are nearly equal, $\partial c_m/\partial t$ and $\partial c_m/\partial z$ may be replaced by $\partial c_m{}^*/\partial t$ and $\partial c_m{}^*/\partial z$, respectively. Consequently, s_m is given as

$$s_m \cong \frac{\partial c_m{}^*}{\partial t} + v\frac{\partial c_m{}^*}{\partial z} \tag{3.2-20}$$

The first term on the right, $\partial c_m{}^*/\partial t$, is particularly easy to deal with because it is a measure of the overall solute accumulation rate (in both phases) in the given volume element. This is explained by the fact that $c_m{}^*$, being an equilibrium concentration, is a constant fraction R of the total solute concentration c in a given region. Within a constant factor, then, $c_m{}^*$ denotes the status of the overall concentration c. Now the mass conservation equation for c is much simpler than for the components c_m or c_s. Since c is the total concentration, and is the sum of c_m and c_s, mass transfer between the phases does not affect it (mass transfer will add to one component what it subtracts from another, leaving the sum $c = c_m + c_s$ unchanged). Hence the mass transfer term is missing from the mass conservation equation for c. In fact the only significant source of increase of c in a given region is the inflow-outflow balance. Inasmuch as the net gain in c by flow all occurs through influx within the mobile phase (since no other element of the column is transporting solute around), the net gain in c is identical to the net gain (due to flow alone) in c_m,

$$\frac{\partial c}{\partial t} = \left(\frac{dc_m}{dt}\right)_{\text{flow}} = -v\frac{\partial c_m}{\partial z} \tag{3.2-21}$$

The last equality of the two is simply a direct expression of equation (3.2-14). Once again $\partial c_m/\partial z$ can be approximated by $\partial c_m^*/\partial z$. Thus the final (and very simple) mass conservation equation for c is

$$\frac{\partial c}{\partial t} \cong -v \frac{\partial c_m^*}{\partial c} \tag{3.2-22}$$

As noted previously, c_m^* is a constant fraction of c (i.e., $c_m^* = Rc$), and thus $\partial c_m^*/\partial t = R\, \partial c/\partial t$ (R is the usual relative migration velocity of the zone). When combined with equation (3.2-22), this gives

$$\frac{\partial c_m^*}{\partial t} \cong -Rv \frac{\partial c_m^*}{\partial z} \tag{3.2-23}$$

This expression can be used to replace the first term on the right of the mass transfer equation, (3.2-20), thus giving the excellent approximation

$$s_m = (1 - R)v\, \partial c_m^*/\partial z \tag{3.2-24}$$

This equation is significantly easier to deal with than its precursor, equation (3.2-20). The dual time- and distance-derivative terms of the earlier equation have been replaced by a single distance-derivative term. This term, $\partial c_m^*/\partial z$, is equal to $R\, \partial c/\partial z$, and thus relates directly to the observed concentration gradient $\partial c/\partial z$ of the chromatographic zone.

Step Three: The Link Formed by s_m. One expression for s_m was found under Step One and is given by equation (3.2-12). Another expression for s_m was found at Step Two and appears in equation (3.2-24). Equating the two expressions for s_m leads to

$$-c_m^*(k_d + k_a)\epsilon_m = (1 - R)v\, \partial c_m^*/\partial z \tag{3.2-25}$$

which can be solved for ϵ_m

$$\epsilon_m = -\frac{(1 - R)v}{k_d + k_a} \frac{1}{c_m^*} \frac{\partial c_m^*}{\partial z} \tag{3.2-26}$$

This equation relates the all-important equilibruim departure term ϵ_m to the velocity and the overall concentration gradient of the zone. All quantities in this equation are subject to experimental interpretation. Note that ϵ_m is positive in the leading part of the zone (where $\partial c_m^*/\partial z$ is negative), and negative behind the center ($\partial c_m^*/\partial z$ positive), as predicted in Section 3.1.

Step Four: Relationship to Plate Height. Our initial discussion of Section 3.1 indicated that equilibrium departure was responsible for a contribution to zone spreading. Since ϵ_m is now available to us in the form of the last equation, it is possible to proceed directly to the calculation of plate height.

Zone spreading originates because at the front of the zone too much solute is transported forward, and at the rear too little is transported forward. This fact is illustrated by calculating the amount of solute J transported forward through a unit cross-sectional area of the column in unit time. The solute flux J is equal to the concentration of solute c_m directly in front of the unit cross section, and which will soon be carried through the cross section, times the volume swept out by the flow in unit time. This volume is equal to the cross-sectional area (unity) times the distance covered by the mobile phase flow in one second (the velocity v). Thus

$$J = c_m v \tag{3.2-27}$$

(The concentration c_m is used here because only the mobile phase material will move through the cross section.) Since $c_m = c_m^*(1 + \epsilon_m)$, we have

$$J = c_m^* v + c_m^* \epsilon_m v = J^* + \Delta J \tag{3.2-28}$$

The first term on the right represents the solute flux at equilibrium and is denoted by J^*. If solute were transported with a flux J^*, the zone would move down the columns without spreading (uniform movement of the zone would result since the flux J^* is directly proportional to the amount of material, c_m^* or c). The spreading influence originates in the ΔJ term. Since ΔJ is proportional to ϵ_m, it is positive ahead of the zone center (thus giving too large a forward movement of solute) and negative to the rear (giving a deficiency in forward movement). Quantitatively, as seen above, $\Delta J = c_m^* \epsilon_m v$, and with ϵ_m substituted from equation (3.2-26), ΔJ is seen to be

$$\Delta J = -\frac{(1 - R)v^2}{k_d + k_a} \frac{\partial c_m^*}{\partial z} \tag{3.2-29}$$

Since $\partial c_m^* / \partial z = R\, \partial c / \partial z$, this becomes

$$\Delta J = -\frac{R(1 - R)v^2}{k_d + k_a} \frac{\partial c}{\partial z} \tag{3.2-30}$$

The analogy to diffusion appears here. During diffusion along a concentration gradient, the flux of material is always $-D\, \partial c / \partial z$, where D is the diffusion coefficient. Parallel to this we can write

$$\Delta J = -D\, \partial c / \partial z \tag{3.2-31}$$

where D is the apparent diffusion coefficient responsible for zone spreading. The value of D is seen to be[2]

$$D = \frac{R(1 - R)v^2}{k_d + k_a} \tag{3.2-32}$$

Digressing for a moment, it is interesting to note that the nonequilibrium approach has brought us back to the rather universal form of the common diffusion equation.[2] This equation means that Gaussian zones will be obtained from narrow initial spikes, but it also means much more than this. It means that any initial zone configuration can be handled by well-known mathematical techniques,[20] and the final profile thus calculated.

The diffusion equation is one aspect generally resulting from the important long-time approximation. It may be recalled from Section 2.2 that the long-time approximation is also associated with Gaussian zones, numerous transfers between phases, the plate model, and near-equilibrium conditions. These concepts are essentially equivalent, so in practice it makes little difference how the problem is approached. Since it is conventional to describe zone spreading in terms of the plate height H, we will proceed with this association.

At the end of Section 2.3 it was seen that the plate height is related to the effective diffusion coefficient D by

$$H = 2D/Rv \tag{3.2-33}$$

Thus we have

$$H = 2(1 - R)v/(k_d + k_a) \tag{3.2-34}$$

which is essentially the final object of this section. For consistency, however, this equation will be converted to a form appearing elsewhere in this book. When numerator and denominator are multiplied by k_d, H becomes

$$H = \frac{2(1 - R)v}{k_d} \frac{k_d}{k_d + k_a} \tag{3.2-35}$$

The ratio $(k_d + k_a)/k_d$, appearing in reciprocal above, is $1 + k_a/k_d$. From equation (3.2-9) k_a/k_d is seen to be c_s^*/c_m^*. Thus the ratio under discussion $1 + c_s^*/c_m^* = (c_m^* + c_s^*)/c_m^* = c/c_m^* = 1/R$. Substituting this into the above in reciprocal form, we have

$$H = 2R(1 - R)v/k_d \tag{3.2-36}$$

This equation is identical with equation (2.13-16) derived from stochastic theory, and with equation (2.7-6) derived from random walk theory. The first rigorous derivation was through the stochastic theory,[21] although the equation has been derived many times and in many ways since then.[2,4,18,22,23] Thus these methods (except for the random walk theory which is highly approximate) are seen to be consistent. However, our object in this section has been more than the derivation of an old equation. This method, expanded somewhat to its generalized form, has led to numerous plate height contributions not obtained by any other theory. Thus acquaintance with the present derivation is the largest single step in using the theory generally. We shall proceed in the next section with one more example, in this case designed to account for mass transfer by diffusion through a uniform layer of stationary phase. First, however, a general relationship will be established between the plate height and ϵ_m. This will be applicable to both examples treated in this chapter.

A General Relationship between H and ϵ_m. The approach just used was designed to show that the solute flux ΔJ is proportional to the overall concentration gradient, and that the effective diffusion concept (along with the use of the theoretical plate parameter H) is valid for zone spreading. If we now accept the effective diffusion concept as a general principle (and it does always prove to be valid under the assumed conditions), we can derive a somewhat more general expression, applicable both here and in the next section, which in a simple form gives H in terms of ϵ_m for any mass transfer process occurring at a solid surface or otherwise within the stationary phase.

The arguments concerning solute flux which precede equation (3.2-28) are quite general, and ΔJ for stationary phase contributions may always be taken, as in that equation, as

$$\Delta J = c_m{}^*\epsilon_m v \tag{3.2-37}$$

Without looking at specific cases, we will write ΔJ in its effective diffusion form (see equation (3.2-31))

$$\Delta J = -D\ \partial c/\partial z \tag{3.2-38}$$

A comparison of these two equations gives

$$D = -\frac{c_m{}^*\epsilon_m v}{\partial c/\partial z} \tag{3.2-39}$$

The plate height contribution H is related to this by $H = 2D/Rv$, equation (3.2-28). Thus

$$H = -\frac{2c_m{}^*\epsilon_m}{R\ \partial c/\partial z} = -\frac{2(Rc)\epsilon_m}{R\ \partial c/\partial z} \tag{3.2-40}$$

where the last equation is obtained by writing $c_m{}^*$ as Rc. On canceling the R's and writing $\partial c/c\ \partial z$ as $\partial \ln c/\partial z$, we have

$$H = \frac{-2\epsilon_m}{\partial \ln c/\partial z} \tag{3.2-41}$$

a very simple expression achieving the desired purpose. This equation should not be construed to mean that H is inversely proportional to $\partial \ln c/\partial z$ or is independent of v. The mobile phase departure term ϵ_m is always of such a form as to cancel the former and to introduce a velocity-proportional term (see, for example, equation (3.2-26)).

3.3 Derivation of H for Mass Transfer Through a Uniform Film by Diffusion

Partition chromatography is based on the selective absorption of solute into bulk† stationary phase, usually held intact on a scaffold of (presumably) inert porous solid. In practice the configuration of the liquid is very complex as it seeks its way into odd-shaped pores and, in general, follows the complex contours of the solid. This matter has been discussed particularly in connection with gas-liquid chromatography.[24] The general complexity of the problem will be dealt with later. In this section we shall assume that the liquid is distributed in the simplest possible configuration; a uniform film of depth d spread on a flat solid surface with one interface exposed to the mobile phase. This idealized model is shown in Figure 3.3-1. Solute molecules enter the liquid through the exposed surface and then explore the liquid, finally leaving it, through molecular diffusion.

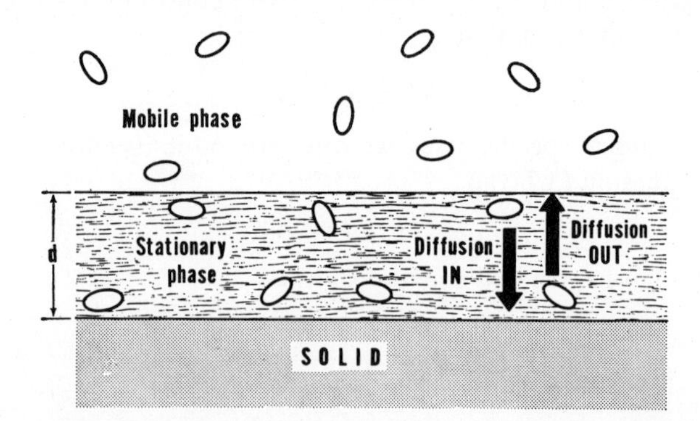

Figure 3.3-1 Schematic illustration of a uniform film of stationary phase in partition chromatography. Molecules gain access to the bulk of liquid by diffusing in from the exposed liquid interface (first arrow). Molecules leave the liquid by means of an outward diffusion process (second arrow).

The nonequilibrium theory connected with diffusional mass transfer is in many ways analogous to the simple kinetics problem treated in Section 3.2. There are a few important differences, however, particularly connected with the necessity for examining the stationary-phase processes in more detail. Nonetheless the development here will proceed in parallel with the last section and, where possible, results obtained there will be applied to the present example.

† The word "bulk" means three-dimensional (Section 1.5).

Nonequilibrium Parameters

The nonequilibrium and concentration parameters associated with partition chromatography are in most cases identical to those used in adsorption chromatography. One still has the overall concentration c and its components c_m and c_s. At equilibrium the component concentrations are $c_m{}^*$ and $c_s{}^*$. Out of equilibrium the fractional departure from these values is indicated by the equilibrium departure terms ϵ_m and ϵ_s. All of these parameters are valid for partition chromatography, but several new and closely related terms are of main importance. The reason why new terms are needed, along with the definitions, merits explanation.

The concentration of adsorbed molecules on a 1-site surface is uniform within any small region of the column. This is not so in partition chromatography. The partitioning (stationary) phase has a finite depth (see Figure 3.3-1), and some of its parts are further removed than others from the fluctuating mobile phase concentration. Solute can reach these deepest regions only by diffusion, and a net transport by diffusion occurs only if concentration gradients exist through the phase. Thus the concentration within a bulk partitioning phase, unlike on a 1-site surface, is not uniform. Since our main object here involves the study of transport of solute in and out of this stationary phase, it is necessary to actually account for the local concentration gradients within the liquid. Thus we define the *local concentration* in the stationary-phase as c_s'. This term is the actual phase concentration, related to unit stationary-phase volume, as opposed to the earlier *overall concentration* referring to unit volume of column material (Section 1.2). Other terms related to the local concentration are also indicated by a prime; for example, the local equilibrium departure term is defined by

$$c_s' = c_s{}^{*\prime}(1 + \epsilon_s') \tag{3.3-1}$$

where $c_s{}^{*\prime}$ is the local equilibrium concentration, obtained when all solute within a given region of column has reached equilibrium. The local concentration is greater than the overall concentration by the ratio of the volumes to which they are related

$$c_s{}^{*\prime} = c_s{}^* \times \frac{\text{total column volume}}{\text{vol. of stat. phase in column}} \tag{3.3-2}$$

and thus $c_s{}^{*\prime}$ is larger than, usually much larger than, $c_s{}^*$. The local departure term ϵ_s' is not necessarily larger than ϵ_s, however, since the ϵ terms represent the fractional departure from equilibrium. The mean of the local departure term, $\overline{\epsilon_s'}$, is in fact equal to ϵ_s.

The other analogous term which must be primed because it refers to the local concentration is the *local mass transfer term* s_s'. This gives the rate of increase of c_s' at some point in the stationary liquid.

Our treatment so far has dealt with local concentrations in the stationary phase, but no such suggestion has been made for the mobile phase. Concentration gradients do exist in the mobile phase (we are not, of course, speaking of the gross concentration gradients associated with the zone profile). Up to this point, however, we have assumed that mass transfer is controlled by kinetics (Section 3.2) or by stationary phase diffusion, and thus the gradients exist in relationship only to these processes. Later, when the mobile phase is considered in detail, local concentration terms will be introduced for this phase.

In parallel with the case of adsorption-desorption kinetics, various equilibrium departure terms are related to one another. Equation (3.2-6), expressing the balance of nonequilibrium between the two phases, is still valid

$$\epsilon_s = -c_m^* \epsilon_m / c_s^* \tag{3.3-3}$$

Since the average of the local departure $\overline{\epsilon_s'}$ equals ϵ_s, we have $\overline{\epsilon_s'} = -c_m^* \epsilon_m / c_s^*$. The quantity c_m^*/c_s^* is simply the ratio of solute in the mobile phase to that in the stationary phase, $R/(1 - R)$. Thus

$$\overline{\epsilon_s'} = -R\epsilon_m/(1 - R) \tag{3.3-4}$$

Theoretical Approach

As before, the theory will proceed on the basis that all ϵ terms are small. The step-by-step derivation, similar to the last section, follows.

Step One: Mass Transfer and Nonequilibrium. This step[6] is considerably different in detail than the analogous step for 1-site surfaces. This is a result of diffusional transport replacing the somewhat simpler first-order kinetics of the last section, and the attendant need to consider local concentration variations in the stationary phase.

We will begin our considerations with the stationary phase where the relevant diffusion is occurring. The mass transfer term, giving the rate of concentration increase due to mass transfer into a small region of the stationary phase, is governed by Fick's second law of diffusion (see Chapter 6)

$$s_s' = D_s \frac{\partial^2 c_s'}{\partial x^2} \tag{3.3-5}$$

where D_s is the diffusion coefficient of solute in the stationary phase and x is the distance through the uniform film measured from its bottom (from the liquid-solid interface). Now since $c_s' = c_s^{*\prime}(1 + \epsilon_s') = c_s^{*\prime} + c_s^{*\prime}\epsilon_s'$, the first derivative of c_s' is

$$\partial c_s'/\partial x = c_s^{*\prime}\, \partial \epsilon_s'/\partial x \tag{3.3-6}$$

since $c_s^{*\prime}$ is constant. The second derivative is

$$\partial^2 c_s'/\partial x^2 = c_s^{*\prime}\, \partial^2 \epsilon_s/\partial x^2 \tag{3.3-7}$$

(Note that derivatives are taken just as in conventional calculus, although we are actually dealing with partial differentials symbolized by $\partial c_s'$ rather than dc_s'. The differences between the two are rather minor, except the "partial" notation reminds us that we are studying the processes in a given region at a fixed time.) Consequently, s_s' is given by[6]

$$s_s' = D_s c_s^{*\prime} \frac{\partial^2 \epsilon_s'}{\partial x^2} \tag{3.3-8}$$

This equation is analogous to the final s_s' equation, (3.2-12), obtained in step one for 1-site surfaces. However the equation given here is a differential equation, and thus in step three, where the final ϵ equation is obtained, an integration will be necessary.

Step Two: Mass Transfer and Flow. In step one an equation was derived relating ϵ_s' (through its second derivative) and s_s'. It is the purpose of step two to relate s_s' to the column-flow processes and the zone profile. Thus s_s' becomes the connecting link between ϵ_s' and the flow processes. It will be found here that s_s' is considerably easier to obtain than was the relevant term s_m for 1-site surfaces. The reason for this is that the equation of mass conservation for the stationary phase is much simpler than for the mobile phase. This is due to the lack of a flow term, absent because nothing flows into any part of the stationary phase. The only source of solute is the mass transfer term, in this case caused by diffusion. Hence the mass conservation equation is

$$\partial c_s'/\partial t = s_s' \tag{3.3-9}$$

This simple expression replaces equation (3.2-18) of Section 3.2.

Equation (3.3-9) can be used as an expression for s_s', but it is necessary to convert the left-hand side into a measurable quantity. First, we assume that the stationary phase concentration is near equilibrium, and hence

$$s_s' \cong \partial c_s^{*\prime}/\partial t \tag{3.3-10}$$

an expression replacing equation (3.2-20) of the previous case (for more detailed arguments regarding these approximate steps, Section 4.7 may be consulted). Since at equilibrium all concentrations are proportional to c, the time rate of change in c is germane to the problem. Equation (3.2-22) showed this quantity to be

$$\frac{\partial c}{\partial t} \cong -v \frac{\partial c_m^*}{\partial z} \tag{3.3-11}$$

The components of this equation can be written in terms of $c_s^{*\prime}$ as follows. First, since $c_s^{*\prime}$ is proportional to c, we have

$$\frac{\partial c}{\partial t} = \frac{c}{c_s^{*\prime}} \frac{\partial c_s^{*\prime}}{\partial t} \tag{3.3-12}$$

Since $c_s^{*\prime}$ and c_m^* are mutually proportional, we have

$$\frac{\partial c_m^*}{\partial z} = \frac{c_m^*}{c_s^{*\prime}} \frac{\partial c_s^{*\prime}}{\partial z} \tag{3.3-13}$$

When these two expressions are substituted back into equation (3.3-11) and the $c_s^{*\prime}$ terms canceled, the result is

$$\frac{\partial c_s^{*\prime}}{\partial t} \cong -v \frac{c_m^*}{c} \frac{\partial c_s^{*\prime}}{\partial z} = -Rv \frac{\partial c_s^{*\prime}}{\partial z} \tag{3.3-14}$$

The last form arises because $c_m^*/c = R$. This equation, except for the type of concentration used ($c_s^{*\prime}$ instead of c_m^*), is identical to equation (3.2-23).

The above equation can be substituted into equation (3.3-10), and thus provides the desired expression for the mass transfer term

$$s_s' = -Rv\, \partial c_s^{*\prime}/\partial z \tag{3.3-15}$$

This expression is similar to equation (3.2-24), but contains $-R$ in place of $1 - R$.

Step Three: The s_s' Link and Integration for ϵ. If the two equations, (3.3-8) and (3.3-15), for s_s' are set equal, the resulting expression is

$$D_s c_s^{*\prime} \frac{\partial^2 \epsilon_s'}{\partial x^2} = -Rv \frac{\partial c_s^{*\prime}}{\partial z} \tag{3.3-16}$$

On rearranging,

$$\frac{\partial^2 \epsilon_s'}{\partial x^2} = -\frac{Rv}{D_s} \frac{\partial c_s^{*\prime}}{c_s^{*\prime}\, \partial z} \tag{3.3-17}$$

Recall that a differential of the form du/u is equivalent to $d \ln u$. The same holds for partial differentials, and thus $\partial c_s^{*\prime}/c_s^{*\prime} = \partial \ln c_s^{*\prime}$. Since

$c_s{}^{*\prime}$ and c are proportional, this also equals $\partial c/c = \partial \ln c$. Hence the last equation becomes

$$\frac{\partial^2 \epsilon_s'}{\partial x^2} = -\frac{Rv}{D_s}\frac{\partial \ln c}{\partial z} \tag{3.3-18}$$

This equation, like equation (3.2-21) for 1-site surfaces, relates the ϵ-containing term to the velocity and concentration profile of the zone. However, this expression is a differential equation and thus needs integrating before ϵ_s' can be obtained explicitly.

The integration of equation (3.3-18) is quite simple since the right-hand side is constant (the right-hand side is a function of downstream coordinate z and time t, but the partial derivative expression $\partial^2 \epsilon_s'/\partial x^2$ implies that the integration involves only x with z and t fixed). A first integration gives

$$\frac{\partial \epsilon_s'}{\partial x} = \left(-\frac{Rv}{D_s}\frac{\partial \ln c}{\partial z}\right)x + g_0 \tag{3.3-19}$$

where g_0 is the constant of integration. A second integration gives

$$\epsilon_s' = \left(-\frac{Rv}{D_s}\frac{\partial \ln c}{\partial z}\right)\frac{x^2}{2} + g_0 x + g_1 \tag{3.3-20}$$

where g_1 is the constant for this integration. It is clear that g_0 and g_1 must be evaluated before ϵ_s' can be regarded as a known quantity. These constants can be obtained through certain *boundary conditions* or in terms of *boundary values*.[25] The boundary conditions applicable here are simply short mathematical expressions of some rather obvious relationships existing in the column.[6] Our first boundary condition expresses the evident fact that no solute is diffusing through the stationary phase-solid interface (*S*-*S*) into the nonpermeable solid. The mathematical translation of this is that the solute flux through the *S*-*S* interface is zero

$$J_{(S\text{-}S\text{ interface})} = 0 \tag{3.3-21}$$

The outward flux, by Fick's first law, is the negative product of D_s' and the concentration gradient existing in the stationary phase right at the interface

$$J_{(S\text{-}S\text{ interface})} = -D_s\left(\frac{\partial c_s'}{\partial x}\right)_{x=0} = 0 \tag{3.3-22}$$

The condition $x = 0$ appended to the last term indicates the location of the *S*-*S* interface. Since $\partial c_s'/\partial x = c_s{}^{*\prime}\,\partial \epsilon_s'/\partial x$ (see equation (3.3-6)), we have

$$-D_s c_s{}^{*\prime}\left(\frac{\partial \epsilon_s'}{\partial x}\right)_{x=0} = 0 \tag{3.3-23}$$

Both D_s and $c_s{}^{*\prime}$ are real and finite. This equation can only be valid, then, if

$$\left(\frac{\partial \epsilon_s'}{\partial x}\right)_{x=0} = 0 \tag{3.3-24}$$

if the gradient of ϵ_s' disappears at $x = 0$. Now equation (3.3-19) gives an alternate expression for $\partial \epsilon_s'/\partial x$. When $x = 0$, this equation reduces to

$$\left(\frac{\partial \epsilon_s'}{\partial x}\right)_{x=0} = g_0 \tag{3.3-25}$$

Thus g_0 must equal zero

$$g_0 = 0 \tag{3.3-26}$$

Our second boundary condition is based on the fact that solute in the interface between the uniform film and the mobile phase is in equilibrium with the mobile phase. (By choice, this problem is concerned only with mass transfer in the uniform film. This is tantamount to assuming that mass transfer rates are infinite elsewhere, and that all other regions are in mutual equilibrium.) This boundary condition is equivalent to the equation

$$\epsilon_m = (\epsilon_s')_{x=d} \tag{3.3-27}$$

where $x = d$ (d = film thickness) indicates that we are referring to the value of ϵ_s' at the interface adjacent to the mobile phase. (This equation is written in terms of ϵ values, since the same fractional departure from equilibrium ϵ will exist under conditions of mutual equilibrium. Neither ϵ is zero, of course, because the system as a whole is thrown out of equilibrium by the gradients in the stationary film.) Another expression for ϵ_s' is found in equation (3.3-20). When $x = d$, this expression, equal to ϵ_m from the above, becomes

$$(\epsilon_s')_{x=d} = \epsilon_m = \left(-\frac{Rv}{D_s}\frac{\partial \ln c}{\partial z}\right)\frac{d^2}{2} + g_1 \tag{3.3-28}$$

where g_0 has been omitted due to equation (3.3-26). The second integration constant g_1 can be obtained directly from this and substituted back into equation (3.3-20) to yield the desired ϵ_s' expression

$$\epsilon_s' = \epsilon_m + \frac{1}{2}\left(-\frac{Rv}{D_s}\frac{\partial \ln c}{\partial z}\right)(x^2 - d^2) \tag{3.3-29}$$

The only term on the right of this equation not subject to direct measurement is ϵ_m. Since we will, in any case, need ϵ_m in the next section for obtaining the plate height, this quantity will be determined here, thus making ϵ_s' a directly calculable nonequilibrium parameter.

For reasons to be explained shortly it is useful to obtain the average value of the local departure term ϵ_s'. This can be done with the above expression for ϵ_s' substituted into the usual expression for average quantities

$$\overline{\epsilon_s'} = \int_0^d \epsilon_s' \, dx \bigg/ \int_0^d dx \tag{3.3-30}$$

Evaluation of these straightforward integrals gives

$$\overline{\epsilon_s'} = \epsilon_m - \left(-\frac{Rv}{D_s}\frac{\partial \ln c}{\partial z}\right)\frac{d^2}{3} \tag{3.3-31}$$

thus yielding a relationship between $\overline{\epsilon_s'}$ and ϵ_m (all other quantities being measurable). Earlier in this section (equation (3.3-4)) another such relationship was presented

$$\overline{\epsilon_s'} = -R\epsilon_m/(1 - R) \tag{3.3-32}$$

If the right-hand sides of these two equations are set equal to one another, ϵ_m can be solved for

$$\epsilon_m = \left(-\frac{Rv}{D_s}\frac{\partial \ln c}{\partial z}\right)\frac{(1 - R)\, d^2}{3} \tag{3.3-33}$$

This, substituted back into equation (3.3-29), yields the hoped for relationship between ϵ_s' and measurable quantities

$$\epsilon_s' = \left(-\frac{Rv}{D_s}\frac{\partial \ln c}{\partial z}\right)\left[\frac{x^2}{2} - \frac{d^2}{6}(1 - 2R)\right] \tag{3.3-34}$$

Step Four: Relationship to Plate Height. This step is made with trivial ease as a result of some preceding developments. It was shown in equation (3.2-41) that the plate height for any mass transfer process in the stationary phase was $H = -2\epsilon_m/(\partial \ln c/\partial z)$. We have already obtained ϵ_m in the form of equation (3.3-33). A direct substitution of this into the H expression yields

$$H = \tfrac{2}{3}R(1 - R)\frac{d^2 v}{D_s} \tag{3.3-35}$$

This is the desired plate height expression. The form of this expression is the same as obtained by the random walk theory, equation (2.8-1). The two values differ by a numerical constant, with the above expression being the rigorously correct one. This equation was first derived by a different method for the film of liquid on the wall of a capillary column.[14] It was obtained later by means of the generalized nonequilibrium theory.[6]

3.4 References

1. J. C. Giddings in *Chromatography*, E. Heftmann, ed., Reinhold, New York, 1961, Chap. 3.
2. J. C. Giddings, *J. Chromatog.*, **2,** 44 (1959).
3. J. N. Wilson, *J. Am. Chem. Soc.*, **62,** 1583 (1940).
4. J. C. Giddings, *J. Chem. Phys.*, **31,** 1462 (1959).
5. J. C. Giddings, *J. Chromatog.*, **3,** 443 (1960).
6. J. C. Giddings, *J. Chromatog.*, **5,** 46 (1961).
7. J. C. Giddings, *J. Phys. Chem.*, **68,** 184 (1964).
8. J. C. Giddings, *J. Chem. Phys.*, **26,** 1210 (1957).
9. J. C. Giddings and J. O. Hirschfelder, *Sixth Symposium on Combustion* (1956), Reinhold, New York, 1957, p. 199.
10. J. O. Hirschfelder, C. F. Curtiss, and R. B. Bird, *Molecular Theory of Gases and Liquids*, Wiley, New York, 1954.
11. J. W. Westhaver, *J. Res. Nat. Bur. Stand.*, **38,** 169 (1947).
12. Sir Geoffrey Taylor, *Proc. Roy. Soc.* (*London*), **A219,** 186 (1953).
13. R. Aris, *Proc. Roy. Soc.* (*London*), **A235,** 67 (1956).
14. M. J. E. Golay in *Gas Chromatography, 1958*, D. H. Desty, ed., Academic Press, New York, 1958, p. 36.
15. M. A. Khan in *Gas Chromatography, 1962*, M. van Swaay, ed., Butterworths, Washington, 1962, p. 3.
16. K. J. Mysels, *J. Chem. Phys.*, **24,** 371 (1956).
17. P. C. Scholten and K. J. Mysels, *J. Chem. Phys.*, **35,** 1845 (1961).
18. T. A. Bak and W. G. Kauman, *Trans. Faraday Soc.*, **55,** 1109 (1959).
19. R. C. L. Bosworth, *Transport Processes in Applied Chemistry*, Wiley, New York, 1956.
20. J. Crank, *The Mathematics of Diffusion*, Clarendon Press, Oxford, 1956.
21. J. C. Giddings, *J. Chem. Phys.*, **26,** 1755 (1957).
22. P. C. Scholten and K. J. Mysels, *Trans. Faraday Soc.*, **56,** 994 (1960).
23. A. Klinkenberg, *Chem. Eng. Sci.*, **15,** 255 (1961).
24. J. C. Giddings, *Anal. Chem.*, **34,** 458 (1962).
25. R. V. Churchhill, *Fourier Series and Boundary Value Problems*, McGraw-Hill, New York, 1941.

Chapter Four
The Generalized Nonequilibrium Theory

4.1 Nature of the General Problem

The specific nonequilibrium theories discussed in Chapter 3 have been extended to include complex mass transfer phenomena of a general type.[1,2,3,4] The need to extend and generalize the theory is very obvious in view of the known complexity of chromatographic materials. The generalized approach suffers, naturally, from appearing to be a little more removed from concrete cases, but it renders an important service in being immediately applicable to a large number of real chromatographic problems. Through this approach a more significant insight can be obtained into the relative role of various processes. A good example of this arises in connection with the tendency of experimentalists to avoid nonuniform (heterogeneous) surfaces in adsorption chromatography. Not only can an interpretation of this experimental preference be given, but it is possible to specify the degree of nonuniformity (in terms of the spread in adsorption energies) which can be tolerated before column efficiency deteriorates.

We have seen already that mass transfer may be controlled by either of two mechanisms: by step-wise (or single-step) kinetic processes or by diffusional transport. The former is applicable primarily to adsorption chromatography and the latter to partition chromatography. However, only a few experimental columns are purely of one type. Hence a mixed mass transfer mechanism must often be considered (Section 4.6).

In Section 4.2 we will consider mass transfer by step-wise kinetics. This will include the effect of any process which causes a sudden change in the state of a solute molecule at a single occurrence. Desorption is one such process. Chemical reaction is another. Diffusion is not usually regarded as such a process since many individual molecular displacements are needed to effect any significant change in a molecule's location.

In Chapter 2 the random walk approach was used to show that zone spreading was caused by a molecular species erratically alternating between regions of different velocities. This phenomenon is not confined solely to classical chromatography. In electrochromatography, or electrophoresis, zone spreading may originate[5] in an isomerization reaction involving a change of ionic charge, $A^+ \rightleftarrows A^{2+} + e^-$, and thus a change in mobility. The faster moving species (probably A^{2+} here) is analogous to a molecule in the mobile phase, while the slower of the two is analogous to a molecule retarded by the stationary phase. More complex kinetics, along with more species, may be involved. The zone spreading due to this process has been termed electrodiffusion,[5] and it is also covered by the general theory given below.[6]

We will now proceed with the theoretical development, following the same sequence of steps used in Sections 3.2 and 3.3. Much of the material in this chapter is relatively mathematical. The reader may wish to peruse only the nonmathematical parts or proceed directly to the summarizing section, 4.8, in order to obtain an idea of the type of problems treated and the practical consequences.

4.2 Step-Wise Kinetics: Adsorption Chromatography

We wish to consider here an arbitrary number of discrete states which can be reached by a solute species through first-order interconversion processes. A state may correspond to a solute molecule or a particular type of site, or to a molecule in the mobile phase, etc. These states will be denoted by A_i, A_j, A_k, etc. The interconversion process may be represented by the kinetic analog[1]

$$A_i \underset{k_{ji}}{\overset{k_{ij}}{\rightleftarrows}} A_j \tag{4.2-1}$$

in which the first-order rate constants, as indicated by the arrows, are k_{ij}, k_{ji}, etc. When a solute molecule is in the ith form its downstream velocity will be v_i. The overall concentration (referred to unit volume of column) of solute in state i is c_i. The sum of all such concentrations is c,

$$c = \sum c_i \tag{4.2-2}$$

When equilibrium is attained in the system, each state will contain the fraction X_i^* of the total number of solute molecules

$$X_i^* = c_i^*/c \tag{4.2-3}$$

Under nonequilibrium conditions each concentration will be related to its equilibrium value c_i^* through the nonequilibrium departure term ϵ_i

$$c_i = c_i^*(1 + \epsilon_i) \tag{4.2-4}$$

As before, each positive displacement from equilibrium is matched by a negative displacement elsewhere. There is consequently a "balance of nonequilibrium" which can be obtained by equating Σc_i with Σc_i^*. This equality leads to

$$\sum c_i^* \epsilon_i = \sum X_i^* \epsilon_i = 0 \tag{4.2-5}$$

thus relating the various ϵ_i terms.

The mean velocity of the zone, $\mathscr{V}$, may be obtained as a weighted average of the individual velocities

$$\mathscr{V} = \sum X_i v_i \tag{4.2-6}$$

The v_i for each state associated with the stationary phase will, of course, be zero, and thus relatively few terms will contribute to this summation. Ordinarily this velocity equals Rv, where $R = X^*$ (mobile phase) and v, as usual, is the mean velocity of the mobile phase or mobile state.

Theoretical Approach

Step One: Mass Transfer and Nonequilibrium. The mass transfer term for state i may be obtained as a sum over all reaction paths by which its concentration may be enlarged or decreased

$$s_i = \sum_j k_{ji} c_j - c_i \sum_j k_{ij} \tag{4.2-7}$$

This equation is a generalized form of equation (3.2-8). When equilibrium is attained, this expression becomes (the omitted summation index j being understood)

$$0 = \sum k_{ji} c_j^* - c_i^* \sum k_{ij} \tag{4.2-8}$$

Now if each concentration in equation (4.2-7) is written in the nonequilibrium form, $c_i = c_i^*(1 + \epsilon_i)$ and $c_j = c_j^*(1 + \epsilon_j)$, s_i is found to be

$$s_i = \underbrace{\sum k_{ji} c_j^* - c_i^* \sum k_{ij}}_{\text{zero}} + \sum k_{ji} c_j^* \epsilon_j - c_i^* \epsilon_i \sum k_{ij} \tag{4.2-9}$$

The part which we have indicated to be zero is identical to equation (4.2-8). Thus s_i is

$$s_i = \sum k_{ji} c_j^* \epsilon_j - c_i^* \epsilon_i \sum k_{ij} \tag{4.2-10}$$

The principle of microscopic reversibility,[7,8] which indicates that equilibrium concentrations are the same irrespective of the interconversion mechanism, gives us the equilibrium relationship

$$k_{ji}c_j^* = k_{ij}c_i^* \tag{4.2-11}$$

When this is substituted into the summation of the first term on the right of equation (4.2-10), we have

$$s_i = \sum k_{ij}c_i^*\epsilon_j - c_i^*\epsilon_i \sum k_{ij} \tag{4.2-12}$$

After the c_i^* is factored out, this becomes

$$s_i = c_i^* \left(\sum k_{ij}\epsilon_j - \epsilon_i \sum k_{ij}\right) \tag{4.2-13}$$

This may be written in the slightly simpler form

$$s_i = c_i^* \sum k_{ij}(\epsilon_j - \epsilon_i) \tag{4.2-14}$$

Either of the last two equations may be used conveniently to relate s_i to the ϵ terms.

Step Two: Mass Transfer and Flow. The equation of mass conservation for solute in state i is very similar to the form derived earlier for 1-site kinetics, equation (3.2-18)

$$\frac{\partial c_i}{\partial t} = s_i - v_i \frac{\partial c_i}{\partial z} \tag{4.2-15}$$

where longitudinal diffusion has again been ignored as a solute transport mechanism. This assumption will be discussed in more detail in Section 4.7. The above equation gives the following s_i expression

$$s_i = \frac{\partial c_i}{\partial t} + v_i \frac{\partial c_i}{\partial z} \tag{4.2-16}$$

Application of the near-equilibrium approximation (also discussed in Section 4.7) gives

$$s_i \cong \frac{\partial c_i^*}{\partial t} + v_i \frac{\partial c_i^*}{\partial z} \tag{3.4-17}$$

The term $\partial c_i^*/\partial t$ can be converted to a distance derivative by noting that the entire concentration profile is advancing with a velocity of essentially $\mathscr{V}$ (slight departures from this at back and front of zone center were noted in Section 3.1). Thus

$$\frac{\partial c_i^*}{\partial t} \cong -\mathscr{V} \frac{\partial c_i^*}{\partial z} \tag{4.2-18}$$

(A more detailed argument for this equation proceeds along the same line used to derive equation (3.3-14).) Hence

$$s_i = (v_i - \mathscr{V})\, \partial c_i^*/\partial z \tag{4.2-19}$$

This equation is in a suitable form for immediate use.

Step Three: The ϵ Values. Equations (4.2-13) and (4.2-19) may be combined to relate the ϵ-containing expression to the flow and concentration-gradient expression

$$c_i^* \left(\sum k_{ij}\epsilon_j - \epsilon_i \sum k_{ij}\right) = (v_i - \mathscr{V})\, \partial c_i^*/\partial z \tag{4.2-20}$$

Since c_i^* is proportional to the overall concentration c, this may be written as

$$\sum k_{ij}\epsilon_j - \epsilon \sum k_{ij} = (v_i - \mathscr{V})\partial \ln c/\partial z \tag{4.2-21}$$

If there are n states—if i (and j) runs from 1 to n—then there are n of these equations (one for each state i) and n unknowns, $\epsilon_1, \epsilon_2, \ldots, \epsilon_n$. However, these n equations do not yield unique ϵ values. This may be seen by writing the left-hand side in the form of equation (4.2-14)

$$\sum k_{ij}(\epsilon_j - \epsilon_i) = (v_i - \mathscr{V})\partial \ln c/\partial z \tag{4.2-22}$$

Suppose we obtain n solutions, $\epsilon_1, \epsilon_2, \ldots, \epsilon_n$, to this set of n equations. It is easy to see that the set of ϵ values, $\epsilon_1'', \epsilon_2'', \ldots, \epsilon_n''$, where each ϵ'' is related to ϵ by $\epsilon_i'' = \epsilon_i + \lambda$ (λ = a constant), also yields a solution to the n equation. This is a result of the left-hand side being proportional to the difference term $\epsilon_j - \epsilon_i$, and this difference is the same no matter what constant value (λ) is added to each ϵ value. Thus the set of n equations in (4.2-21) and (4.2-22) gives ϵ values only within an additive constant; some other equation is needed to fix their absolute value. This other equation is the balance-of-nonequilibrium equation, (4.2-5). Hence the n equations used to specify equilibrium departure in each state are, for one, this equation, and for the others, any $n - 1$ of the remaining equations expressed in (4.2-21),

$$\begin{aligned} \sum k_{ij}\epsilon_j - \epsilon_i \sum k_{ij} &= E_i \qquad (n-1 \text{ equations}) \\ \sum X_i^* \epsilon_i &= 0 \qquad (\text{one equation}) \end{aligned} \tag{4.2-23}$$

where E_i is a shorthand notation for the left-hand side of equation (4.2-2)

$$E_i = (v_i - \mathscr{V})\partial \ln c/\partial z \tag{4.2-24}$$

The E_i values are known parameters, being obtainable from the velocity components and the concentration gradient as seen here.

The n solutions to the n linear equations given above can be obtained from Cramer's rule.[9] For ϵ_i, as an example, we have the ratio of the two n by n determinants

$$\epsilon_i = \frac{\begin{vmatrix} -\sum k_{1j} & k_{12} & k_{13} & \cdots & E_1 & \cdots & k_{1n} \\ k_{21} & -\sum k_{2j} & k_{23} & \cdots & E_2 & \cdots & k_{2n} \\ k_{31} & k_{32} & -\sum k_{3j} & \cdots & E_3 & \cdots & k_{3n} \\ \cdot & & & & & & \\ \cdot & & & & & & \\ \cdot & & & & & & \\ X_1^* & X_2^* & X_3^* & \cdots & 0 & \cdots & X_n^* \end{vmatrix}}{\begin{vmatrix} -\sum k_{1j} & k_{12} & k_{13} & \cdots & k_{1i} & \cdots & k_{1n} \\ k_{21} & -\sum k_{2j} & k_{23} & \cdots & k_{2i} & \cdots & k_{2n} \\ k_{31} & k_{32} & -\sum k_{3j} & \cdots & k_{3i} & \cdots & k_{3n} \\ \cdot & & & & & & \\ \cdot & & & & & & \\ \cdot & & & & & & \\ X_1^* & X_2^* & X_3^* & \cdots & X_i^* & \cdots & X_n^* \end{vmatrix}} \tag{4.2-25}$$

The other ϵ values can be similarly expressed. In most actual cases of interest a large number of the rate constants k_{ij} vanish, and the determinants reduce to fairly simple forms.

Step Four: The Plate Height Expression. The solute flux through a given cross section of column is given by the sum of the flux carried by solute in each state

$$J = \sum c_i v_i = \sum c_i^*(1 + \epsilon_i)v_i \tag{4.2-26}$$

Since the equilibrium fraction of solute in state i is given by $X_i^* = c_i^*/c$, this can be written as

$$J = c \sum X_i^*(1 + \epsilon_i)v_i \tag{4.2-27}$$

When expanded into its two components, this becomes

$$J = c \sum X_i^* v_i + c \sum X_i^* \epsilon_i v_i \tag{4.2-28}$$

The component which leads to zone spreading is, as before

$$\Delta J = c \sum X_i^* \epsilon_i v_i \tag{4.2-29}$$

The summation here is simply a series of terms, each containing one of the ϵ values. Now it may be observed from equation (4.2-25) that ϵ is proportional to a determinant, a whole column of which is composed of

E values (the lone zero is of no effect here). Furthermore each E, as seen by equation (4.2-24), is proportional to $\partial \ln c/\partial z$. Since it is a property of determinants[10] that any term, such as $\partial \ln c/\partial z$, common to any column (or row) is a factor of the entire determinant, then each ϵ is proportional to $\partial \ln c/\partial z$. The ΔJ expression, then, is seen to be proportional to $\partial \ln c/\partial z$ and also to c. The product $c\, \partial \ln c/\partial z$, of course, is the concentration gradient $\partial c/\partial z$. This is a general proof, then, that the flux is proportional to the concentration gradient, and consequently that this source of zone spreading is an effective diffusion process with coefficient D.

Quantitatively ΔJ may be written as $-D\, \partial c/\partial z$, and D thus obtained as

$$D = -\frac{\sum X_i^* \epsilon_i v_i}{\partial \ln c/\partial z} \tag{4-2.30}$$

The plate height contribution is related to D through the mean zone velocity $\mathscr{V}$ (Section 2.3), $H = 2D/\mathscr{V}$, and thus

$$H = -\frac{2 \sum X_i^* \epsilon_i v_i}{\mathscr{V} \partial \ln c/\partial z} \tag{4.2-31}$$

A measurable value for this can be calculated once the ϵ_i values have been obtained.

Applications

The preceding formulation will be applied to a number of special cases which have chromatographic implications. We will first treat 1-site adsorption kinetics to illustrate with the simplest case the procedure to be followed, and to show that this general formulation is consistent with the special derivation of Section 3.2.

1-Site Adsorption Kinetics. Adsorption and desorption in a 1-site system may be represented by the simple kinetic analog[1]

$$\underset{\binom{\text{mobile}}{\text{phase}}}{A_1} \underset{k_{21}}{\overset{k_{12}}{\rightleftarrows}} \underset{\binom{\text{stationary}}{\text{phase}}}{A_2} \tag{4.2-32}$$

Where k_{12} and k_{21} are recognized as the adsorption and desorption rate constants k_a and k_d, respectively. The solute fraction X_1^* is obviously R and X_2^* is $1 - R$. The zone velocity $\mathscr{V}$ is Rv. The component velocities are $v_1 = v$ and $v_2 = 0$.

The set of n equations, (4.2-23), reduces to two equations of the following type

$$\begin{aligned} -k_{12}\epsilon_1 + k_{12}\epsilon_2 &= E_1 \\ X_1^*\epsilon_1 + X_2^*\epsilon_2 &= 0 \end{aligned} \tag{4.2-33}$$

Both ϵ_1 and ϵ_2 can be obtained readily from this set. However we may note from equation (4.2-31) that the only ϵ values of importance are those connected with a finite velocity. Therefore only the direct calculation of ϵ_1 is necessary. By analogy with equation (4.2-25) this is

$$\epsilon_1 = \frac{\begin{vmatrix} E_1 & k_{12} \\ 0 & X_2{}^* \end{vmatrix}}{\begin{vmatrix} -k_{12} & k_{12} \\ X_1{}^* & X_2{}^* \end{vmatrix}} = -\frac{E_1 X_2{}^*}{k_{12}(X_1{}^* + X_2{}^*)} \tag{4.2-34}$$

In view of the fact that

$$X_2{}^*/(X_1{}^* + X_2{}^*) = 1 - R \quad \text{and} \quad E_1 = (1 - R)v\, \partial \ln c/\partial z,$$

from equation (4.2-24), ϵ_1 is

$$\epsilon_1 = -\frac{(1 - R)^2 v}{k_{12}} \frac{\partial \ln c}{\partial z} \tag{4.2-35}$$

A direct substitution of this into equation (4.2-31), noting only that $X_1{}^* = R$ and $\mathscr{V} = Rv$, gives

$$H = 2(1 - R)^2 v/k_{12} \tag{4.2-36}$$

Since $R/(1 - R) = k_{21}/k_{12}$, we have

$$H = 2R(1 - R)v/k_{21} \tag{4.2-37}$$

Inasmuch as k_{21} is the desorption rate constant k_d, this equation is identical to that obtained fom the specialized nonequilibrium theory, equation (3.2-36).

Multisite Adsorption. The obvious generalization of 1-site adsorption is multisite adsorption[2] (2-site adsorption has been treated as a special case[1]). We will postulate that the adsorbing surface contains an arbitrary number $n - 1$ of different kinds of sites, each with its own particular kinetic constants for adsorption and desorption. For simplicity we will stipulate that transfer of solute from one site to that of a different kind is not possible.

The kinetic analog for this type of adsorption process is

$$\begin{array}{c} A_1 \rightleftarrows A_2 \\ A_1 \rightleftarrows A_3 \\ \cdot \\ \cdot \\ \cdot \\ A_1 \rightleftarrows A_n \end{array} \tag{4.2-38}$$

where A_1 represents the mobile phase (and thus $v_1 = v$ and $X_1{}^* = R$) and A_2 through A_n represent the $n - 1$ different types of sites. The velocity of solute attached to these sites is obviously zero and thus $v_i = 0$ for $i = 2$ to n. The mean zone velocity $\mathscr{V}$ is $Rv = X_1{}^*v$.

The n equations expressed in (4.2-23) may be written as (note that $i = 1$ has been chosen as the one excluded from the set)

$$\begin{aligned}
&k_{21}\epsilon_1 - k_{21}\epsilon_2 &&= E_2\\
&k_{31}\epsilon_1 \qquad\qquad - k_{31}\epsilon_3 &&= E_3\\
&k_{41}\epsilon_1 \qquad\qquad\qquad\qquad - k_{41}\epsilon_4 &&= E_4\\
&\quad\vdots &&\quad\vdots\\
&X_1{}^*\epsilon_1 + X_2{}^*\epsilon_2 + X_3{}^*\epsilon_3 + X_4{}^*\epsilon_4 + \ldots &&= 0
\end{aligned} \tag{4.2-39}$$

Only ϵ_1 need be obtained again because state one (the mobile state), alone, is associated with a finite velocity. From Cramer's rule

$$\epsilon_1 = \frac{\begin{vmatrix} E_2 & -k_{21} & 0 & 0 & 0 & & 0\\ E_3 & 0 & -k_{31} & 0 & 0 & & 0\\ E_4 & 0 & 0 & -k_{41} & 0 & & 0\\ E_5 & 0 & 0 & 0 & -k_{51} & & 0\\ \vdots & & & & & & \vdots\\ 0 & X_2{}^* & X_3{}^* & X_4{}^* & X_5{}^* & \ldots & X_n{}^* \end{vmatrix}}{\begin{vmatrix} k_{21} & -k_{21} & 0 & 0 & 0 & & 0\\ k_{31} & 0 & -k_{31} & 0 & 0 & & 0\\ k_{41} & 0 & 0 & -k_{41} & 0 & & 0\\ k_{51} & 0 & 0 & 0 & -k_{51} & & 0\\ \vdots & & & & & & \vdots\\ X_1{}^* & X_2{}^* & X_3{}^* & X_4{}^* & X_5{}^* & \ldots & X_n{}^* \end{vmatrix}} \tag{4.2-40}$$

An expansion by minors[11,12] yields the result

$$\epsilon_1 = E \sum_{i=2}^{n} X_i{}^*/k_{i1} \tag{4.2-41}$$

where

$$E = -Rv\,\partial \ln c/\partial z \\ = E_2 = E_3 = E_4, \text{ etc.} \tag{4.2-42}$$

The equality of E_2, E_3, E_4, etc., can be seen to result from equation (4.2-24) in which each of the corresponding v_i terms is zero. A direct substitution of the last two equations into the general plate height expression, (4.2-31), yields

$$H = 2Rv \sum_{i=2}^{n} X_i^*/k_{i1} \tag{4.2-43}$$

This is a surprisingly simple result for such an extended mechanism of adsorption-desorption. When only one type of site is active, the summation reduces to $X_2^*/k_{21} = (1 - R)/k_{21}$ and the foregoing expression becomes identical with equation (4.2-37) for 1-site kinetics.

While it is difficult to simplify equation (4.2-43) any further, it is possible to relate the equation to a more physically meaningful parameter. Consider the fact that a desorption rate, such as k_{i1}, is one divided by the average desorption time, t_{di}, for solute leaving that particular type of site (see Section 2.7). The summation in equation (4.2-43) adds each such desorption time, weighted according to its fractional contribution to the total solute reservoir. It is thus related to an overall mean desorption time. The summation proceeds only over the stationary phase solute—only over the fraction $1 - R$ of solute—and it is thus actually equal only to the fraction $1 - R$ of the mean desorption time $\bar{t}_d$

$$\sum_{i=2}^{n} X_i^*/k_{i1} = (1 - R)\bar{t}_d \tag{4.2-44}$$

(A more rigorous derivation of this equation can be found in the original paper.[2]) Thus the plate height is

$$H = 2R(1 - R)v\bar{t}_d \tag{4.2-45}$$

The quantity $\bar{t}_d$ should be accessible to kinetic measurement as the mean time required to desorb an equilibrium distribution of initially adsorbed molecules. The magnitude of this quantity will be discussed in detail in Chapter 6.

Consecutive Processes: Adsorption in Partition Chromatography. While a 1-site surface is an oversimplification in adsorption chromatography, a simple diffusion model is probably erroneous in many cases of partition chromatography. A great number of chromatographic systems generally regarded as purely of the partition type will likely involve some adsorption on the solid support material which holds the liquid in place. The model considered here will bear on the importance of such adsorption effects, as well as on other problems.

The kinetic analog for a 2-step consecutive process is

$$A_1 \rightleftarrows A_2 \rightleftarrows A_3 \tag{4.2-46}$$

This is an analog for adsorption, providing we assume that one transition is between the mobile and the stationary phase and that the other is between the stationary phase and the adsorbed state. However, for the moment the problem will be treated in a general fashion with the arbitrary velocities v_1, v_2, and v_3 attached to the states.

The three ϵ equations chosen from the set shown in (4.2-23) are

$$\begin{aligned} -k_{12}\epsilon_1 + k_{12}\epsilon_2 \qquad\qquad &= E_1 \\ k_{32}\epsilon_2 - k_{32}\epsilon_3 &= E_3 \\ X_1{}^*\epsilon_1 + X_2{}^*\epsilon_2 + X_3{}^*\epsilon_3 &= 0 \end{aligned} \tag{4.2-47}$$

It is necessary to obtain all three ϵ values in order to apply the plate height equation, (4.2-31), since all velocities can be finite. However, this can easily be done using the same method as applied to the previous cases. It can be shown that the final result is[2]

$$H = \frac{2X_1{}^*(v_1 - \mathscr{V})^2}{\mathscr{V}k_{12}} + \frac{2X_3{}^*(v_3 - \mathscr{V})^2}{\mathscr{V}k_{32}} \tag{4.2-48}$$

This can be reduced to the study of adsorption disturbances in partition chromatography by assuming that A_1 is solute in the mobile phase, A_2 solute in the stationary phase, and A_3 adsorbed solute. It is clear that $X_1{}^* = R$, $X_2{}^* + X_3{}^* = 1 - R$, $v_1 = v$, $v_2 = 0$, and $v_3 = 0$. In addition, the mean zone velocity is $\mathscr{V} = Rv$. The plate height expression for this case becomes

$$H = \frac{2(1 - R)^2 v}{k_{12}} + \frac{2X_3{}^*Rv}{k_{32}} \tag{4.2-49}$$

The fact that this (as well as the previous expression) is a sum of terms, each related solely to the rate of one of the reversible reactions, is of interest. The first term, related to $A_1 \rightleftarrows A_2$, is equal to equation (4.2-36) for 1-site kinetics. This suggests that complex problems may be broken into their simple components. This is a simple statement of the *additive law*. We will deal at some length with this concept in several places later (see Sections 4.6 and 4.7) since it is one key to simplifying the theoretical treatment of chromatography.

Adsorption of Complex Molecules. Large, complex molecules may be presumed to adsorb in a step-wise procedure, with one end or segment attaching as a preliminary step to total adsorption. This case can be treated with any desired degree of sophistication using the generalized nonequilibrium theory. However, it is adequate here to illustrate the nature of the problem with simple bifunctional adsorption. We will presume that the solute molecule has two active ends, either of which may first combine with the surface. The kinetics analog for this system is

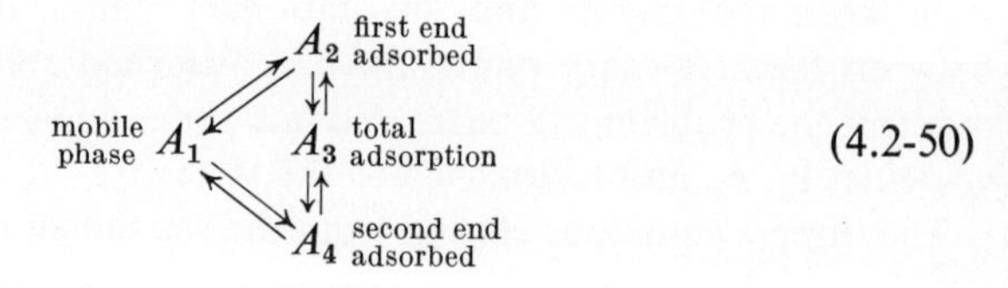

The nature of each state is explained in this sequence. In three of the states (A_2, A_3, A_4) the solute is stationary and the corresponding velocity is thus zero.

Four of the equations from (4.2-23) which may be chosen to represent this kinetic analog are

$$\begin{aligned}
k_{21}\epsilon_1 - (k_{21} + k_{23})\epsilon_2 + k_{23}\epsilon_3 &= E_2 \\
+ k_{32}\epsilon_2 - (k_{32} + k_{34})\epsilon_3 + k_{34}\epsilon_4 &= E_3 \\
k_{41}\epsilon_1 + k_{43}\epsilon_3 - (k_{43} + k_{41})\epsilon_4 &= E_4 \\
X_1^*\epsilon_1 + X_2^*\epsilon_2 + X_3^*\epsilon_3 + X_4^*\epsilon_4 &= 0
\end{aligned} \tag{4.2-51}$$

Only ϵ_1 needs be determined here because it alone is associated with a finite velocity. The ϵ_1 expression, as well as the rigorous answer, is too long to reproduce here. However, we find that by making the reasonable assumption that the two intermediate states A_2 and A_4 have short lifetimes, the plate height reduces to the relatively simple form[2]

$$H = \frac{2(1 - R)^2 v}{k_{12}f_2 + k_{14}f_4} \tag{4.2-52}$$

where f_2 is the fraction of solute molecules in state 2 which proceed to state 3. An analogous description applies to f_4. It is interesting to note that the sum $k_{12}f_2 + k_{14}f_4$ is the total rate for reaching state 3; it is the overall adsorption rate. With this interpretation H is identical to the one-site expression, (4.2-36), with the sum replacing the adsorption rate k_{12}. The fact that all these cases are consistent with one another gives us more confidence in the theoretical procedure.

Reversible Chemical Changes. Reversible chemical reactions were postulated in Section 2.13 as being the source of double zones and other distorted profiles. When these reactions proceed rapidly (such that the near-equilibrium assumption is valid), distortion will cease but there may still be a resultant zone spreading. This problem requires the consideration of four processes; the chemical change which occurs in both mobile and stationary phase and the adsorption-desorption processes connected with each chemical species. The kinetic analog is

$$\begin{array}{ccc} A_1 & \rightleftarrows & A_2 \\ \downarrow\uparrow & & \uparrow\downarrow \\ A_4 & \rightleftarrows & A_3 \end{array} \quad \text{chemical change} \updownarrow \qquad \text{phase change} \longleftrightarrow \tag{4.2-53}$$

where A_2 and A_3 may be regarded as the mobile solute, each with velocity v. Chemical change occurs between the species (A_1, A_2) and (A_3, A_4.)

The detailed mathematics, too cumbersome to repeat here, proceeds much as before. If we assume that the phase change processes are rapid by comparison to the chemical changes, the normally complex plate height expression reduces to[2]

$$H = \frac{2v(X_1^*X_3^* - X_2^*X_4^*)^2}{R(k_{14}X_1^* + k_{23}X_2^*)} + \frac{2vX_1^*X_2^*}{R(X_1^* + X_2^*)(k_{12} + k_{21})} + \frac{2vX_3^*X_4^*}{R(X_3^* + X_4^*)(k_{34} + k_{43})} \tag{4.2-54}$$

It can be shown that the first term accounts for the chemical change (and thus contains only the chemical reaction rates k_{14} and k_{23}) and that the last two terms account for the phase change (or adsorption-desorption) process. Thus once again the composite expression is a sum of terms, each applying to simpler groups of processes.

4.3 Theory of Diffusional Mass Transfer

We will consider here the role of diffusion through the microscopic masses of material which compose the stationary and mobile phases.[3] Diffusional mass transfer is particularly applicable to partition chromatography. It should be noted, however, that even in adsorption chromatography the mobile phase is three dimensional, and that access to each point in the mobile phase can therefore be achieved only through diffusion.

Significant Parameters

We may in general be interested in diffusion through any phase i. Usually, however, there will be only a single stationary phase and a single mobile phase so that i may be denoted by s (stationary) or m (mobile). Our attention must focus on the detailed processes occurring within each phase (as explained in Section 3.3). Hence our main parameters will be those which describe a very small element of a given phase rather than describing a small region of the overall column. As in Section 3.3, those parameters (excluding distance coordinates) related to a small element of a given phase are indicated by primes, and are denoted by the word *local* (as opposed to *overall*). Thus the local concentration in phase i is c_i'. At equilibrium it is $c_i^{*\prime}$. The equilibrium departure term ϵ_i' is given at each point by $c_i' = c_i^{*\prime}(1 + \epsilon_i')$. The mass transfer term, giving the local rate of accumulation of solute in phase i is s_i'. In addition, we must define the local downstream velocity in phase i as v_i'. This is necessary because the velocity of the mobile phase v_m varies from point to point within any given flow channel. The stationary phase velocity v_s' is of course, zero.

It is interesting to note that the average of some *local* quantities, e.g., $\overline{v_i'}$ or $\overline{\epsilon_i'}$, are simply the *overall* quantities v_i or ϵ_i. The reason for this is perhaps clear in view of the fact that the word overall, as used here, implies reference to a phase's gross characteristics, that is, average values. The exception found with concentration is explained by equation (3.3-2).

Theoretical Approach

The balance-of-nonequilibrium expression, always an integral part of the nonequilibrium theory, is obtained as usual by equating the sum total of solute as it actually exists with that found at equilibrium

$$\sum \int c_i' \, dV_i = \sum \int c_i^{*\prime} \, dV_i \tag{4.3-1}$$

Where V_i is the volume of phase i per unit volume of column and dV_i is the differential element of V_i. Using $c_i' = c_i^{*\prime}(1 + \epsilon_i')$, this expression becomes

$$\sum_i c_i^{*\prime} \int \epsilon_i' \, dV_i = 0 \tag{4.3-2}$$

where $c_i^{*\prime}$ has been taken outside the integral because of its constancy. The integral, $\int \epsilon_i' \, dV_i$ equals $\overline{\epsilon_i'} V_i$, and thus

$$\sum_i c_i^{*\prime} \overline{\epsilon_i'} V_i = 0 \tag{4.3-3}$$

We may also write this in terms of overall column parameters by writing

$\overline{\epsilon_i'} = \epsilon_i$ and by noting that $c_i{*}'V_i$ is equal to the total solute content of phase i in a unit volume of the column, $c_i{*}$. Consequently

$$\sum c_i{*}\epsilon_i = \sum X_i{*}\epsilon_i = 0 \tag{4.3-4}$$

In this form the equation is identical to equation (3.4-5).

The mean velocity of the zone is given by the weighted average

$$\mathscr{V} = \frac{\sum \int c_i{*}'v_i' \, dV_i}{\sum \int c_i{*}' \, dV_i} \tag{4.3-5}$$

A somewhat less complicated expression is

$$\mathscr{V} = \sum X_i{*}\overline{v_i'} \tag{4.3-6}$$

Since the average of the local velocity equals the overall, this expression is identical to equation (4.2-6), $\mathscr{V} = \sum X_i{*}v_i$.

Step One: Mass Transfer and Nonequilibrium. The rate of accumulation of solute within a phase is given by the three-dimensional form of Fick's second law as

$$s_i' = D_i \nabla^2 c_i' \tag{4.3-7}$$

where ∇^2 is the Laplacian operator.† Since $c_i{*}'$ is a constant, $\nabla^2 c_i' = c_i{*}' \nabla^2 \epsilon_i'$. Hence this expression becomes

$$s_i' = D_i c_i{*}' \nabla^2 \epsilon_i' \tag{4.3-8}$$

This equation accomplishes the desired task of relating mass transfer and nonequilibrium. The basic diffusion equation, (4.3-7), is based on the assumption that the diffusion coefficient D_i is constant. This should be valid for any of the homogeneous phases found in chromatographic work.

Step Two: Mass Transfer and Flow. The equations required here are essentially identical to those derived in the last section. The only modification involves replacing overall parameters by local parameters. Thus the initial mass conservation equation is like equation (4.2-15)

$$\frac{\partial c_i'}{\partial t} = s_i' - v_i' \frac{\partial c_i'}{\partial z} \tag{4.3-9}$$

but gives a much more detailed picture because of its use of local (primed) quantities. The procedure from here is identical and the final equation is in the same form as equation (4.2-19)

$$s_i' = (v_i' - \mathscr{V}) \, \partial c_i{*}'/\partial z \tag{4.3-10}$$

† In rectangular coordinates ∇^2 is given by $\partial^2/\partial x^2 + \partial^2/\partial y^2 + \partial^2/\partial z^2$.

Step Three: Method for Obtaining ϵ_i. The two mass transfer expressions, (4.3-8) and (4.3-10), may be combined to yield an equality relating ϵ_i and the zone's flow properties

$$D_i c_i^{*\prime} \nabla^2 \epsilon_i' = (v_i - \mathscr{V})\, \partial c_i^{*\prime}/\partial z \tag{4.3-11}$$

On rearrangement this equation becomes

$$\nabla^2 \epsilon_i' = \frac{(v_i' - \mathscr{V})}{D_i} \frac{\partial \ln c}{\partial z} \tag{4.3-12}$$

This is the fundamental differential equation for ϵ_i'.[3] It can be integrated to obtain ϵ_i through various mathematical methods, both analytical and numerical. In some cases, fortunately, the integration is very simple (the uniform film of Section 3.3 is an example of this). Equation (4.3-12) is equivalent to Poisson's equation in mathematical physics,† an equation which describes the distribution of electrical potential (in place of ϵ_i') in a charged medium.

The complete specification of ϵ_i', once it is obtained by integration, depends on evaluating the integration constants through the use of certain restraining relationships or boundary conditions.‡ These have been divided into the following types:[3]

(*1*) The balance-of-nonequilibrium relationship given in equations (4.3-2) to (4.3-4).

(*2*) The boundary condition at a closed interface, i.e., an interface through which no mass transfer occurs because one of the adjacent phases is a solid. Since there is no solute transport through such an interface and thus no concentration gradients in the vicinity, this boundary condition is

$$\partial \epsilon_i'/\partial w = 0 \tag{4.3-13}$$

to be evaluated at the interface with w the distance normal to the interface.

(*3*) The boundary condition at a connecting interface, i.e., an interface (such as between stationary and mobile phases) through which mass transfer proceeds readily. The solute transport must be equal on the two sides of the interface, and thus

$$D_i(\partial \epsilon_i'/\partial w) = D_j(c_j^{*\prime}/c_i^{*\prime})(\partial \epsilon_j'/\partial w) \tag{4.3-14}$$

(*4*) The boundary condition expressing equilibrium at an interface

$$\epsilon_i' = \epsilon_j' \tag{4.3-15}$$

† Methods of solution are discussed in ref. 13.

‡ An excellent discussion of the general role of boundary conditions is found in ref. 14.

where, as before, the ϵ values must be evaluated in the immediate vicinity of the interface. While the interfacial-equilibrium assumption is generally valid, there may be a few systems where slow transfer through the interface vitiates this boundary condition.

(5) The condition at planes, lines, and points of symmetry which expresses the fact that there is no solute transport through such places

$$\partial \epsilon_i'/\partial w = 0 \tag{4.3-16}$$

This condition is a result of the fact that transport through a plane of symmetry would in fact destroy the symmetry by making the two symmetrical parts unequal, i.e., one would gain solute and the other would lose it.

There may be other conditions applicable to specific cases (see Section 4.6). However these five conditions have been found more than adequate for most of the diffusional mass transfer problems so far encountered.

Step Four: The Calculation of Plate Height. The quantity of solute passing through a unit area normal to flow in unit time is given by the solute flux

$$J = \sum \int c_i' v_i' \, dA_i \tag{4.3-17}$$

where A_i is the fraction of the cross-sectional area occupied by phase i. On writing c_i' in terms of equilibrium departure, this expression becomes

$$J = \sum c_i^{*\prime} \int v_i' \, dA_i + \sum c_i^{*\prime} \int \epsilon_i' v_i' \, dA_i \tag{4.3-18}$$

where the second term, containing ϵ_i', is the flux ΔJ responsible for zone spreading

$$\Delta J = \sum c_i^{*\prime} \int \epsilon_i' v_i' \, dA_i \tag{4.3-19}$$

It is useful to factor the concentration units out of this expression. First we note that

$$c_i^{*\prime}/c_i^* = 1/A_i \tag{4.3-20}$$

(This is consistent with equation (3.3-2) since relative areas are proportional to relative volumes and thus A_i is equal to the fractional column volume V_i occupied by phase i.) However $c_i^* = X_i^* c$, and thus

$$c_i^{*\prime} = X_i^* c / A_i \tag{4.3-21}$$

When this is substituted back into equation (4.3-19), there results

$$\Delta J = c \sum X_i^* (1/A_i) \int \epsilon_i' v_i' \, dA_i \tag{4.3-22}$$

This equation can be equated to the expression for zone diffusion, $-D\, \partial c/\partial z$, thus yielding

$$D = -\frac{\sum X_i^*(1/A_i) \int \epsilon_i' v_i'\, dA_i}{\partial \ln c/\partial z} \tag{4.3-23}$$

The plate height, $H = 2D/\mathscr{V}$, equals

$$H = -\frac{2 \sum X_i^*(1/A_i) \int \epsilon_i' v_i'\, dA_i}{\mathscr{V}\, \partial \ln c/\partial z} \tag{4.3-24}$$

Thus when the ϵ values have been obtained, this equation yields the plate height in terms of these values and in terms of the concentration gradients, velocities, physical dimensions and the equilibrium fractions in the system.

4.4 Diffusion in the Stationary Phase

The general theory of the last section has many applications to both stationary and mobile phases. We shall consider only the former in the next several pages.

The typical chromatographic system, to be considered here, has a single mobile phase and a single stationary phase. Adsorption of any kind will be ignored here since it is outside the scope of this section. The rate of mass transfer will be limited by diffusion through the bulk stationary phase, and concentration gradients will thus be found in this phase but not in the mobile phase. Although we are dealing with the stationary phase, mobile phase parameters are not entirely irrelevant as seen by the fact that the plate height, equation (4.3-24), involves a summation over both phases. The mobile phase departure term ϵ_m is particularly important as indicated by equation (3.2-41). This equation is still valid as seen by the following reasoning. The summation in equation (4.3-24) is effectively related only to the mobile phase since the local velocity v_i' is zero elsewhere. In addition, $\mathscr{V} = Rv$. Taking only the mobile phase contribution to the summation and setting $X_m^* = R$ as well as

$$\int \epsilon_m' v_m'\, dA_m = \epsilon_m v A_m$$

(or $\epsilon_m' v A_m$), H is found to be

$$H = \frac{-2\epsilon_m}{\partial \ln c/\partial z} \tag{4.4-1}$$

the same as equation (3.2-41). The evaluation of the integral was accomplished easily in view of the fact that ϵ_m' is constant under the assumed conditions, and by equating the average local velocity, $\overline{v_m'}$, to the overall mobile phase velocity v (throughout this book v is equivalent to v_m). This equation shows that the evaluation of a mobile phase parameter ϵ_m is critical to the study of stationary phase mass transfer.

Our general line of attack, then, is to obtain the stationary phase departure term through the integration of equation (4.3-12), and relate this to ϵ_m through the boundary and other conditions applicable to this differential equation. Realistically, however, the direct integration of this equation is not expected to be easily accomplished for the complex distribution of stationary phase found on real columns. (The stationary phase, in following the contours of the solid support to which it adheres, will present a total picture of near-infinite complexity.) The direct integration over the entire stationary phase is skirted through the use of the so-called general combination law.

General Combination Law for the Stationary Phase[4]

The general combination law shows how the entire, complex mass of the stationary phase can be divided into simple units, each of which can be treated singly in a straightforward manner. This is in keeping with the spirit of Section 4.3, where is was shown by examples that many complex kinetic sequences separated into their component parts.

We will assume that the total bulk of the stationary phase can be broken into small *units*, where a unit is defined by the following properties of its enclosing boundary. Such a bounding surface must have, first, an open area of contact with the mobile phase through which solute exchange takes place. We stipulate that the remaining area of the unit's boundary must be closed to solute exchange. Closure may be accomplished by a solid impermeable interface or by symmetry conditions, etc. A mathematical expression of closure (identical to the type 2 boundary condition) is

$$\partial \epsilon_s'/\partial w = 0 \qquad \text{(closed surface)} \qquad (4.4\text{-}2)$$

The concentration (and thus nonequilibrium) gradient along the normal, w, must be zero at the bounding surface.

The bulk of the stationary phase in partition chromatography presumably occupies, as a result of surface tension, the separate pores and cavities of the solid support material. These small masses of stationary phase are usually somewhat isolated from one another, being connected, perhaps, only through a thin film of liquid. This is illustrated in Figure 4.4-1. To a good approximation then, each cavity filled with stationary

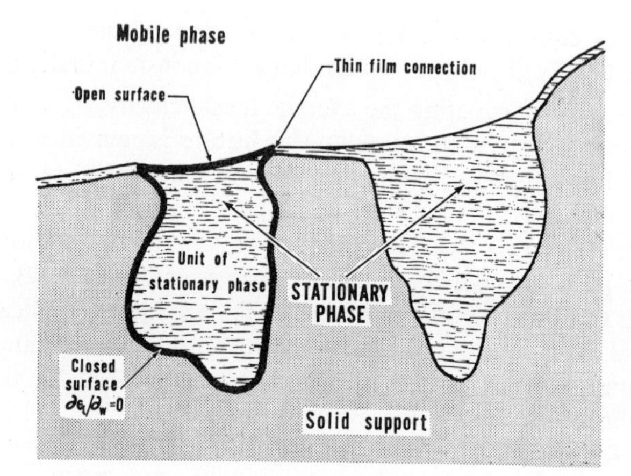

Figure 4.4-1 An illustration of a *unit* of stationary phase within the internal pore structure of a solid support particle. Both closed and open surfaces are present. The tenuous, thin-film connection has a negligible effect. This is the smallest mass of stationary phase which may be treated independently.

phase is a unit. The containing wall of the cavity is the closed surface and the exchanging interface is the open surface. The tenuous links with other cavities are ordinarily negligible insofar as solute transport is concerned.

Whatever configuration the separate units finally attain, they will be the smallest entity which can be treated independently. The theoretical treatment of this problem is as follows.

First, we define a new nonequilibrium parameter θ

$$\theta = \frac{(\epsilon_s' - \epsilon_m) D_s}{Rv \, \partial \ln c/\partial z} \tag{4.4-3}$$

At any point where ϵ_s' may be zero, θ acquires the value θ_0 where

$$\theta_0 = \frac{-\epsilon_m Ds}{Rv \, \partial \ln c/\partial z} \tag{4.4-4}$$

Since from equation (4.4-1) $H = -2\epsilon_m/\partial \ln c/\partial z$, the plate height in terms of θ_0 can be written

$$H = 2Rv\theta_0/D_s \qquad (4.4\text{-}5$$

This equation will be used shortly.

The conservation-of-nonequilibrium expression, equation (4.3-4), reduces to the following for this simple two-phase system

$$\epsilon_m R = -\overline{\epsilon_s'}(1 - R) \tag{4.4-6}$$

The quantity $\overline{\epsilon_s'}$ may also be obtained by solving for ϵ_s' from equation (4.4-3) and forming an average

$$\overline{\epsilon_s'} = \epsilon_m + \bar{\theta}(Rv/D_s)\partial \ln c/\partial z \tag{4.4-7}$$

where $\bar{\theta}$ is the mean value of θ averaged over the total mass of the stationary phase in a small region (for simplicity, a unit volume) of the column. In view of the relationship between θ_0 and ϵ_m expressed in (4.4-4), equation (4.4-7) becomes

$$\overline{\epsilon_s'} = (\bar{\theta} - \theta_0)(Rv/D_s)\partial \ln c/\partial z \tag{4.4-8}$$

and equation (4.4-6) becomes

$$\overline{\epsilon_s'} = \frac{R\theta_0}{1 - R}\frac{Rv}{D_s}\frac{\partial \ln c}{\partial z} \tag{4.4-9}$$

When these are equated and identical terms canceled, the two expressions yield

$$\theta_0 R = (\bar{\theta} - \theta_0)(1 - R) \tag{4.4-10}$$

or

$$\theta_0 = \bar{\theta}(1 - R) \tag{4.4-11}$$

In view of equation (4.4-5) we have

$$H = 2R(1 - R)v\bar{\theta}/D_l \tag{4.4-12}$$

Since $\bar{\theta}$ is an average over the total stationary phase volume within a unit volume of the column, it may be written as

$$\bar{\theta} = \sum_j (V_j/V_s)\bar{\theta}_j \tag{4.4-13}$$

where V_j is the volume of unit j, V_s the stationary phase volume per unit column volume, and the summation is over all the units which the stationary phase divides itself into. The volume in unit j relative to the entire stationary phase volume considered is seen to be the ratio V_j/V_s, and the average value of θ within unit j is $\bar{\theta}_j$. The plate height is now in the desired form

$$H = \frac{2R(1 - R)v}{D_s}\sum_j \left(\frac{V_j}{V_s}\right)\bar{\theta}_j \tag{4.4-14}$$

and is made up of a sum of terms, each term being concerned only with a given unit of the stationary phase. This is the main part of the proof sought, and the equation states the principal idea of the combination law.

It is now necessary to show how the unit-averaged θ's ($\bar{\theta}_j$, etc.) are obtained. The following will show that they are obtained independently for each unit, and will thus complete the proof that H is composed of a sum of independent terms, one for each unit.

The basic differential equation for the ϵ's, (4.3-12), reduces to the following for the case of the stationary phase

$$\nabla^2 \epsilon_s' = -(Rv/D_s)\partial \ln c/\partial z \tag{4.4-15}$$

since $\mathscr{V} = Rv$ and $v_s = 0$. In view of the definition of θ, equation (4.4-3), we have

$$\nabla^2 \theta = \frac{D_s \nabla^2 \epsilon_s'}{Rv \, \partial \ln c/\partial z} \tag{4.4-16}$$

Substituting $\nabla^2 \epsilon_s'$ into this from above, we have the basic differential equation for θ

$$\nabla^2 \theta = -1 \tag{4.4-17}$$

an equation still applying to the entire mass of stationary phase. At all closed surfaces the boundary condition (type 2) is, as expressed in (4.4-2), $\partial \epsilon_s'/\partial w = 0$. Since θ is linear in ϵ_s', the corresponding closed-surface boundary condition for θ is identical

$$\partial \theta/\partial w = 0 \qquad \text{(closed surface)} \tag{4.4-18}$$

The other boundary condition (type 4, equation (4.3-15)) expresses equilibrium at the open surface, i.e., $\epsilon_s' = \epsilon_m$, where ϵ_s' is to be evaluated directly at the open surface. Equation (4.4-3) shows that θ is zero when this equality holds

$$\theta = 0 \qquad \text{(open surface)} \tag{4.4-19}$$

The quantity θ is completely defined by the differential equation, (4.4-17), and the two subsequent boundary conditions which, between them, cover the entire surface of the liquid mass. However, if we look at a single unit we notice that the information is just as specific—there is the same differential equation with precise boundary conditions applying over the enclosing surface. Hence the value of θ within each unit is specified in terms of a universal (differential) equation and the geometry of that unit—the exact location of its open and closed surfaces. There are, therefore, a set of $\overline{\theta}_j$ values, each of which can be calculated independently for its particular unit through the equations

$$\nabla^2 \theta_j = -1, \qquad \partial \theta_j/\partial w = 0 \quad \text{(closed)}, \qquad \theta_j = 0 \quad \text{(open)} \tag{4.4-20}$$

Thus the plate height, equation (4.4-14), is a sum of independent terms as postulated.

It is possible to remain completely general and still indicate, in a somewhat more detailed way, the nature of the plate height. To do this we note that each unit j of stationary phase has a characteristic distance d_j usually equal (by convention) to the maximum depth of the unit from its open surface. A dimensionless distance can be defined as $\hat{x} = x/d_j$, and a dimensionless θ_j by $\hat{\theta}_j = \theta_j/d_j^2$. Use of the dimensional distance (or distances) in the Laplacian (thus changing it from ∇^2 to $\hat{\nabla}^2$) yields the same equation and boundary conditions for $\hat{\theta}_j$ as for θ_j

$$\hat{\nabla}^2\hat{\theta}_j = -1, \qquad \partial\hat{\theta}_j/\partial w = 0 \quad \text{(closed)}, \qquad \hat{\theta}_j = 0 \quad \text{(open)} \tag{4.4-21}$$

The plate height equation in terms of $\hat{\theta}_j$ differs from equation (4.4-14) by having a d_j appear explicitly in the contributing terms

$$H = \frac{2R(1-R)v}{D_s} \sum_j \bar{\hat{\theta}}_j \left(\frac{V_j}{V_s}\right) d_j^2 \tag{4.4-22}$$

This can be written as

$$H = \sum q_j \left(\frac{V_j}{V_s}\right) R(1-R) \frac{d_j^2 v}{D_s} \tag{4.4-23}$$

which is probably the most useful form of the general combination law and shows more clearly that each plate height contribution is proportional to the square of the depth d_j of the stationary phase and to the velocity v, and inversely proportional to the diffusion coefficient for the stationary phase, D_s. The dimensionless quantity q_j (or simply q) is called the *configuration factor* (since it depends on the configuration of the stationary phase units), a term which is usually within an order of magnitude of unity. A comparison of the last two equations shows that its value is given by

$$q_j = 2\bar{\hat{\theta}}_j \tag{4.4-24}$$

The term can thus be calculated for any solid support pore structure, where the pores have a known configuration, by means of the basic $\hat{\theta}_j$ equations, (4.4-21).

If there is only one kind of unit composing the total stationary phase (as, for example, in glass bead columns in gas chromatography where the liquid units all occupy the "pores" around the bead contact points), the plate height reduces to

$$H = qR(1-R)d^2v/D_s \tag{4.4-25}$$

This approximation may be made for most stationary phases where the detailed variation in structure from unit to unit is not understood.

Most chromatographic equations have been obtained with the assumption that all the column's stationary phase has an identical configuration, and thus a variety of H terms, of the general form of equation (4.4-25), have been derived. The general combination law as shown in equation (4.4-23) demonstrates that where several configurations exist

$$H = \sum (V_j/V_s)H_j \tag{4.4-26}$$

The final plate height is the average of the component plate heights where the average is weighted by the stationary phase volume. The fact that this is a volume-averaged plate height is significant because it shows the relative contribution of terms and indicates that some of these, involving small amounts of stationary phase, are negligible.

Application to the Uniform Film Model. The simplest possible distribution of stationary phase, as discussed in Section 3.3, is in the form of a uniform film covering the surface of a solid support. The unit of stationary phase may be considered as the entire film or any patches thereof. This example will help illustrate the use of the generalized nonequilibrium theory.

For a uniform film the basic differential equation in (4.4-21) takes the form

$$\partial^2\hat{\theta}/\partial\hat{x}^2 = -1 \tag{4.4-27}$$

which integrates simply to

$$\partial\hat{\theta}/\partial\hat{x} = -\hat{x} + g_0 \tag{4.4-28}$$

and

$$\hat{\theta} = -\hat{x}^2/2 + g_0\hat{x} + g_1 \tag{4.4-29}$$

where g_0 and g_1 are integration constants. The boundary condition at the closed surface, $\hat{x} = 0$, is shown by (4.4-21) to be $\partial\hat{\theta}/\partial w = 0$. Applied to equation (4.4-28) this gives $g_0 = 0$. At the open surface, $\hat{x} = 1$, $\hat{\theta} = 0$. Equation (4.4-29), with g_0 already equal to zero, shows that $g_1 = \frac{1}{2}$. Thus

$$\hat{\theta} = -\hat{x}^2/2 + \tfrac{1}{2} \tag{4.4-30}$$

The average value of this is

$$\bar{\hat{\theta}} = \int_0^1 \hat{\theta}\, d\hat{x} = \tfrac{1}{3} \tag{4.4-31}$$

The configuration factor, equal to $2\bar{\hat{\theta}}$ as shown by equation (4.4-24), is thus $q = \frac{2}{3}$. The plate height, equation (4.4-25), thus equals

$$H = \tfrac{2}{3}R(1 - R)\frac{d^2v}{D_s} \tag{4.4-32}$$

an expression identical to equation (3.3-35).

Application to the Circular Rod (Paper Chromatography) Model. The stationary phase in paper chromatography is the permeable part of the swollen paper fibers. These may be approximated by narrow cylinders, and thus the unit stationary phase may be assumed to have a cylindrical structure. The basic differential equation for $\hat{\theta}$, now in cylindrical coordinates, is

$$\frac{1}{\hat{r}}\frac{\partial}{\partial \hat{r}}\left(\hat{r}\frac{\partial \hat{\theta}}{\partial \hat{r}}\right) = -1 \tag{4.4-33}$$

where the angular and axial coordinates fail to appear because of symmetry and because end effects are negligible in a long rod. In this case $\hat{r}$ is the dimensionless radius measured from the rod center.

The integration of (4.4-33) yields

$$\partial \hat{\theta}/\partial \hat{r} = -\hat{r}/2 + g_0/\hat{r} \tag{4.4-34}$$

and

$$\hat{\theta} = -\hat{r}^2/4 + g_0 \ln \hat{r} + g_1 \tag{4.4-35}$$

The closed surface boundary condition, $\partial \hat{\theta}/\partial \hat{r} = 0$ at $\hat{r} = 0$, is in this case a result of symmetry; solute transport through the rod center must be zero. (It is also assumed, of course, that $\partial \hat{\theta}/\partial \hat{r}$ is a continuous function of $\hat{r}$.) This condition once again gives $g_0 = 0$. At the open surface, $\hat{r} = 1$, $\hat{\theta} = 0$, and thus $g_1 = -\frac{1}{4}$. Consequently

$$\hat{\theta} = -\hat{r}^2/4 + \tfrac{1}{4} \tag{4.4-36}$$

The average value of $\hat{\theta}$ is

$$\bar{\hat{\theta}} = \int_0^1 2\pi\hat{\theta}\hat{r}\, d\hat{r} \bigg/ \int_0^1 2\pi\hat{r}\, d\hat{r} = \tfrac{1}{8} \tag{4.4-37}$$

and thus the configuration factor, twice this, is $q = \frac{1}{4}$. The plate height contribution, (4.4-25), is therefore

$$H = \tfrac{1}{4}R(1 - R)\frac{d^2 v}{D_s} = \tfrac{1}{16}R(1 - R)\frac{d_p{}^2 v}{D_s} \tag{4.4-38}$$

where d (or $d_p/2$) is the fiber radius. This equation was first derived by use of the generalized nonequilibrium theory.[3]

Application to the Spherical Bead (Ion-Exchange) Model. The unit of stationary phase in ion-exchange chromatography is the spherical ion-exchange bead. The basic differential equation for $\hat{\theta}$, now in polar coordinates, is

$$\frac{1}{\hat{r}^2}\frac{\partial}{\partial \hat{r}}\left(\hat{r}^2\frac{\partial \hat{\theta}}{\partial \hat{r}}\right) = -1 \tag{4.4-39}$$

where the angular coordinates fail to appear because of symmetry. In this case $\hat{r}$ is the dimensionless radius measured from the bead center.

The integration of (3.5-62) yields

$$\partial \hat{\theta}/\partial \hat{r} = -\hat{r}/3 + g_0/\hat{r}^2 \tag{4.4-40}$$

and

$$\hat{\theta} = -\hat{r}^2/6 - g_0/\hat{r} + g_1 \tag{4.4-41}$$

Application of the boundary conditions, as before, yields $g_0 = 0$ and $g_1 = \frac{1}{6}$. Thus $\hat{\theta}$ is

$$\hat{\theta} = -\hat{r}^2/6 + \tfrac{1}{6} \tag{4.4-42}$$

The average value of $\hat{\theta}$ is

$$\bar{\hat{\theta}} = \int_0^1 4\pi\hat{\theta}\hat{r}^2\,d\hat{r} \Big/ \int_0^1 4\pi\hat{r}^2\,d\hat{r} = \tfrac{1}{15} \tag{4.4-43}$$

The configuration factor is twice this, $q = \frac{2}{15}$, and the plate height is consequently

$$H = \tfrac{2}{15}R(1 - R)\frac{d^2v}{D_s} = \tfrac{1}{30}R(1 - R)\frac{d_p^{\,2}v}{D_s} \tag{4.4-44}$$

where d (or $d_p/2$) is the bead radius. The mass transfer coefficient, forming an alternate approach to this equation, was first derived by Glueckauf[15] and later by Bogue[16] under the assumption of frontal analysis and that the zone profile was Gaussian, respectively. The equation was derived by Giddings[3] using the generalized nonequilibrium theory.

General Theory of Narrow Pores

If it is assumed that the stationary phase is drawn into narrow pores by surface tension, a number of chromatographic systems can be covered. If, more generally, we simply assume that the unit of stationary phase is a narrow spike of some kind, we can derive numerous practical equations, including the last two. The ion-exchange bead, for instance, may be divided (as a result of symmetry) into tapered slivers as indicated in Figure 4.4-2. The uniform film may also be divided into narrow units as

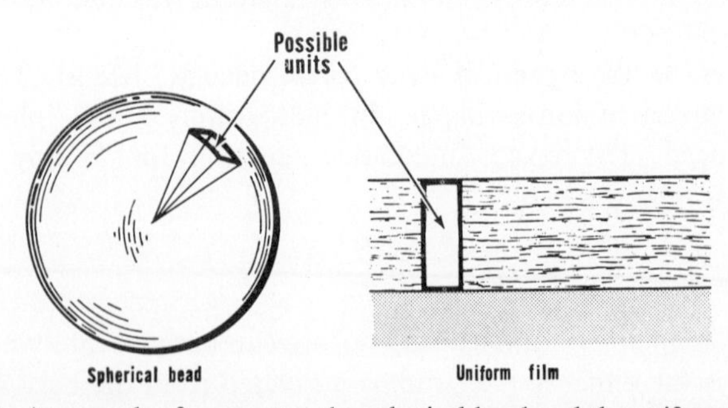

Figure 4.4-2 As a result of symmetry, the spherical bead and the uniform film can be divided into narrow units of the type shown. Each of the stationary phase units may be treated independently as discussed in the text.

shown in the figure. From a theoretical point of view it doesn't matter whether stationary phase can be divided into narrow units through symmetry or whether this occurs naturally through the occupancy of narrow pores.

In order to treat the narrow pore case we must recast the basic differential equation, $\hat{\nabla}^2\hat{\theta} = -1$. Consider the narrow pore (or unit of stationary phase) shown in Figure 4.4-3. The word "narrow," as used here, implies that all gradients are directed along the main axis of the pore; the lateral dimensions are by comparison so limited that concentration and non-equilibrium gradients related thereto are negligible. As before, we will measure the distance from the bottom of the pore in terms of the dimensionless distance, $\hat{x} = x/d$, where x is the actual distance from the deepest point. The normal cross-sectional area of the pore a is some function of the depth into the pore.

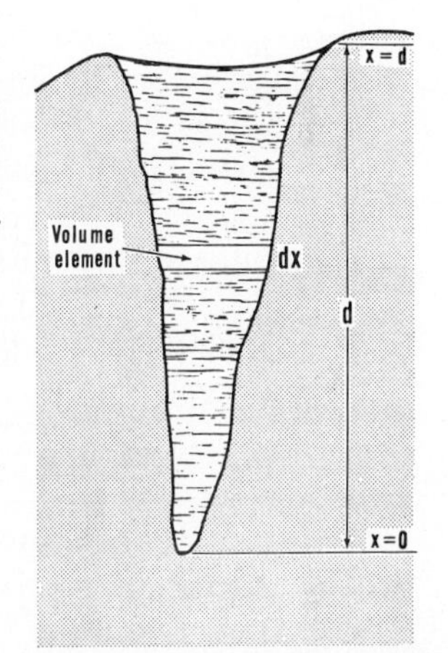

Figure 4.4-3 A narrow pore (or *unit* of stationary phase) of depth d. Gauss's theorem may be applied to the volume element as shown in the text.

A small volume element existing within the pore is shown in Figure 4.4-3. According to Gauss's theorem[17]

$$\int \nabla^2 \theta \, d(\text{volume}) = \int (\text{grad}\, \theta)_n \, d(\text{area}) \quad (4.4\text{-}45)$$

where the first integral covers the entire volume of the element and the second the enclosing surface area. The quantity $(\text{grad}\, \theta)_n$ is the outwardly directed normal component of the vector gradient of θ. The integral over area covers only the two flat surfaces (at x and at $x + dx$) since the gradient of θ normal to the outside ring is zero. Since all gradients and the enclosing surfaces are aligned with the x axis, $(\text{grad}\, \theta)_n$ is simply $\partial\theta/\partial x$ (or $-\partial\theta/\partial x$ for the lower surface). Hence the integral on the right hand side is

$$\int \frac{\partial \theta}{\partial x}\, da = \left(a \frac{\partial \theta}{\partial x}\right)_{x+dx} - \left(a \frac{\partial \theta}{\partial x}\right)_{x} = \frac{\partial}{\partial x}\, a \frac{\partial \theta}{\partial x}\, dx \quad (4.4\text{-}46)$$

Now since $\nabla^2\theta = -1$ (see equation (4.4-17)), the integral on the left is simply the negative of the elements volume, $-a\, dx$. Equating these two expressions, we have

$$\frac{1}{a} \frac{\partial}{\partial x}\, a \frac{\partial \theta}{\partial x} = -1 \quad (4.4\text{-}47)$$

or, in reduced coordinates,

$$\frac{1}{a}\frac{\partial}{\partial \hat{x}} a \frac{\partial \hat{\theta}}{\partial \hat{x}} = -1 \tag{4.4-48}$$

The pore cross-sectional area may be written either in reduced form ($\hat{a} = a/d^2$) or not; the factor d^2 cancels since one area term is in the numerator and one in the denominator. Equation (4.4-48) is the basic differential equation for $\hat{\theta}$, replacing the one in equation (4.4-21) for the narrow pore case considered here. The boundary conditions remain identical as expressed in (4.4-21).

Applications of Narrow Pores. Numerous applications can be found for the narrow pore equations just discussed. The uniform film, the rod, and the spherical forms of stationary phase are special cases as indicated in connection with Figure 4.4-2. For instance, the small spike shown in connection with the spherical (ion-exchange) bead in the figure has a cross-sectional area a which increases in proportion to r^2. If this r^2 dependence is used in place of a in the last equation, and the coordinate $\hat{x}$ replaced by $\hat{r}$, we arrive at the basic differential equation for the bead, equation (4.4-39).

The cross-sectional area of many pores or units of stationary phase will be given either precisely or approximately by the tapered-pore model

$$a = a_0 \hat{x}^n \tag{4.4-49}$$

where n is known as the taper factor. The differential equation, (4.4-48), becomes

$$\frac{1}{\hat{x}^n}\frac{\partial}{\partial \hat{x}} \hat{x}^n \frac{\partial \hat{\theta}}{\partial \hat{x}} = -1 \tag{4.4-50}$$

The integration with respect to x is achieved in a straightforward fashion

$$\frac{\partial \hat{\theta}}{\partial \hat{x}} = -\frac{\hat{x}}{n+1} + \frac{g}{\hat{x}^n} \tag{4.4-51}$$

$$\hat{\theta} = -\frac{\hat{x}^2}{2n+2} + \frac{g_0(\hat{x})^{1-n}}{1-n} + g_1 \tag{4.4-52}$$

If we make the restriction $n \geq 0$, g_0 is zero as shown by the usual closed-surface boundary condition and $g_1 = 1/(2n + 2)$. Thus

$$\hat{\theta} = \frac{1 - \hat{x}^2}{2n+2} \tag{4.4-53}$$

The mean value is

$$\bar{\hat{\theta}} = \int_0^1 a\hat{\theta}\, d\hat{x} \bigg/ \int_0^1 a\, d\hat{x} = 1/(n+1)(n+3) \tag{4.4-54}$$

and the configuration factor, twice this, is

$$q = 2/(n + 1)(n + 3) \tag{4.4-55}$$

The plate height may consequently be expressed as[18]

$$H = \frac{2}{(n + 1)(n + 3)} R(1 - R) \frac{d^2 v}{D_s} \tag{4.4-56}$$

This equation, as it stands or as a component of the combination law, (4.4-23), is probably a good approximation to real laboratory systems involving complex pore structures. In addition it has the following specialized applications:[18]

$n = 0$. This is equivalent to the uniform film model. The value of q is $\frac{2}{3}$, in agreement with equation (4.4-32).

$n = 1$. This is equivalent to the cylindrical rod (paper chromatography) model. It is seen that $q = \frac{1}{4}$. This agrees with equation (4.4-38).

$n = 2$. This value of the taper factor is applicable to spherical (ion-exchange) beads. We find $q = \frac{2}{15}$, in agreement with equation (4.4-44).

$n = 3$. This is applicable to the stationary liquid which accumulates around the contact points of glass beads when the latter are used in gas-liquid chromatography. The value of q is $\frac{1}{12}$.

The narrow pore model is not, of course, limited to the tapered-pore model of equation (4.4-49). Any narrow pore can be treated numerically. Truncated tapered pores have been treated analytically,[19] but the resultant plate height is a rather complicated expression.

Limiting Forms for Wide Pores. Some pores will undoubtedly resemble shallow depressions more than they will deep, narrow pores. Thus we will treat here pores whose overall width is considerably greater than the average depth. This problem is rather easy to solve, and by obtaining both wide and narrow pore limits, it will be possible to estimate the effect of mass transfer in intermediate pores. This is particularly useful since the latter would otherwise require a numerical procedure for an exact solution.

A wide pore is shown schematically in Figure 4.4-4. The width is such that very little mass transfer occurs by means of lateral diffusion. (This is true as long as the angle, α, remains small over the containing surface.) Close access to the open surface is responsible for the great bulk of mass transfer. Thus planes drawn perpendicular to the open surface become, to a good approximation, closed surfaces. A series of such planes is shown in the figure. Because they act as closed surfaces they serve to divide the liquid into *units*. These units have the advantage of simplicity; to a good approximation they are "tapered" units with a taper factor

$n = 0$. The general combination law, (3.5-23), for this collection of units may be written

$$H = \tfrac{2}{3}R(1 - R)\frac{v}{D_s}\sum\left(\frac{V_j}{V_s}\right)(\text{depth})^2 \tag{4.4-57}$$

The summation term is simply the square of the depth averaged over the stationary phase volume. It may be written as

$$\sum\frac{V_j}{V_s}(\text{depth})^2 = \int_0^d (d - x)^3 \frac{da}{dx}\,dx \Big/ \int_0^d (d - x)\frac{da}{dx}\,dx \tag{4.4-58}$$

where $d - x$ is the variable depth and d is the maximum depth. The integral in the denominator is simply the total volume in the wide pore.

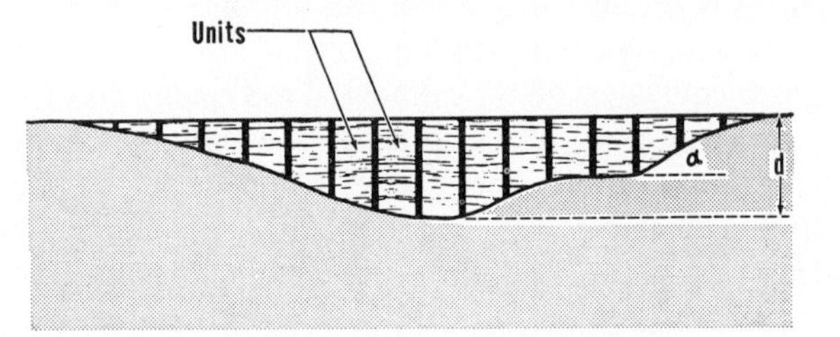

Figure 4.4-4 Schematic diagram of a wide pore filled with stationary phase. To a good approximation, with α small everywhere, the stationary phase may be divided into the simple *units* shown.

In the particular case of the tapered pore, where $a = a_0x^n$ (this can be applied to wide as well as narrow pores), both integrals of the last expression can be integrated by parts. The resulting equation is

$$H = \frac{4}{(n + 2)(n + 3)} R(1 - R)\frac{d^2v}{D_s} \tag{4.4-59}$$

that is, $q = 4/(n + 2)(n + 3)$. This compares to the configuration factor, $q = 2/(n + 1)(n + 3)$, given for narrow pores by equation (3.5-55). The ratio of q values (or plate heights) is

$$\frac{q(\text{wide pore})}{q(\text{narrow pore})} = \frac{2n + 2}{n + 2} \tag{4.4-60}$$

This ratio is given as a function of n in Table 4.4-1.

This table shows that the wide and narrow pore extremes are not vastly different for any of the taper factors in the spectrum. Many porous solids will probably support stationary phase in pores with an effective taper factor in the vicinity of unity. The divergence of the extremes is thus only about 30%. If the pore is intermediate between wide and narrow,

Table 4.4-1 Values of the q Ratio from Equation (4.4–60)

n	q Ratio
0	1
1	4/3
2	3/2
3	8/5
4	5/3
.	.
.	.
.	.
∞	2

it will be bracketed by the two extremes, and either one will be in error no more than about 15%. It should be pointed out, however, that not all pores of chromatographic interest are in the somewhat vague intermediate category; the contact point pores of a low-load glass-bead column in gas chromatography may be regarded as truly narrow pores.

4.5 Diffusion in the Mobile Phase

The problems connected with a sound and rigorous treatment of mass transfer in the mobile phase are rather formidable. No simplifying method comparable to the combination law for the stationary phase has been found applicable to the mobile phase. In addition the mass transfer terms which are calculated must be coupled with the eddy diffusion phenomenon as explained in Section 2.10 (see also Section 4.7).

The reasons for the excessive complexity of mobile phase nonequilibrium are not hard to find. This phase not only inherits all the problems of geometrical disarray which characterize the stationary phase, it also possesses an intricate flow-velocity pattern which must be quantitatively understood before rigorous calculations can be made. These two complexities are certainly not unrelated since the disorder of flow is a result of geometrical disorder. Nonetheless the conversion of the latter (once it is comprehended) to the former must be accomplished through extremely tedious hydrodynamic treatments. This problem is avoided with the stationary phase since its whole mass is at rest.

Approximate models for mobile phase nonequilibrium have been proposed, and these will be discussed briefly. Complete success (as measured by the ability to treat a problem rigorously and quantitatively) has been obtained in only three extreme cases; the capillary column of gas chromatography, certain transcolumn effects in large columns and the transparticle effect. Unfortunately, for most ordinary analytical columns, these phenomena are not always highly significant.

It will be recalled from Section 2.8 that five mass transfer mechanisms could be distinguished (with all due regard for overlapping effects) in the mobile phase. These mechanisms, illustrated in Figure 2.8-1, are: **(1)** transchannel diffusion, **(2)** transparticle diffusion, **(3)** short-range interchannel diffusion, **(4)** long-range interchannel diffusion, and **(5)** transcolumn diffusion. Approximate values of the mass transfer term ω (where $H = \omega d_p^2 v / D_m$) were calculated in that section. It was shown that **(1)**, which is analogous to the effect in capillary columns, is negligible, with a contribution less than 1% of total. Mechanism **(2)**, when applicable, is the order of 10%. Mechanisms **(3)** and **(4)** probably constitute the bulk of the plate height in regards to mobile phase mass transfer. However, mechanism **(5)**, particularly in large columns but perhaps also in some analytical columns, makes a significant contribution.

The theoretical status of the five mass transfer mechanisms is as follows:

1. Transchannel diffusion (involving diffusion across a single channel) is analogous to the effect in capillary columns.[20,21] The equations for capillary columns can thus be applied to an approximation, although this approach does not allow for the fluctuating diameter of the channel nor the rather complex flow pattern within its boundaries. The plate height can probably be calculated within a factor of 2 or 3.

2. Transparticle diffusion effects can be calculated with more confidence since the velocity of any mobile phase within the support particles is essentially zero (this is a result of the fine channel system, with consequent low permeability, which characterizes most chromatographic support particles). The main approximation concerns particle geometry and the obstructive factor for diffusion. With this difficulty the plate height is probably calculable only within 50% or so. The transparticle effect does not exist, of course, where the mobile phase does not (paper and ion-exchange chromatography) penetrate the supporting solid.

3 and **4.** Interchannel diffusion phenomena can also be treated approximately.[21] It is necessary to estimate the nature of the velocity inequalities within the packing (see Section 2.8) and the distances over which these inequalities exist. Effect **(3)** can probably be calculated within a factor of 2 or 3, but effect **(4)** is less certain. (Although of comparable magnitude, effect **(4)** is probably the least important of the two because of the coupling phenomenon; see Figure 2.10-4.)

5. Transcolumn diffusion effects can probably be calculated with more accuracy than any of the other effects. It is still necessary to know the velocity profile, but presumably this is measurable. If not, it can be estimated in terms of column parameters and particle size distribution. Two separate phenomena have been described with particular accuracy: the velocity inequalities associated with coiled or bent columns, and the general case of velocity inequalities which show radial symmetry.[22]

For most of the above cases the numerical results are not decidedly superior to the calculations of the random walk approach. The non-equilibrium approach is much more informative, since it gives a rigorous solution for any assumed model, thus indicating the propriety of the model (the random walk approach leaves one uncertain as to whether the error is in the model or in the calculations related thereto). For our purposes here, however, it is only necessary to discuss a few particular cases.

Mobile Phase Equations

As before, we will assume a simple two-phase system. The average velocity in the mobile phase is v and the local velocity is v_m'. The fraction of solute in the mobile phase at equilibrium is R. Thus the mean zone velocity $\mathscr{V}$ is Rv. The differential equation, (4.3-12), for the local departure term thus reduces to

$$\nabla^2 \epsilon_m' = \frac{v_m' - Rv}{D_m} \frac{\partial \ln c}{\partial z} \tag{4-5.1}$$

and the expression for plate height, equation (4.3-24), becomes

$$H = -\frac{\dfrac{2}{A_m} \displaystyle\int \epsilon_m' v_m' \, dA_m}{v \, \partial \ln c / \partial z} \tag{4.5-2}$$

Since the integral is equal to the mobile phase cross-sectional area A_m times the average of the product $\epsilon_m' v_m'$, this expression is equivalent to the simple form

$$H = -\frac{2\overline{\epsilon_m' v_m'}}{v \, \partial \ln c / \partial z} \tag{4.5-3}$$

an equation only slightly more complex than that for the stationary phase, (4.4-1).

The plate height calculated from the above equation for the mobile phase is always (except for the transcolumn effect) approximately proportional to the particle size d_p squared and to the velocity v, and inversely proportional to the mobile phase diffusion coefficient D_m

$$H = \omega d_p^2 v / D_m \tag{4.5-4}$$

where ω is essentially constant. Arguments indicating the validity of this expression for a particular effect were given in Section 2.8, and led to the equivalent equation, (2.8-9). A rigorous proof of this form can be obtained from the nonequilibrium expression just given. It is necessary to assume that all diffusion and flow processes locate themselves on a scale directly proportional to the particle diameter d_p (see Figure 2.8-1), and that the relative orientation of particles, packing density, etc., remains the same despite changes in d_p, v, or D_m.

First, we define the dimensionless variables

$$\hat{\epsilon}_m' = \epsilon_m' D_m / v d_p^2 \; \partial \ln c / \partial z, \; \hat{x} = x / d_p \tag{4.5-5}$$

where the coordinate x is symbolic of all lateral coordinates and thus $\hat{\nabla}^2 = d_p^2 \nabla^2$. With these definitions equation (4.5-1) can be written as

$$\hat{\nabla}^2 \hat{\epsilon}_m' = (v_m'/v) - R \tag{4.5-6}$$

The relative velocity v_m'/v remains constant despite changes in v and D_m. Since the flow processes were assumed to occur on a scale proportional to d_p, this ratio is a function of the dimensionless coordinate, x/d_p. Thus the right hand side is a function of this coordinate only

$$\hat{\nabla}^2 \hat{\epsilon}_m' = f(\hat{x}) \tag{4.5-7}$$

On integration this equation will yield a solution which, also, is solely a function of the dimensionless coordinate

$$\hat{\epsilon}_m' \equiv \hat{\epsilon}_m'(\hat{x}) \tag{4.5-8}$$

(The boundary conditions leading to (4.5-8) also expand on a scale proportional to d_p.) Now when ϵ_m' in equation (4.5-3) is replaced by $\hat{\epsilon}_m'$ through the use of equation (4.5-5), we obtain

$$H = -2(\overline{\hat{\epsilon}_m' v_m'/v}) \, v d_p^2 / D_m \tag{4.5-9}$$

Since both $\hat{\epsilon}_m'$ and the relative velocity v_m'/v are functions only of $\hat{x}$, the averaging procedure yields a constant. Thus we may write

$$\omega = -2(\overline{\hat{\epsilon}_m' v_m'/v}) \tag{4.5-10}$$

in which ω is constant despite changes in v, d_p, or D_m. Consequently equation (4.5-9) reduces to $H = \omega d_p^2 v / D_m$, as postulated in equation (4.5-4).

The Capillary Column Equation

The capillary column equation provides a classic example of mobile phase mass transfer where local velocity differences are of paramount importance. While the results are of little direct value to packed column studies, they provide a useful illustration of the mathematical procedure to be used and they are significant for capillary column technology. The derivation is valid for both gas and liquid mobile phases although the former has received nearly exclusive attention experimentally.

Laminar flow in a round capillary tube is accompanied by a parabolic flow profile in which the local velocity is

$$v_m' = 2v(1 - r^2/r_c^2) \qquad (4.5\text{-}11)$$

where r is the distance from the tube center and r_c is the tube radius. The basic differential equation (4.5-1), acquires the following form after this velocity expression has been substituted for v_m'

$$\frac{1}{r}\frac{\partial}{\partial r}r\frac{\partial \epsilon_m'}{\partial r} = \frac{v}{D_m}\frac{\partial \ln c}{\partial z}\left[(2 - R) - \frac{2r^2}{r_c^2}\right] \qquad (4.5\text{-}12)$$

The Laplacian expression on the left, $\nabla^2\epsilon_m'$, has been cast into cylindrical coordinates (the same form appears for cylindrical rods, equation (4.4-33), in the study of paper chromatography). The integration with respect to r is straightforward

$$\epsilon_m' = \frac{v}{D_m}\frac{\partial \ln c}{\partial z}\left[\frac{(2 - R)r^2}{4} - \frac{r^4}{8r_c^2}\right] + g_0 \ln r + g_1 \qquad (4.5\text{-}13)$$

The symmetry boundary condition, equation (4.3-16) (expressing the fact that $\partial\epsilon_m'/\partial r$ must vanish at $r = 0$), leads to the usual result, $g_0 = 0$. The interfacial-equilibrium condition, equation (4.3-15), can be applied to equation (4.5-13) to yield the following relationship between ϵ_s and g_1 (in keeping with our previous procedure, nonequilibrium gradients in one phase are considered in conjunction with the lack of gradients in the other phase; the constancy in the departure term of the latter means that its local value, in this case ϵ_s', can be replaced by the overall value, ϵ_s)

$$\epsilon_s = \frac{vr_c^2}{D_m}\frac{\partial \ln c}{\partial z}\left(\frac{3}{8} - \frac{R}{4}\right) + g_1 \qquad (4.5\text{-}14)$$

The balance-of-nonequilibrium expression, in the form of

$$\epsilon_m R = -\epsilon_s(1 - R),$$

provides another relationship between ϵ_s and g_1, thus allowing a solution for these two terms. The proper relationship is obtained in view of the fact that $\epsilon_m = \overline{\epsilon_m'}$, and that the latter can be acquired by averaging equation (4.5-13) over the cross-sectional area. The resultant solution for g_1 is

$$g_1 = -\frac{vr_c^2}{24D_m}\frac{\partial \ln c}{\partial z}(3R^2 - 10R + 9) \qquad (4.5\text{-}15)$$

With this and the fact that $g_0 = 0$, the final ϵ_m' expression following equation (4.5-13) is

$$\epsilon_m' = \frac{v}{4D_m}\frac{\partial \ln c}{\partial z}\left[(2 - R)r^2 - \frac{r^4}{2r_0^2} - \frac{r_c^2}{6}(3R^2 - 10R + 9)\right] \qquad (4.5\text{-}16)$$

a result in agreement with an earlier derivation.

As equation (4.5-3) shows, the plate height expression requires an averaging (over the tube cross section) of the product $\epsilon_m' v_m'$. Both of these terms are now available, equations (4.5-11) and (4.5-16), and following the usual averaging procedure (involving some rather detailed but routine steps) the plate height is obtained as

$$H = (6R^2 - 16R + 11)r_c^2 v/24D_m \qquad (4.5\text{-}17)$$

Had the velocity been assumed uniform over the column cross section, this result would be[1]

$$H = (1 - R)^2 r_c^2 v/4D_m \qquad (4.5\text{-}18)$$

The significant difference between these expressions arises in the detailed consideration of the local velocity profile in (4.5-17). Numerically, the latter yields a plate height $\frac{11}{6}$ times larger than the uniform flow case at total retention ($R = 0$) and infinitely greater (since equation (4.5-18) goes to zero) when there is no retention ($R = 1$).

A form equivalent to (4.5-17) was first derived by Golay using an approach suggested by the telegrapher's equation.[23] It has since been derived in different ways,[3,24] including the generalized nonequilibrium method just outlined.[3] It will be noticed that the final product (4.5-17), despite first appearances, is still consistent with the general form of equation (4.5-4) if we consider the fact that diffusion is occurring on a scale proportional to the tube radius, r_c, rather than d_p.

Effect of Transcolumn Velocity Variations

The capillary column equation, just derived, is approximately applicable to the phenomenon of smallest scale leading to mobile phase nonequilibrium—namely, to transchannel diffusion. On the other end of the spectrum we must deal with nonequilibrium over entire column dimensions, particularly important with large columns. Such nonequilibrium is generally caused by variations in the mobile phase velocity over the column cross section.

Several authors have dealt with particular cases of transcolumn velocity profiles.[25,26] Below we indicate a general approach which allows for any form of velocity profile so long as it remains symmetric around the center axis of the column. This broad approach has also been used together with the generalized nonequilibrium theory in obtaining the plate height equation in the presence of a variable cross-sectional retention as well as a variable velocity.[22] This will be discussed later.

For now we may assume a very general type of symmetric profile in which the mobile phase velocity v is expanded in the series[22]

$$v = \tilde{v} \sum G_n(r/r_c)^n \tag{4.5-19}$$

where $\tilde{v}$ is the mean velocity averaged over the column cross section, r_c is the column radius, r the variable distance along the radial axis, and the G_n's are the expansion coefficients. If the actual velocity profile is not of some unusual form, it may be approximated to any desired degree of accuracy by taking sufficient terms in the above equation. This is a consequence of Taylor's theorem or its special case, Maclaurin's.

The differential equation for equilibrium departure may be written in a form slightly different from equation (4.5-1), i.e.

$$\nabla^2\epsilon = \frac{v - \tilde{v}}{\gamma D_m} \frac{\partial \ln c}{\partial z} \tag{4.5-20}$$

where ϵ is the overall (instead of local) equilibrium departure in a given region of the column. It measures the total departure from equilibrium in the given region, including both stationary and mobile phases. Another "overall" parameter which replaces its local counter-part is γD_m. The factor γ, discussed in connection with longitudinal diffusion in Sections 2.6, and 6.4, accounts for the obstruction to lateral diffusion caused by the fixed particles. It is assumed, of course, that the medium is isotropic, and that longitudinal and lateral obstructive factors are the same.

When the v from equation (4.5-19) is substituted into this, and $\nabla^2\epsilon$ written in cylindrical coordinates (with angular and axial variations assumed negligible), an integration can be made with respect to r. The boundary conditions are analogous to those used in connection with capillary columns. Following the usual procedure the plate height is obtained as[22]

$$H = \frac{\tilde{v}r_c^{\,2}}{\gamma D_m} \sum\sum \frac{G_n G_m}{(n+2)} \left[\frac{2}{(m+4)} - \frac{4}{(n+2)(n+m+4)} - \frac{n(n+6)}{(m+2)(n+2)(n+4)} \right] \tag{4.5-21}$$

and can thus be obtained generally in terms of the G_n coefficients.

An interesting and special case of the above, simple enough to be of practical value, occurs when the velocity variation can be approximated as a simple quadratic form. In this case the plate height is simply[22]

$$H = \left(\frac{\nu - 1}{\nu + 1}\right)^2 \frac{\tilde{v} r_c^{\,2}}{24\gamma D_m} \tag{4.5-22}$$

where ν is the ratio of the extreme outside velocity to the center velocity.

As mentioned earlier, the nonequilibrium theory can be expanded to include several transcolumn effects in addition to the velocity variation. Thus if we consider the fact that relative retention (measured by R) may vary over the cross section, and that structural differences exist in the column packing thus giving a variation in porosity p and in γ, the basic differential equation becomes[22]

$$p\nabla \cdot \gamma\nabla \frac{\epsilon}{p} = \left[v - \frac{\tilde{v}}{R(\widetilde{1/R})}\right] \frac{1}{D_m} \frac{\partial \ln c}{\partial z} \tag{4.5-23}$$

where $\widetilde{1/R}$ is the cross-sectional average of $1/R$. This equation reduces to (4.5-20) as a special case. While a great number of solutions have not yet been sought for this equation, the plate height has been calculated for a general variation in both v and R. The literature should be consulted for more details on this problem.[22]

The Nature of Interchannel Diffusion

Interchannel diffusion (effects **3** and **4**) is the most difficult mass transfer phenomenon to account for. We must postulate, first, a model which reproduces the main features of the erratic flow profile within chromatographic materials. The generalized nonequilibrium theory may then be applied to the model in order to derive a plate height expression. As pointed out earlier, the rigorous nonequilibrium approach has a real advantage over approximate theories such as that based on the random walk. Because the theory is exact, discrepancies between theory and experiment can be attributed to failings in the model, and remedial steps can be taken. At the present time, however, experimental parameters have not been measured in sufficient detail to require refinements in the crude model discussed below.

A certain degree of "order" can be discerned in the otherwise random structure of column-packing material (Section 5.2). After every several particle diameters an unusually large void space, caused by the poor fit of adjacent particles, occurs.[21,27] This space offers a relatively unrestricted path for flow, and it may be presumed that the mobile phase velocity speeds up through such regions. The velocity differential between the voids and the restricted channels develops a nonequilibrium with respect to the two, and leads to a plate height contribution.

The model which has been developed[21] to account for this short-range effect is shown in Figure 4.5-1. The "outer annulus" corresponds to tightly packed particles while the "inner core" is the high velocity region

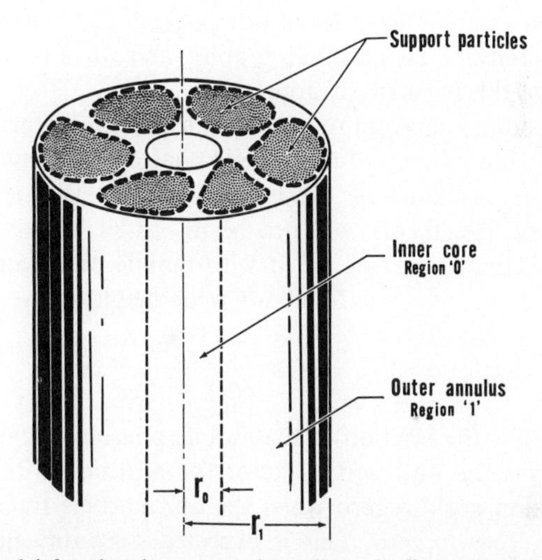

Figure 4.5-1 A rough model for the short-range interchannel effect. The inner core corresponds to the void (bridged) spaces which allow the relatively free passage of mobile phase.

analogous to the open void spaces. Approximate estimates can be made for the relative dimensions of these two regions and the relative velocities found therein. The arguments and the detailed calculations are too long to reproduce here. The approximations obtained for the universal ω parameter of equation (3.5-87) are

$$\omega \cong 0.70(1 - 0.15R) \tag{4.5-24}$$

for porous solids permeated with the mobile phase, and

$$\omega \cong 0.62(1 - 0.3R) \tag{4.5-25}$$

for nonporous support particles.[21] These are the same order of magnitude[21] (although somewhat lower because this is only part of the total mobile phase contribution) as the experimentally determined[20,28-32] values of ω.

Although this model is rather crude, it has not been possible to do even this well for long-range interchannel diffusion. However if the speculation indicated by Figure 2.10-4 is correct the latter effect, shown as the h_4 curve, is a negligible part of the whole.

Transparticle Diffusion

This is the single remaining phenomenon of the five not yet related to nonequilibrium concepts. This process occurs when the support particle is porous and its channels are filled with mobile phase (the stationary phase is presumably in small side pockets and pores). It is most important in the study of gas chromatography, and was first recognized by workers in that field, particularly Jones.[33]

The nonequilibrium theory of transparticle diffusion is relatively simple because the velocity of the mobile phase is uniformly zero. This is caused by the restrictions of the extremely fine channels within the support particles. The theory will not be discussed in detail here, but it can be shown[18] that a particle laden with mobile phase and approximated by a sphere contributes the following plate height term

$$H = \frac{2(1 - \Phi R)^2}{60(1 - \Phi)\gamma_p} \frac{d_p{}^2 v}{D_m} \tag{4.5-26}$$

where Φ is the fraction of the mobile phase occupying interparticle space and γ_p is the obstruction factor for diffusion within the particles. This expression goes to zero when the intraparticle fraction of mobile phase, $1 - \Phi$, goes to zero. This is expected since diffusion through the intraparticle mobile phase leads to the exchange of solute through the particles. The absence of such mobile phase would lead to a zero rate of mass transfer and, consequently, a very ineffective column with H calculated as infinity. (The actual value would not be infinity since the lack of mass transfer would render the near-equilibrium hypothesis inoperative.)

The ω parameter of equation (4.5-4) can be identified with the first part of the above equation. Its value is somewhat uncertain, but it probably resides somewhere in the neighborhood 0.1. This is the same magnitude as predicted by the random walk model, Table 2.10-1.

4.6 Some Mixed Mass Transfer Mechanisms

In striving to mold the nonequilibrium theory to real chromatographic systems, we must proceed considerably beyond the simple mass transfer mechanisms of Chapter 3. In this section we will indicate several approaches to be used when mass transfer occurs by a combined process of diffusion and step-wise kinetics.

The obvious approach to complex mass transfer processes is to consider, simultaneously, the nonequilibrium in every part of the system. This approach is rather complex, but it does provide a rigorous solution to the problem. Hence we will illustrate this approach in relation to a very simple example immediately below. A second approach, which may be followed in most circumstances, involves the use of the *additive assumption.* This greatly simplifies the calculation of nonequilibrium and plate-height parameters.

The Total Nonequilibrium Approach: Simple Adsorption from a "Flat" Channel

Probably the simplest model of the chromatographic process which still involves mass transfer by both diffusion and step-wise kinetics is simple adsorption from a "flat" channel. The adsorbing surface will be a 1-site surface with adsorption and desorption rate constants k_a and k_d. The channel will be that existing between two parallel surfaces a distance $2d$ apart (these will represent the solid particles which form the boundaries to the flow channels within the packing). The flow profile will be assumed uniform—the local velocity v_m' will be constant throughout the channel and equal to v. The quantity x will be the distance out from the channel center.

Step One: Mass Transfer and Nonequilibrium. In this case we have two such parameters, one for the mobile phase and one for the stationary phase. The former is a three-dimensional phase involving only diffusion, and the mass transfer term is consequently the same as obtained in the last section, equation (4.3-8)

$$s_m' = D_m c_m{}^{*\prime} \nabla^2 \epsilon_m' \tag{4.6-1}$$

The stationary phase involves only step-wise kinetics. The mass transfer term is therefore obtainable as a special case of equation (4.2-14)

$$s_s = c_s{}^* k_d[(\epsilon_m')_{x=d} - \epsilon_s] \tag{4.6-2}$$

where $(\epsilon_m')_{x=d}$ is the value of ϵ_m' at the adsorbing wall, $x = d$. Note that concentrations and nonequilibrium terms in the mobile phase are local values as indicated by a prime, whereas stationary phase quantities are in terms of overall values.

Step Two: Mass Transfer and Flow. The equations here are nearly identical for diffusion and step-wise kinetics. Thus equation (4.3-10) for diffusion reduces to

$$s_m' = (1 - R)v\, \partial c_m{}^{*\prime}/\partial z \tag{4.6-3}$$

and equation (4.2-19) for step-wise kinetics gives the stationary phase term

$$s_s = -Rv\, \partial c_s{}^*/\partial z \tag{4.6-4}$$

Step Three. This involves a combination of the last two steps to obtain the departure terms. The combination of mobile phase terms, (4.6-1) and (4.6-3), yields

$$\nabla^2 \epsilon_m' = \frac{\partial^2 \epsilon_m'}{\partial x^2} = \frac{(1-R)v}{D_m} \frac{\partial \ln c}{\partial z} \tag{4.6-5}$$

while the combination of (4.6-2) and (4.6-4) gives

$$(\epsilon_m')_{x=d} - \epsilon_s = -\frac{Rv}{k_d} \frac{\partial \ln c}{\partial z} \tag{4.6-6}$$

Integration of the first of these gives

$$\epsilon_m' = \frac{(1-R)v}{D_m} \frac{\partial \ln c}{\partial z} \frac{x^2}{2} + g_0 x + g_1 \tag{4.6-7}$$

The last two simultaneous equations in ϵ_s and ϵ_m' incorporate the four unknowns; ϵ_m', ϵ_s, g_0, and g_1. Once these are obtained, both the parameter ϵ_s and the continuous variable ϵ_m' will be specified. Four relationships can be used to establish the value of the four quantities. First, the symmetry condition can be applied to the center of the channel, $\partial \epsilon_m'/\partial x = 0$ at $x = 0$. This yields $g_0 = 0$. Second, the balance-of-nonequilibrium condition applies as

$$R\overline{\epsilon_m'} = -(1-R)\epsilon_s \tag{4.6-8}$$

Finally, the two equations, (4.6-6) and (4.6-7), give the final relationships between the unknowns, making four relationships in all. The four unknowns obtained from these relationships are

$$g_0 = 0 \tag{4.6-9}$$

$$g_1 = v \frac{\partial \ln c}{\partial z} \left[\frac{(1-R)(2R-3)\, d^2}{6D_m} + \frac{(R-1)R}{k_d} \right] \tag{4.6-10}$$

$$\epsilon_s = v \frac{\partial \ln c}{\partial z} \left[\frac{R(1-R)\, d^2}{3D_m} + \frac{R^2}{k_d} \right] \tag{4.6-11}$$

$$\epsilon_m' = v \frac{\partial \ln c}{\partial z} \left[\frac{(1-R)x^2}{2D_m} + \frac{(1-R)(2R-3)\, d^2}{6D_m} + \frac{(R-1)R}{k_d} \right] \tag{4.6-12}$$

The mean value of ϵ_m' is

$$\overline{\epsilon_m'} = -(1-R)v \frac{\partial \ln c}{\partial z} \left[\frac{(1-R)\, d^2}{3D_m} + \frac{R}{k_d} \right] \tag{4.6-13}$$

an equation which will be used below.

Step Four. This leads to the final plate height equation. The solute flux which leads to zone dispersion originates entirely in the mobile phase.

We are therefore justified in using the equation of Section 4.3, dealing with diffusional transport, to calculate H. Both equations (4.3-24) and (4.5-3) show that the plate height for the present example is

$$H = -\frac{2\overline{\epsilon_m' v_m'}}{v\, \partial \ln c/\partial z} \tag{4.6-14}$$

Since, however, v_m' is constant and equal to v, H becomes

$$H = \frac{-2\overline{\epsilon_m'}}{\partial \ln c/\partial z} \tag{4.6-15}$$

Substitution of (4.6-13) into this yields

$$H = \tfrac{2}{3}(1 - R)^2 \frac{d^2 v}{D_m} + \frac{2R(1 - R)v}{k_d} \tag{4.6-16}$$

This equation is a sum of the two terms. The first term contains D_m and is thus related only to mobile phase mass transfer. The second contains only k_d and is related to the stationary phase process. On closer inspection it is found that each term is the same as obtained when considering each process singly; the first term has been reported in the literature for the same type of mobile phase[3] and the second term is identical to that shown in equation (3.2-36) for 1-site surfaces. This equation strongly suggests that complex processes may be broken into their constituent parts. This point was also emphasized in Section 4.2. This matter will be discussed in the following paragraphs.

Additive Law for Multiple Processes

The rule which specifies how complex, multiple processes may be separated into their constituent parts is the *additive law*.[3] This law applies directly to the plate height (or, equally, to the effective diffusion coefficient). Consider a group of mass transfer processes involving both step-wise kinetics (with rate constants $k_1, k_2, \ldots k_m$) and several diffusion processes (with diffusion coefficients $D_1, D_2, \ldots D_n$). The plate height is obviously a function of these rate parameters

$$H \equiv H(k_1, k_2, \ldots k_m; D_1, D_2, \ldots D_n) \tag{4.6-17}$$

The additive law states that H may be broken into the following component terms[3]

$$H = H(k_1, \infty) + H(k_2, \infty) + \ldots + H(D_1, \infty) + H(D_2, \infty) + \ldots \tag{4.6-18}$$

where, for example, $H(k_2, \infty)$ means that all rate parameters except k_2

are considered as infinity. Equation (4.6-16) represents a special case of this law, and may be written as

$$H(D_m, k_d) = H(D_m, \infty) + H(\infty, k_d) \tag{4.6-19}$$

It is important to note that each component term must be calculated with full allowance made for the equilibrium concentration in each phase or state. Thus the additive law simplifies the rate processes, but it does not do away entirely with the necessity for considering the total system with each calculation.

No general proof has yet been given for the additive law. It is obeyed in many special cases, but exceptions can be found. The general validity of this law will be discussed further in Section 4.7.

Application to Simultaneous Partition-Adsorption

The presence of adsorption in partition chromatography was discussed in Section 4.2. The kinetic analog used for this system, given in equation (4.2-46), is applicable in a slightly modified form

$$\begin{array}{ccccc} D_m & & D_s & & \\ A_m & \underset{k_{sm}}{\overset{k_{ms}}{\rightleftarrows}} & A_s & \underset{k_{as}}{\overset{k_{sa}}{\rightleftarrows}} & A_a \\ \text{mobile phase} & & \text{bulk stat. phase} & & \text{adsorbed state} \end{array} \tag{4.6-20}$$

There are four rate processes which may influence mass transfer in this system. This is illustrated in Figure 4.6-1. There are two diffusion processes (not considered in the earlier treatment), found in the mobile and

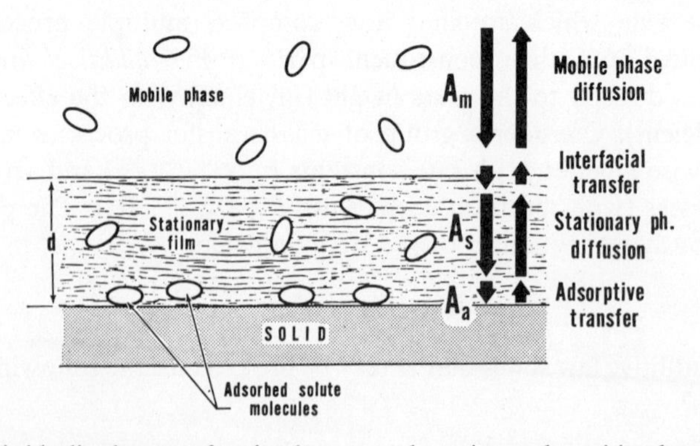

Figure 4.6-1 This idealized system for simultaneous adsorption and partition has four mass transfer processes, each indicated by a double arrow. Solute may reside in three states as indicated by A_m, A_s, and A_a.

stationary phases. There is a single-step kinetic process for the adsorptive exchange and also for transfer across the interface between mobile and stationary phases (the double arrows indicate a forward and a backward reaction, but this double process can be characterized with one rate constant if an equilibrium constant is used). Below we will consider the application of the additive law to this complex system. The bulk stationary phase will be assumed to form a uniform film covering the adsorbing surface.

The separation of terms by the additive law is indicated in the following equation

$$H = H(D_m, \infty) + H(D_s, \infty) + H(k_{sm}, \infty) + H(k_{as}, \infty) \quad (4.6\text{-}21)$$

An analysis of the component terms follows.

***H*(D_m, ∞).** This term is of precisely the same form as found in Section 4.5 where adsorption was not considered. (Adsorption may change the R value, but at a given value of R this contribution is independent of the amount of adsorption.) To demonstrate, consider the pair of mobile phase equations, (4.5-1) and (4.5-3). The first of the two is the basic differential equation for ϵ_m' and involves only mobile phase properties (v_m' and D_m) aside from R and the concentration gradient. All of the boundary and other conditions applicable to this equation are equally unresponsive to the details of the stationary phase except as it influences the equilibrium parameter R. The interfacial equilibrium condition may seem to violate this rule, but it does not in fact do so. Thus if we stipulate that ϵ_m' (at interface) $= \epsilon_s$, there would appear to be a dependence on a stationary phase nonequilibrium property ϵ_s. However, the balance-of-nonequilibrium condition specifies that ϵ_s must equal $-R\overline{\epsilon_m'}/(1 - R)$, and thus, except through R, the stationary phase has no effect. The balance-of-nonequilibrium expression is valid in this form even if there are several retention mechanisms as illustrated here by partition and adsorption. Since the rates related thereto are effectively infinite for this calculation, each part has the same ϵ value, ϵ_s, and the total unbalance with respect to equilibrium is always the same: the product of ϵ_s and the total fraction existing within the stationary phase at equilibrium, $1 - R$. Since ϵ_m' is unaffected by stationary phase kinetics, the plate height, equation (4.5-3), is also unchanged. In view of these considerations the mobile phase contribution may be written in its conventional form, equation (4.5-4)

$$H(D_m, \infty) = \omega d_p^2 v / D_m \quad (4.6\text{-}22)$$

***H*(D_s, ∞).** This term cannot be obtained so simply as that for diffusion in the mobile phase. The amount of underlying adsorption does make a difference here, even at constant R, as can be seen by means of the following physical argument. Each solute molecule which adsorbs on the underlying surface must gain access to that surface by diffusion through the bulk stationary phase. Thus the rate at which mass transfer occurs between the mobile and the adsorbed state is determined in large part by the intervening diffusion rate of the stationary phase. This illustrates our earlier argument that the role of a given rate process (for stationary phase diffusion) may be affected by the relative equilibrium distribution of solute in the other phases and locations (for the adsorbed state).

Since the earlier equation, (3.3-35), for a uniform stationary film is now invalid, we must start anew with the derivation of this particular plate height term.

The basic differential equation for ϵ_s' is the same as derived generally in equation (4.3-12) and shown specifically in equation (3.3-18)

$$\frac{\partial^2 \epsilon_s'}{\partial x^2} = -\frac{Rv}{D_s}\frac{\partial \ln c}{\partial z} \tag{4.6-23}$$

Integration of this gives (as shown by equation (3.3-20))

$$\epsilon_s' = \left(-\frac{Rv}{D_s}\frac{\partial \ln c}{\partial z}\right)\frac{x^2}{2} + g_0 x + g_1 \tag{4.6-24}$$

where, as before, x is the distance from the bottom of the liquid film where the adsorption takes place. Altogether in this system there are five unknown quantities: ϵ_s', g_0, g_1, ϵ_a, ϵ_m. Along with equation (4.6-24), other relationships between these quantities are the two expressing interfacial equilibrium (these are rigorous expressions since all rates but D_s are assumed infinite)

$$(\epsilon_s')_{x=0} = \epsilon_a \tag{4.6-25}$$

$$(\epsilon_s')_{x=d} = \epsilon_m, \tag{4.6-26}$$

and the balance of nonequilibrium expression

$$X_m^* \epsilon_m + X_s^* \overline{\epsilon_s'} + X_a^* \epsilon_a = 0 \tag{4.6-27}$$

This gives four expressions in all. One more must apparently be obtained outside of our usual procedure. This last one is acquired by writing the mass transfer term for the adsorbed state in terms of the flux of solute by diffusion through the stationary phase

$$D_s\left(\frac{\partial c_s'}{\partial x}\right)_{x=0} A = s_a = -Rv\frac{\partial c_a^*}{\partial z} \tag{4.6-28}$$

where A is the area of the adsorbing surface (and also the area of the stationary phase film) per unit volume of the column. If we use the equations $c_s' = c_s^{*\prime}(1 + \epsilon_s')$, $A = c_s^*/c_s^{*\prime}d$, and $c_a^*/c_s^* = X_a^*/X_s^*$, this expression becomes

$$\left(\frac{\partial \epsilon_s'}{\partial x}\right)_{x=0} = -\frac{X_a^*}{X_s^*}\frac{Rv\,d}{D_s}\frac{\partial \ln c}{\partial z} \tag{4.6-29}$$

This gives the fifth and last equation for determining the five unknowns. The process of actually obtaining these unknowns involves manipulations too lengthy to reproduce here. The methods for this are quite standard and no major difficulties are involved. We find that the quantity of greatest direct interest is

$$\epsilon_m = -\frac{Rv\,d^2}{D_s}\frac{\partial \ln c}{\partial z}\left(\frac{X_s^*}{3} + X_a^* + \frac{X_a^{*2}}{X_s^*}\right) \tag{4.6-30}$$

Since the plate height for stationary-phase processes is $H = -2\epsilon_m/\partial \ln c/\partial z$, equation (3.2-41), we have[18]

$$H = \tfrac{2}{3}R\frac{d^2v}{D_s}\left(X_s^* + 3X_a^* + 3\frac{X_a^{*2}}{X_s^*}\right) \tag{4.6-31}$$

The total retained fraction, $1 - R$, is equal to the sum

$$1 - R = X_s^* + X_a^* \tag{4.6-32}$$

Solving for X_s^* and using this in equation (4.6-31), we have

$$H(D_s, \infty) = \tfrac{2}{3}R(1 - R)\frac{d^2v}{D_s}\left[1 + 2\phi + \frac{3\phi^2}{1 - \phi}\right] \tag{4.6-33}$$

where

$$\phi = X_a^*/(1 - R) \tag{4.6-34}$$

when there is no adsorption, $X_a^* = 0$ and thus $\phi = 0$, the quantity in square brackets equals unity and the plate height is identical to that found for a uniform film in pure partition chromatography, equation (3.3-35). In the presence of adsorption the quantity in square brackets represents the relative increase in plate height for stationary phase diffusion

$$\frac{H(\text{with adsorption})}{H(\text{without adsorption})} = \left[1 + 2\phi + \frac{3\phi^2}{1 - \phi}\right] \tag{4.6-35}$$

This ratio increases strongly with ϕ, as shown in Figure 4.6-2, indicating that a large ϕ might be a real detriment to practical systems. Since ϕ is the fraction of retained solute which is actually adsorbed, this indicates that adsorption should not be allowed to become excessive in systems which are mainly of the partition type.

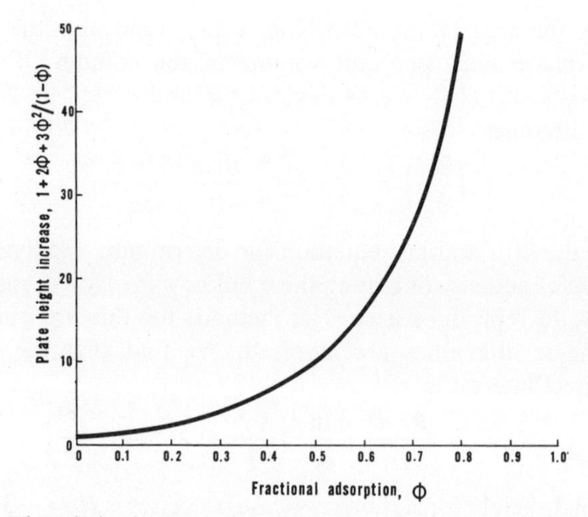

Figure 4.6-2 The relative increase of plate height caused by adsorption in a partition-type system (see equation (4.6-34)). At high fractional adsorption the plate height becomes excessive. The plate height term corresponds to diffusion in the stationary film, only, and is not related to the actual kinetics of adsorption.

$H(k_{sm}, \infty)$. This term corresponds to the limited rate of transfer through the interface separating the mobile phase and the stationary film. The problem has been considered by several authors in gas-liquid chromatography.[24,34,35] Although this term represents part of a complex system, it is actually equivalent to the simple term for 1-site adsorption presented in Section 2.3. The reason for the identity between the two is that adsorption, in each case, transfers solute from the mobile to the stationary phase, while desorption has the opposite effect. Since the calculation of $H(k_{sm}, \infty)$ is made on the basis that all other rates are infinite, this becomes the rather universal problem of transferring solute between mobile and stationary phases (the relative amount in all cases governed by the equilibrium ratios, R and $1 - R$) by means of simple, first-order reaction steps. Thus the plate height contribution is identical to equation (3.2-36), except that the desorption rate constant is replaced by k_{sm}

$$H(k_{sm}, \infty) = 2R(1 - R)v/k_{sm} \tag{4.6-36}$$

This expression is also equivalent to the first part of equation (4.2-49) where simultaneous adsorption and partition was discussed.

$H(k_{as}, \infty)$. This term is not identical to that for the 1-site problem since this particular adsorption transfers solute between two parts of the stationary phase. It can, however, be equated to the term obtained in our earlier discussion of adsorption-partition processes. If k_{32} is written as

k_{as} and $X_3{}^*$ as $X_a{}^*$ in the last term of equation (4.2-49) we have

$$H(k_{as}, \infty) = 2RX_a{}^*v/k_{as} = 2R(1 - R)\phi v/k_{as} \qquad (4.6\text{-}37)$$

The justification for using this expression from an earlier treatment where diffusion was ignored is based on the fact that $H(k_{as}, \infty)$ is, by definition, a term in which nothing but k_{as} has any effect on mass transfer and nonequilibrium.

The addition of equations (4.6-22), (4.6-33), (4.6-36), and (4.6-37) gives an expression for the plate height in accordance with the additive law of equation (4.6-21)

$$H = \omega \frac{d_p{}^2 v}{D_m} + \tfrac{2}{3}R(1 - R)\frac{d^2 v}{D_s}\left[1 + 2\phi + \frac{3\phi^2}{1 - \phi}\right] + 2R(1 - R)v(1/k_{sm} + \phi/k_{as}) \qquad (4.6\text{-}38)$$

where, as before, ϕ is the fraction of the retained solute which is adsorbed (see equation (4.6-34)).

As an alternative the adsorption-partition problem could have been treated without the breakdown afforded by the additive law. The intermediate equations would be, in this case, exceedingly cumbersome.

Adsorption at the Interface Between Mobile and Stationary Phases

One other adsorption effect which may accompany partition chromatography is that occurring between the mobile and stationary phases. Adsorption of this type has been observed for gas-liquid chromatography,[36,37] and has been related to the Gibbs adsorption equation.[38] Such a phenomenon may also occur in liquid chromatography where solute molecules may have a polar group immersed in the most polar phase with a nonpolar end left behind in the relatively nonpolar medium.

Consider the very general adsorption-partition problem indicated by the following kinetic analog

$$\underset{\text{mobile phase}}{\overset{D_m}{A_m}} \rightleftarrows \underset{\text{first adsorbed state}}{A_\alpha} \rightleftarrows \underset{\text{bulk stat. phase}}{\overset{D_s}{A_s}} \rightleftarrows \underset{\text{second adsorbed state}}{A_a} \qquad (4.6\text{-}39)$$

There are 5 important rate processes, 2 of diffusion and 3 of step-wise kinetics. The additive law can be expressed as

$$H = H(D_m, \infty) + H(D_s, \infty) + H(k_{\alpha m}, \infty) + H(k_{\alpha s}, \infty) + H(k_{as}, \infty) \qquad (4.6\text{-}40)$$

This problem can be treated in a fashion analogous to the preceding case in which A_α was not present. The detailed derivation will not be given here.

Proceeding through the above terms we find that the first, $H(D_m, \infty)$, is the same as before, equation (4.6-22). The second term is also identical to the form expressed in equation (4.6-31) (but is not the same as equation (4.6-33) since the intervening equation is not applicable). The third and fourth terms are found to be[18]

$$H(k_{\alpha m}, \infty) = 2R(1 - R)^2 v / X_\alpha^* k_{\alpha m} \tag{4.6-41}$$

$$H(k_{\alpha s}, \infty) = 2R(X_s^* + X_a^*)^2 v / X_\alpha^* k_{\alpha s} \tag{4.6-42}$$

The fifth term is the same as before, and is given by the first part of equation (4.6-37).

Adsorption Kinetics in Ion-Exchange Chromatography

Although diffusion through ion-exchange beads is generally the controlling step of solute exchange,[39] we may calculate the plate height for those cases in which mass transfer is limited both by diffusion and by adsorption[40] at particular sites within the bead. The kinetic analog is much the same as the partition-adsorption case, $A_m \rightleftarrows A_s \rightleftarrows A_a$, corresponding to the reversible transition between mobile, stationary, and adsorbed solute. The model is different, however, since adsorption sites are distributed evenly throughout the bulk stationary phase rather than at the bottom of a "pool" or "film" of this phase. See Figure 4.6-3.

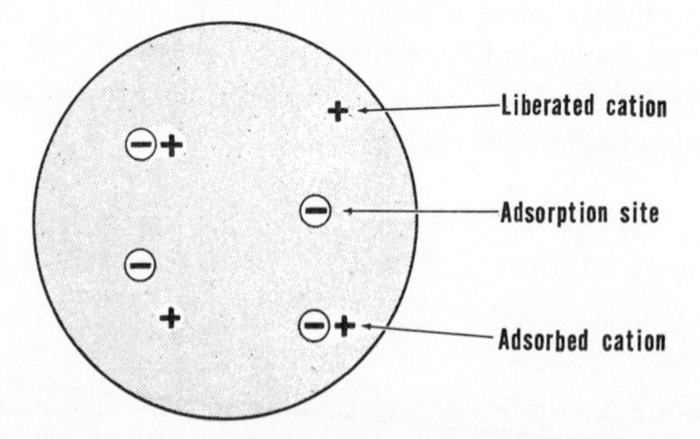

Figure 4.6-3 A schematic diagram of an ion-exchange bead showing anionic adsorption sites along with adsorbed and liberated cations. The rate of cation adsorption depends on (1) the rate of diffusion of the cation to the site, and (2) the adsorption rate. These two cases are covered by equations (4.6-43) and (4.6-44).

It can be shown[19] that the foregoing model leads to the following expression for the stationary phase diffusion process

$$H(D_s, \infty) = \frac{R(1 - R)}{30(1 - \phi)} \frac{d_p^2 v}{D_s} \tag{4.6-43}$$

where, as in equation (4.6-34), ϕ is the relative adsorption parameter. The adsorption process leads to the terms

$$H(k_{as}, \infty) = 2RX_a^* v/k_{as} = 2R(1 - R)\phi v/k_{as} \tag{4.6-44}$$

which is identical to equation (4.6-37). This identity is reasonable since, if diffusion rates are taken as infinite, all sites are equally accessible to the solute molecules in the system. The matter of accessibility itself is a part of the stationary phase diffusion term, $H(D_s, \infty)$, and it is the latter which changes as the sites become distributed differently.

4.7 Extensions, Limits, and Assumptions

The generalized nonequilibrium theory is not a total and complete theory of the dynamics of chromatography. At least in its present stage (see below), it does not account for the phenomenon of erratic stream flow as made manifest in "eddy" diffusion. It does not apply to slow rate processes within a chromatographic column. Finally, the theoretical calculations must be based on some model which gives the dimensions, velocity, and relative positions of all parts of the mobile and stationary phases, and which pinpoints all adsorption processes. The acquisition of a realistic model is itself a major theoretical and experimental hurdle. In recognition of the important role played by "structural" models, we may think of the entire approach to meaningful plate height calculations as the "structure-nonequilibrium" approach.[41] One of the most important extensions of the present theory, then, is related to the detailed characterization of the system's architecture and basic processes irrespective of the niceties of zone migration and nonequilibrium. Such factors will be considered mainly in the subsequent volumes where specific systems will be considered.

Other extensions of the generalized nonequilibrium theory are given below.

Extension to Eddy Diffusion

The basis of eddy diffusion was discussed in Sections 2.9 and 2.10. In Section 2.10 it was shown that eddy diffusion acts to exchange solute between fast and slow streamlines in the mobile phase—not by conventional mass transfer of the type considered here but as a result of sudden random changes in each stream-path's velocity. These abrupt velocity changes undoubtedly complicate the picture of mobile phase nonequilibrium. It would be useful to have a more direct theoretical access to this phenomenon than is afforded by the coupling theory. Such has not yet been worked out in detail, but the basic approach can be stated here.

The most significant aspect of eddy diffusion is illustrated in Figure 4.7-1—the erratic velocity changes along any given streampath due to

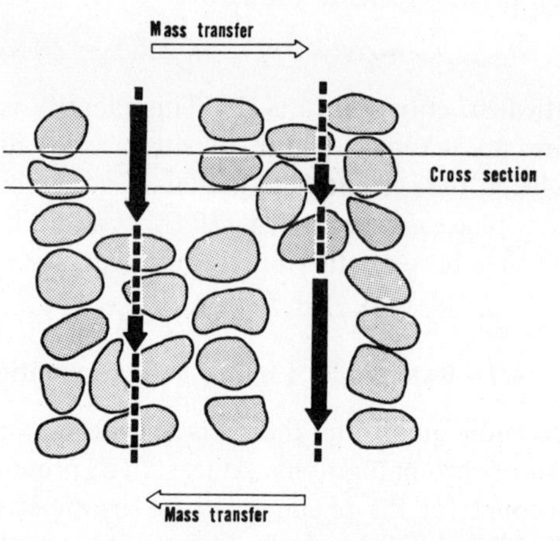

Figure 4.7-1 The source of eddy diffusion is the erratic variation in velocity (indicated by the length of the heavy arrows) along streampaths (dashed lines). For simplicity the streampaths are shown as straight parallel lines. The direction of mass transfer, assuming that this part of the column is engulfed by the leading half of the zone, is shown (see text).

variations in "channel" constrictions (lateral velocity components are ignored here). Let us suppose that this small volume element is in the leading half of the zone, and then investigate its nonequilibrium characteristics. The qualitative picture of nonequilibrium given in Section 3.1 indicates that the main source of solute in any given cross section (as indicated in the figure) is found in the region where flow velocity is greatest (on the left). Thus there is a mass transfer from left to right as the whole cross section strives to reach equilibrium. This picture changes rapidly as we proceed down through the volume element. In the lower half of the element the right-hand channel becomes the main source of new solute, and mass transfer occurs from right to left. This change in the direction of mass transfer means that the magnitude of the nonequilibrium departure (which may be regarded as the driving force for mass transfer) flip-flops rapidly along any given streampath. This demands a different mathematical approach to the nonequilibrium theory of the mobile phase. In essence what we have done until now is neglect the rapid variations in velocity along a streampath in the formal theory and then we have accounted for them in the coupling approach of Section 2.10.

The main difference from the previous theory arises in step two where mass transfer and flow are related. Equation (4.3-9) shows that the mass transfer term for the mobile phase is

$$s_m' = \frac{\partial c_m'}{\partial t} + v_m' \frac{\partial c_m'}{\partial z} \tag{4.7-1}$$

since

$$c_m' = c_m^{*\prime}(1 + \epsilon_m') \tag{4.7-2}$$

we may write the first term on the right as

$$\frac{\partial c_m'}{\partial t} = c_m^{*\prime} \frac{\partial \epsilon_m'}{\partial t} + (1 + \epsilon_m') \frac{\partial c_m^{*\prime}}{\partial t} \tag{4.7-3}$$

The equilibrium departure term ϵ_m' appears twice in this expression. Now since ϵ_m' does not change rapidly with time at a given point (its fluctuations are about as frequent as those of c_m^*), and since it is small anyway, the first expression on the right is negligible compared to the second. The second expression is composed of two terms through the $1 + \epsilon_m'$ quantity. Since ϵ_m' is small compared to unity, the entire equation reduces in good approximation to

$$\partial c_m'/\partial t = \partial c_m^{*\prime}/\partial t = -Rv\, \partial c_m^{*\prime}/\partial z \tag{4.7-4}$$

where, as usual, Rv is the overall zone velocity.

The second term on the right of equation (4.7-1) contains the term $\partial c_m'/\partial z$. Using (4.7-2), this is found to equal

$$\frac{\partial c_m'}{\partial z} = c_m^{*\prime} \frac{\partial \epsilon_m'}{\partial z} + (1 + \epsilon_m') \frac{\partial c_m^{*\prime}}{\partial z} \tag{4.7-5}$$

Again ϵ_m' is small compared to unity. In the first term, however, the smallness of ϵ_m' is counteracted by its very rapid "flip-flop" variation with distance z as discussed earlier. Thus $\partial \epsilon_m'/\partial z$ is not necessarily a negligible quantity, and must be retained in the equation. However ϵ_m' itself is still small and in the second term it may be ignored by comparison to unity. Thus

$$\frac{\partial c_m'}{\partial z} = c_m^{*\prime} \frac{\partial \epsilon_m'}{\partial z} + \frac{\partial c_m^{*\prime}}{\partial z} \tag{4.7-6}$$

When (4.7-4) and (4.7-6) are substituted back into equation (4.7-1) we have

$$s_m' = (v_m' - Rv)\frac{\partial c_m{}^{*\prime}}{\partial z} + v_m' c_m{}^{*\prime}\frac{\partial \epsilon_m'}{\partial z} \tag{4.7-7}$$

Since from equation (4.3-8) we have $s_m' = D_m c_m{}^{*\prime}\,\nabla^2\epsilon_m'$, the basic differential equation for ϵ_m' becomes

$$\nabla^2\epsilon_m' = \frac{(v_m' - Rv)}{D_m}\frac{\partial \ln c}{\partial z} + \frac{v_m'}{D_m}\frac{\partial \epsilon_m'}{\partial z} \tag{4.7-8}$$

This replaces equation (4.3-12), and differs from the latter by containing the last term on the right. Insofar as the nonequilibrium theory is concerned, this latter term may be regarded as the addition needed to allow for eddy diffusion. The full implications of this equation have not yet been explored.

Extension to Nonlinear Sorption Isotherms

The generalized nonequilibrium theory can be extended to the nonlinear sorption of solutes. When we know the rate laws governing sorption and desorption, the procedure is much the same as outlined earlier. We end up with the following fundamental equation for zone movement

$$\frac{\partial c}{\partial t} = -\mathscr{V}\frac{\partial c}{\partial z} + D\frac{\partial^2 c}{\partial z^2} \tag{4.7-9}$$

which is identical to the basic migration equation of Section 2.3, equation (2.3-9). In this case, however, the mean zone velocity $\mathscr{V}$ and the effective diffusion coefficient D are functions of solute concentration. The solution of equation (4.7-9) does not, therefore, generally yield a Gaussian or related profile. The problem of dealing with equation (4.7-9) is probably the most difficult aspect of nonlinear problems. The usual treatment of nonlinearity, of course, proceeds on the assumption that there is total equilibrium. The foregoing approach is designed to account for the rate processes which occur in combination with nonlinearity.

The Assumptions of the Generalized Nonequilibrium Theory

Various assumptions have been used throughout this chapter to facilitate and simplify the theoretical approach. In the following paragraphs some indication will be given regarding the basis of these assumptions and the range of their applicability. In this section we will pursue the meaning of these assumptions in a semi-quantitative manner. A more rigorous analysis would involve more detail than we wish to consider at this point.

Near-Equilibrium Assumption

Several important steps in the nonequilibrium approach are made possible by means of this assumption. Mathematically this assumption means that

$$\epsilon_i \quad \text{or} \quad \epsilon_i' \ll 1 \tag{4.7-10}$$

Strictly speaking, this inequality should hold at all points and at all times. Practically, however, this is not necessary and is probably unrealistic. Any partition column will have deep or otherwise inaccessible pockets of stationary phase which will not equilibrate rapidly. Any adsorption column will have sites which, for a number of possible reasons, will fail to combine rapidly with solute. A few such exceptions can make little difference to theory or practice since only a small fraction of the solute will be in any way involved. Our main concern is the behavior of the bulk of the solute peak. An index of this behavior is the average equilibrium departure term. This is an order-of-magnitude index only, since numerous fluctuations from the mean will occur within the column.

We will first consider the average (or overall) mobile phase departure term ϵ_m. We saw in equation (3.2-41) that all stationary phase processes lead to the plate height

$$H = \frac{-2\epsilon_m}{\partial \ln c/\partial z} \tag{4.7-11}$$

Mobile phase contributions are of a similar nature as shown by equation (4.5-3). (In the latter case velocity also enters into the consideration of averages.)

The value of ϵ_m may be estimated by use of the last equation

$$\epsilon_m = -\frac{H}{2}\frac{\partial \ln c}{\partial z} \tag{4.7-12}$$

The order of magnitude of the slope term $\partial \ln c/\partial z$ may be estimated by assuming a Gaussian zone

$$c = c_{\max} \exp(-\Delta z^2/2\sigma^2) \tag{4.7-13}$$

where $c_{\max}$ is the overall concentration at the center of the zone and Δz is the distance from this center. By differentiation

$$\partial \ln c/\partial z = -\Delta z/\sigma^2 \tag{4.7-14}$$

If this and the plate height expression $H = \sigma^2/L$ are substituted back into equation (4.7-12), we find

$$\epsilon_m = \Delta z/2L \tag{4.7-15}$$

(This expression holds only if the total zone spreading indicated by equation (4.7-13) is caused by the mechanism in question. Otherwise ϵ_m is even less than indicated because the two σ^2 values, which canceled one another, are not really equivalent.) We are generally only interested in the properties of the zone within those regions where its largest bulk is located—regions for which Δz does not greatly exceed 2σ. Thus most of the zone has an ϵ_m value ranging between zero (at the zone center, $\Delta z = 0$) and $\sim\sigma/L$. This quantity equals $1/\sqrt{N}$ where N is the number of plates which have been passed through. Thus to an approximation[42]

$$\epsilon_m(\max) \sim 1/\sqrt{N} \ll 1 \tag{4.7-16}$$

(In reality the left-hand term should read "the maximum in the absolute value of ϵ_m," since negative discursions of equal magnitude occur in the rear of the zone.) The above equation shows that the near-equilibrium assumption is valid for the bulk of the mobile phase after several plates have been passed and the zone has had an opportunity to establish its normal pattern.

The balance-of-nonequilibrium condition, $\epsilon_m R = -\epsilon_s(1 - R)$, shows that the stationary phase departure ϵ_s is usually the same order of magnitude as ϵ_m. It would be considerably larger than ϵ_m only for high R values (exceeding, say, 0.9). Such R values are uncommon as a great deal of resolution is lost as R approaches unity (see Chapter 7).

The conclusion reached by equation (4.7-16) is also expected on an intuitive basis. The theoretical plate is defined (see Section 2.3) essentially as an equilibration stage. The time required for passage through this stage should thus correspond to the "equilibration" time or "relaxation" time for the rate processes. The departure from equilibrium generally becomes small after the passage of several relaxation periods (or plates), thus confirming the results of equation (4.7-16). The relationship of the theoretical plate to the relaxation time has been discussed quantitatively.[42]

The Additive Law

The additive law is not a unique feature of the nonequilibrium approach. It is a general law concerning the combination of component plate height terms to yield the overall plate height contribution. It plays an important role in conjunction with the generalized nonequilibrium theory as seen in Section 4.6.

By way of review the additive law indicates that a plate height H determined by a multiplicity of processes (with "rate" parameters k_1, $k_2, \ldots k_m$; $D_1, D_2, \ldots D_n$) can be broken into a sum of terms (see equation (4.6-18) for an equivalent expression of this law)

$$H = \sum_{i=1}^{m} H(k_i, \infty) + \sum_{i=1}^{n} H(D_i, \infty) \tag{4.7-17}$$

where each term is the plate height calculated with its particular rate parameter intact and with all other rate parameters at infinity.

The range of validity of the additive law has not yet been precisely defined. By actual example it has been shown that the law is valid in some cases but not in others. In fact it has been shown that there are special cases for which the additive law is obviously incorrect. At the present time the most satisfactory approach is to employ the criterion used by Giddings[3] for rejecting the additive law, and assuming that it is approximately valid in all other cases. This criterion, somewhat generalized here, states that the additive law is incorrect when there are two competitive mechanisms (involving different rate parameters) for a given mass transfer process. It is obvious in this case that the additive law errs by underestimating H; the use of an "infinite" rate for one process while considering the finite nature of the other process clearly leads to a too high mass transfer rate and thus a too low plate height. Fortunately, such competing mechanisms can generally be recognized easily.

As an example, consider the kinetic analog given in equation (4.2-50) to illustrate the adsorption of complex molecules. There are two competitive mechanisms whereby total adsorption can be achieved—one for each of the two active ends which may adsorb first. The additive law would yield a zero plate height whereas the actual calculation, equation (4.2-52), shows this to be finite.

A second example was used in connection with the original discussion[3] of additivity; two diffusion paths, one through the mobile phase and one through the stationary phase, provided competitive routes for mass transfer between separate regions. The plate height equation derived for this case did not show additivity.

Fortunately cases such as those above are easy to recognize and remedy. When any mass transfer process occurs by competitive mechanisms, all of the involved mechanisms should be considered as a separate package. Additivity can then be used in relating this package to all other mechanisms, but cannot be used within the package itself.

While it is quite clear that competitive mechanisms destroy additivity, it is not so obvious that all noncompetitive mechanisms obey the additive law. A large number of examples support this assumption, and lead us to believe that it may generally be used as a working hypothesis. Some of the examples used in previous sections are:

1. The general case of multisite adsorption as given by equation (4.2-43).

2. The case of a two-step consecutive process as shown by equation (4.2-48).

3. The general combination law for the stationary phase, equation (4.4-23), is a special case of the additive law.

4. A mixed mechanism (involving both step-wise kinetics and diffusion), adsorption from a "flat" channel, as shown by equation (4.6-16).

5. Another mixed mechanism involving adsorption in ion-exchange chromatography, equation (4.6-43) and (4.6-44), was first treated with the total nonequilibrium approach.

Examples which appear in the literature are the following:

6. The full Golay equation for gas-liquid chromatography in capillary columns.[23]

7. The case of two-site adsorption.[1]

8. Gas-solid chromatography involving adsorption from a "flat" channel with parabolic flow (a mixed mechanism).[43]

9. Partition into a uniform film from a "flat" channel with uniform flow in gas chromatography.[3]

10. Gas-liquid chromatography in an ideal capillary with interfacial resistance (a mixed mechanism).[24]

If we may assume that additivity depends on noncompetitive mechanisms, then it would appear that most real chromatographic columns obey the additive law to a good approximation. There will usually be slight disturbances which prevent an exact application. For instance, in partition chromatography the interface between the mobile and the stationary phases may not all be used equally in mass transfer; the relative use, in fact, may depend on the diffusion rates in the two phases (thus giving competitive processes). Some preliminary calculations made in the author's laboratory[44] for a complex system resembling ion exchange showed that a 10% error may sometimes exist in additivity. This is somewhat uncertain because the extensive numerical program needed for such calculations has not yet been completed.

The Neglect of Longitudinal Diffusion Terms

If we include longitudinal diffusion in the mass transfer term of the last several sections, we obtain an expression of the form

$$\frac{\partial c_i}{\partial t} = s_i - v_i \frac{\partial c_i}{\partial z} + D_i \frac{\partial^2 c_i}{\partial z^2} \tag{4.7-18}$$

The last term has been ignored in our general treatment of nonequilibrium. This matter was discussed intuitively in Section 3.2 and has been discussed mathematically in the literature.[1] The approach here will be intermediate between the two.

Without focusing on a particular phase or state, i, the justification for ignoring the diffusion term depends in general on the magnitude of the ratio of the third term to the second term on the right

$$\alpha = D \frac{\partial^2 c}{\partial z^2} \Big/ Rv \frac{\partial c}{\partial z} \tag{4.7-19}$$

where Rv, the overall zone velocity, has replaced the specific value v_i. The same change has been made in c_i and D_i. Now assuming a Gaussian concentration zone as in equation (4.7-13), and taking the first and second derivatives, this ratio becomes

$$\alpha = -\frac{(\Delta z^2 - \sigma^2)}{\Delta z \sigma^2} \frac{D}{R_v} \tag{4.7-20}$$

Writing $D = \sigma^2/2t$ (and remembering that this σ^2 is less than above since this is the component due to longitudinal diffusion) and $Rv = L/t$, we have

$$|\alpha| < \frac{(\Delta z^2 - \sigma^2)}{2L\,\Delta z} \tag{4.7-21}$$

where the inequality stems from the use of the two different σ^2 values. Throughout most of the zone where Δz and σ are the same order of magnitude the absolute value of α is approximately σ/L, or less. This is the same ratio as shown in equation (4.7-16)

$$|\alpha| \sim 1/\sqrt{N} \tag{4.7-22}$$

and is generally small compared to unity. Under these circumstances the diffusion term is negligible as postulated. There are, as seen by equation (4.7-21), particular points in the zone where this ratio fluctuates violently. When $\Delta z = \sigma$, the ratio is zero. This is expected since there is no accumulation by diffusion at such a point of inflection. At $\Delta z = 0$, the zone maximum, the ratio goes to infinity. This does not alter our general conclusion, however, since diffusion constitutes such a small term, here as elsewhere, that it may be approximated by zero.

4.8 Summary of Practical Results

The generalized nonequilibrium theory is the main vehicle for translating the physical and chemical parameters of chromatography into plate height or, indirectly, into resolution. Therefore, if we take the view that a scientific discipline will advance faster if its basic processes are better understood, the generalized nonequilibrium theory becomes a highly practical tool. In this section an attempt will be made to summarize some of the more obvious practical implications of the generalized nonequilibrium theory. This will be done without mathematical developments (some applicable equations will be presented, but none derived).

The Role of Models

The generalized nonequilibrium theory applies rigorously to the model chosen to represent the chromatographic system. In a strict sense the theory is applicable to the model, not to chromatography itself. Hence one of the foremost problems in applying this theory is in choosing models which realistically represent the chromatographic system. Each model should express the relative location and dimensions of the stationary and mobile phases; it should approximate the velocity profile of the mobile phase; and it should specify the points where adsorption occurs and the relative adsorptive activity. These specifications may be difficult to obtain exactly (see below), but they lie at the very heart of the chromatographic process and control its quantitative aspects.

It is important to emphasize again that, even with approximate models, this approach is superior to the random walk and other similar approaches. The latter is approximate as a theory, and hence a double error usually occurs, one (as usual) in the envisioned model and one in the theory. Furthermore, discrepancies of theory and experiment cannot easily be pinpointed to one of the two. With the generalized nonequilibrium theory all significant errors reside with the model, and thus the comparison of theory and experiment becomes a weapon for finding more realistic models for the chromatographic process.

The Structure-Nonequilibrium Approach

The fundamental calculation of plate height involves a dual approach; first the model must be established and then the nonequilibrium theory applied. This dual procedure has been indicated by the descriptive term, "structure-nonequilibrium" approach.[24] The word "structure" implies all the specifications discussed earlier which must be known before a realistic model can be constructed. In this chapter the nonequilibrium part of the dual approach has been emphasized and essentially solved. Since the advent of the generalized nonequilibrium theory, the acquisition of structural details is now probably the most difficult aspect of the basic study of dynamic processes in chromatography, and represents the area where most needs to be done. Some phases of this problem will be discussed in the following chapters. Detailed considerations of structure will be made in the later volumes where particular chromatographic systems will be treated at length.

The plate height contribution obtained from the nonequilibrium theory is always proportional to flow velocity; $H = Cv$, where the coefficient C depends on the structural model assumed. Typical C values will be discussed next, followed by a study of the nature of variations in this important coefficient.

Magnitude and Significance of Typical C Terms

The preceding sections have expressed the various mass transfer or nonequilibrium (C) terms as a function of diffusivity, particle diameter, stationary phase dimensions, rate constants, etc. The full implications of this study will be presented in the subsequent volumes. A number of general conclusions can be reached, however, which concern the practical applications of chromatography.

It has been known since the early days of chromatographic theory that a separation could be achieved more effectively if mass transfer were somehow made more rapid (and thus the C terms smaller). First, resolution is improved by the reduction of C terms. Second, the speed of analysis can be increased in proportion to the reduction of the sum total C. Thus the parameters which affect the C's, and the relative magnitude of different C terms, become primary considerations for any system which is not already performing adequately.

It was shown in Section 2.11 that a well-designed column would exhibit an underlying plate height curve, due to mobile phase effects only, which at its minimum would be about two particle diameters (the lower curve of Figure 2.11-1 shows this contribution in terms of reduced plate height, $h = H/d_p$, versus reduced velocity, $\nu = d_p v/D_m$). Combined with this nearly fixed curve are the stationary phase C terms. The latter may vary over wide limits; in some cases C may be so large as to prevent a desired resolution and in other cases it may be negligible. There is, of course, little point in trying to improve this term if it is already negligible, especially since the compromises involved may lead to difficulties in some other direction. Thus, we must know how large the stationary phase C term may grow before it becomes excessive and destroys resolution. In Section 2.11 it was shown that this problem depends on the nature of the mobile phase, since maximum resolution occurs at widely separated velocities in gas and liquid chromatography. Specifically it was shown that the stationary phase mass transfer parameter Ω must be about unity or less. Since by equation (2.11-5)

$$\Omega = CD_m/d_p^2 \tag{4.8-1}$$

the stationary phase C term should be equal or less than

$$C \leq d_p^2/D_m \tag{4.8-2}$$

in any high resolution column.

In partition chromatography the general combination law (Section 4.4 and equation (4.4-23)) shows that the plate height and thus the C value may be written as a sum of precisely defined terms, one for each of the assorted "units" of stationary phase which is absorbing solute. If we assume that there is some typical or average configuration for the stationary phase, then the sum of terms may be approximated by the single term (equation (4.4-25))

$$H = Cv = qR(1 - R)\, d^2v/D_s \tag{4.8-3}$$

where d is the depth or thickness of the stationary phase and q is the configuration factor which depends on the geometry of the phase (if stationary phase is drawn into interior pores by surface tension, its geometry will be fixed by the pore boundaries). As a general rule q will range between unity and one-tenth; a typical value is $q = 0.25$. Using this and the expression for C obtained from equation (4.8-3), the inequality of equation (4.8-2) becomes

$$0.25R(1 - R)\, d^2/D_s \leq d_p^2/D_m \tag{4.8-4}$$

or, rearranged

$$d/d_p \leq \sqrt{(D_s/D_m)/0.25R(1 - R)} \tag{4.8-5}$$

This shows that the ratio of stationary phase dimensions to the particle diameter should be confined within a certain range with a maximum value as shown on the right-hand side. The component, $1/0.25R(1 - R)$, on the right-hand side will normally vary from about 16 ($R \sim 0.5$) to 400 ($R \sim 0.01$ or 0.99), and its square root will thus vary between 4 and 20. Without much error we may assume that this quantity is typically 10. The last equation thus becomes

$$d/d_p \leq 10\sqrt{D_s/D_m} \tag{4.8-6}$$

giving the maximum d/d_p ratio in terms of the relative diffusion coefficient found in the stationary and mobile phases, respectively.

In gas chromatography the ratio of diffusion coefficients, liquid to gas, is about 10^{-5}, giving for the above ratio

$$d/d_p \leq 0.03 \qquad \text{(gas chromatography)} \tag{4.8-7}$$

Assuming that the solid support is 80–100 mesh ($d_p \sim 0.016$ cm) we find

$$d \leq 5 \times 10^{-4}\ \text{cm} = 5\ \mu \qquad \text{(gas chromatography)} \tag{4.8-8}$$

This maximum depth of stationary phase, in view of the simplifying assumptions made, is probably correct to within a factor of two or three. It shows clearly and in a general way that resolution will suffer if the units of liquid stationary phase, which accumulate within the pores of the solid, are allowed to become larger than about 5 μ. This is roughly in agreement with practice. There is now a growing belief that the Johns-Manville solid support, Chromosorb W, with many pores over 5 μ diameter, inhibits resolution to some extent while Chromosorb P, with pores of roughly 1 μ diameter, is nearly as efficient as any known system. The experimental results of DeFord, Loyd, and Ayers[45] show the quantitative relationship of the stationary and mobile phase terms for the latter support.

In liquid chromatography D_s and D_m may be the same order of magnitude (note exceptions below), and thus d/d_p, equation (4.8-6), may be as large as 10. Such a large value will never be required since the units of stationary phase are never larger than the particle diameter (e.g., paper and ion-exchange chromatography). In this extreme case d corresponds roughly to the particle radius since it is the maximum depth of penetration into the stationary phase, and thus $d/d_p = \frac{1}{2}$, well below the maximum of 10. It would thus appear that there is no size limit for the units of stationary phase insofar as stationary phase mass transfer is concerned, providing $D_s \sim D_m$.

In many cases we may expect $D_m > D_s$, thus requiring a reconsideration of the above conclusion. In paper chromatography, for instance, diffusion within the fibers must occur by means of a very tortuous path which avoids the crystalline regions of cellulose.[46] This will decrease D_s to a certain extent and make the earlier conclusion somewhat questionable. In addition D tends to be less for the more polar liquids of the stationary phase (Section 6.3). However, the reduction in D_s would need to be a factor of 10^2 or so to exceed the limit imposed by equation (4.8-6), and this large a change seems unlikely. (It has been shown experimentally that the abnormally small value, $D_s = 10^{-8}$ cm^2/sec, would be required to make stationary phase mass transfer a significant factor in paper chromatography.[47])

In ion-exchange chromatography the ratio of D_s/D_m probably varies between about unity and 0.01. As shown by equation (4.8-6) this still puts no serious restrictions on the nature of stationary phase mass transfer.

The above arguments, using the criterion $\Omega < 1$, relate to the limitations on stationary phase configuration such that a near maximum resolution may be achieved. However, it is often desirable (particularly in liquid chromatography where the optimum flow velocity is very slow) to seek higher speed in analysis at the sacrifice of some resolution for a given length of column. Because of the coupling phenomenon (Section 2.10) and the subsequent disappearance of the mobile phase C terms at high velocities, the stationary phase C term becomes even more critical. The exact criterion then depends on the analysis speed desired; it may be that the much more demanding criterion, $d/d_p \leq D_s/D_m$, replaces equation (4.8-6) (the two differ by a factor of 10). The problem is sufficiently complicated that a simple criterion of this form is not always applicable. The general problem is discussed further in Chapter 7.

For adsorption chromatography it is difficult to establish a concrete criterion which can be used to determine the role of stationary phase mass transfer. The inequality of equation (4.8-2) is still the basic requirement, but the C of adsorption chromatography depends on various rate constants which are sometimes more difficult to pin down than the corresponding diffusion processes of partition chromatography. As a general rule physical adsorption rather than chemical adsorption is responsible for retention. This simplifies the considerations because there is usually no activation energy for such adsorption (i.e., a solute molecule can combine with the surface without needing an initially high energy to do so). It has been shown[48] (see Section 6.5 for the details of this argument) that in gas chromatography a 1-site surface with desorption rate constant k_d has an extremely low C term that easily satisfies (by several orders of magnitude) the requirement of equation (4.8-2). A multisite or heterogeneous surface, which equation (4.2-45) shows to have

$$C_k = 2R(1 - R)\bar{t}_d \tag{4.8-9}$$

has a considerably larger mass transfer effect because $\bar{t}_d$ (the mean desorption time) is increased by heterogeneity. The precise magnitude of C depends on the degree of heterogeneity. This property is difficult to measure, and while it may be speculated that this C usually obeys the criterion of equation (4.8-2), there may be significant exceptions. Much more work needs to be done in this area.

Variations in C

The foregoing discussion has indicated the general order of magnitude of typical C terms. The difference between failure and success in chromatographic separations often depends, however, on unobvious details which cause variations in C, either within or outside the typical range just discussed. Some of the significant C variations will be summarized here.

As indicated earlier, we can usually depend on obtaining a minimum plate height of about two or three particle diameters as the underlying contribution of the mobile phase. However the mobile phase plate height curve (as shown approximately in Figure 2.11-1) is not fixed rigidly at this level. There is little experimental evidence bearing on this matter, but that which exists† shows that considerable variations (by a factor of three or four) are possible even within similar systems—here gas chromatographic columns of analytical dimensions.‡ The theory of this chapter indicates that large variations in plate height or C are possible. In large columns, for example, the C contribution grows in proportion to column diameter squared (see equations (4.6-21) and (4.5-22)) provided that a similar flow profile is maintained over the column cross section. The effect is verified by the experimental loss in resolution generally encountered in large preparative-scale columns. Perhaps of more interest than this is the possibility for lowering (instead of raising) the normal mobile phase contribution in chromatographic systems. Probably the main source of the mobile phase effect is the unequal flow in neighboring channels (i.e., the interchannel effects). In theory it should be possible to assemble a uniform packing structure, without particle bridging, in which interchannel interactions are eliminated.[21] The results of Sections 4.5 and 2.8 show that the mobile phase C could probably be lowered by a factor of 10 or more through this manipulation. This approach offers an exciting potential for a major gain in chromatographic resolution, and should be pursued experimentally. More attempts should also be made to use columns with a diameter not much larger than that of the contained particles.

Once the mobile phase contribution to the plate height has been established, the key to successful operation resides with the stationary phase. In all cases it is imperative that solute exchange rapidly between the mobile and stationary phases in order to minimize the equilibrium departure. In partition chromatography this means that diffusion rates should be large and that the stationary phase should be dispersed into fine "droplets" with a high interfacial area (the necessity for a large interface and rapid equilibration was recognized by Martin and Synge while developing the technique of paper chromatography[51]). In adsorption chromatography it is necessary that the rates of adsorption and desorption be kept high. Sources of variations in the C term for these two methods (partition and adsorption) will be discussed below.

In partition chromatography we may normally assume that the stationary phase is distributed in some simple and uniform manner corresponding to the "average" configuration of the real stationary phase. We thus arrive at the simple descriptive equation given by (4.8-3). In

† A summary of most existing data is found in reference 21.

‡ When the column radius to bead radius is not much larger than unity, a reduction in plate height to less than one particle diameter is found.[49,50]

substituting an "average" or "typical" configuration for the real one, it is necessary to consider the fact that the "thickest" portions of stationary phase have an inordinate effect on the final C term. The accumulation of any significant amount of stationary phase into abnormally deep pockets may cause this part of the stationary phase to dominate the entire plate height effect, and may thus cause a sharp variation upward in plate height. This effect is spelled out quantitatively by the *general combination law*, given here in a form slightly altered from equation (4.4-23)

$$H = C_s v = R(1 - R)(v/D_s) \sum q_j (V_j/V_s)\, d_j^2 \tag{4.8-10}$$

The summation term, extending over all the diverse forms (or "units") of stationary phase which may be present, contains a configuration factor q_j for each unit of stationary phase; the relative volume (V_j/V_s) in a particular unit; and the depth of the unit squared d_j^2. The factor d_j^2 is responsible for the excessive influence of thick portions of the stationary phase. For example, if 85% of the bulk stationary phase were in the form of a uniform film of depth d_0, and only 15% of this phase were in pockets of depth $10d_0$, the C term would be increased four times above normal by the presence of the 15% minority. Thus any attempt to establish an "average" configuration must take account of the large effect of deep pockets of stationary phase. By the same token such deep pockets cause a deterioration in column performance completely out of keeping with the amount of stationary phase involved. In a very real sense the factor $(V_j/V_s)\, d_j^2$ makes it more imperative to decrease the maximum depth of stationary phase than to decrease the average depth. A considerable quantity of stationary phase may exist, for example, as a very thin film with little benefit except that the quantity so involved may otherwise have appeared in deep pockets.

The most significant variations in C probably result from the "depth" term just considered. Other variations, particularly in q_j, may occur, and these are described with remarkable simplicity by the general combination law of equation (4.8-10). As indicated in Section 4.4, the general combination law reduces the vast complexity of the total stationary phase into simple component parts, each part related to elementary "units" of stationary phase. The summation in equation (4.8-10) is made with respect to each of the elementary units. The precise value of the configuration factor q_j depends on the geometry of the units—whether stationary phase is drawn into narrow or wide pores in the solid support, or whether it exists in the form of an ion exchange bead, etc. The theory indicates, for example, that stationary phase held in deep narrow pores whose cross-sectional area a increases with distance x from the bottom in accord with the relationship

$$a = a_0 x^n \tag{4.8-11}$$

has a configuration factor (see equation (4.4-55)) related to the pore taper n by

$$q = 2/(n + 1)(n + 3) \qquad (4.8\text{-}12)$$

For stationary phase in wide pores it was found (equation 4.4-59) that

$$q = 4/(n + 2)(n + 3) \qquad (4.8\text{-}13)$$

These two extremes differ very little in numerical value (Table 4.4-1), and thus the configuration factor for realistic situations will lie in the rather limited range between. These equations show that q decreases as the taper factor becomes larger. Thus a direct relationship exists between q (and C) and the shape of the pore. These variations may not be as extreme as those existing in the depth and relative volume terms, but a 10-fold variation can result as n changes from 0 to 3. Furthermore, pores with a negative taper ("ink-bottle" pores), which can be penetrated only through a narrow neck, would lead to additional large variations (increase) in q and C. Pore shape, when pores act as the vessels for stationary phase, is a rather important factor affecting chromatographic performance.

It is interesting to note that stationary phase in a symmetrical configuration, such as in an ion-exchange bead or a paper fiber, may be divided by symmetry into "narrow" units with equation (4.8-12) applicable. It was shown earlier that a uniform film (with "translational" symmetry) is equivalent to $n = 0$ and has $q = \frac{2}{3}$; a rod-shaped paper fiber is described by $n = 1$ and $q = \frac{1}{4}$; and ion-exchange bead is represented by $n = 2$ and $q = \frac{2}{15}$; the stationary phase accumulated around glass bead contact points is characterized by $n = 3$ and $q = \frac{1}{12}$.

Let us account for the ink-bottle pores (with a narrow neck) that cause increases in C. The physical basis of this effect is that a restricted access will considerably slow the mass transfer of solute into and out of the stationary phase contained in the pore. This is a general phenomenon of some note; a relatively free access must always exist to those regions where solute tends to accumulate in significant amounts. The following examples treated earlier in this chapter demonstrate this principle quantitatively, and show that considerable increases in C may result from restricted access.

1. It is shown by equation (4.6-33) that the plate height for diffusion in a uniform film is increased as follows by the presence of an underlying surface of adsorption

$$\frac{C_s\,(\text{underlying adsorp.})}{C_s\,(\text{no adsorption})} = \left[1 + 2\phi + \frac{3\phi^2}{1 - \phi}\right] \qquad (4.8\text{-}14)$$

where ϕ is the fraction of retained (or immobilized) solute which is adsorbed on the surface. The reason for the increase is that all adsorbed solute must gain access to the underlying surface by diffusion through the film. When $\phi = 1$, as an extreme but illustrative example, equilibrium is shifted such that all retained solute is adsorbed on the surface and none absorbed in the film. Transport through the film can occur only by having a considerable quantity of diffusing solute in the film, a condition lacking in this particular example. Even for less extreme cases (e.g., $\phi = \frac{1}{2}$), access is strained by the large amount of solute moving to and away from the surface, and the C term is increased correspondingly. We are speaking here only of the C term related to diffusion and thus to access; the C term for adsorption kinetics is another matter, treated in Section 4.2.

2. In ion-exchange chromatography, where solute may be adsorbed at particular sites or diffusing freely within the bulk ion-exchange material, the freely diffusing solute provides the only access to adsorbed solute. Equation (4.6-43) shows that the C term is

$$C_s = \frac{R(1-R)}{30(1-\phi)} \frac{d_p^{\,2}}{D_s} \tag{4.8-15}$$

where, again, ϕ is the fraction of immobile solute actually adsorbed. This quantity goes to infinity as ϕ approaches unity. This, as before, indicates the important effect of "strained" access to adsorption sites.

3. Equation (4.5-26) shows that diffusion through intraparticle mobile phase (i.e., transparticle diffusion) is governed by

$$C = \frac{2(1-\Phi R)^2}{60(1-\Phi)\gamma_p} \frac{d_p^{\,2}}{D_m} \tag{4.8-16}$$

where Φ is the fraction of the mobile phase occupying interparticle space. As Φ approaches unity, none of the mobile phase is left to provide access to adsorption sites or absorption regions in the particle's interior, and the mobile phase C term, C_m, naturally goes to infinity (this conclusion holds as long as there is no competing mechanism for gaining access).

The above examples show that chromatographic systems should be designed to provide ready access to solute exchange between points which exhibit an affinity for solute. Otherwise considerable increases may be observed in various C terms.

Some of the discussion above is concerned with a mixed (adsorption-partition) mechanism of retention. Pure partition chromatography is probably a rare phenomenon. Some adsorption will undoubtedly occur on most underlying solid supports which serve as a scaffold for the bulk partitioning phase. Thus mixed retention may be regarded as another major division of chromatography, a division discussed in some detail in Section 4.6. Little effort, experimental or theoretical, has yet been applied in acquiring numerical values for C with mixed retention. In most cases, following the additive law for plate heights (or C terms) presented in Section 4.6, the resultant C can be broken into two contributions, one for the partition mechanism and one for the adsorption mechanism. If adsorption can occur only after passing through a partitioning film, then the partitioning contribution increases above its normal value by the "access" factor of equation (4.8-14). For strong adsorption this may cause serious variations upward in C. The magnitude of the adsorption term will be similar to that found for pure adsorption chromatography, discussed below.

As shown by the expression $C_k = 2R(1 - R)\bar{t}_d$, equation (4.8-9), the C for adsorption chromatography increases in proportion to the mean desorption time $\bar{t}_d$ for solute molecules dissociating from the surface. The rate parameter $\bar{t}_d$ can be treated theoretically and should also be accessible to experimental measurement. In fact $\bar{t}_d$ should provide the most direct possible approach to the kinetics of desorption; it should be much simpler to obtain than its component terms, the individual desorption rate constants for the various kinds of sites (see equation (3.4-44)).

Detailed variations in $\bar{t}_d$ will be discussed in Chapter 6. Without going into details here it can be shown[48] that for gas-solid chromatography the C term is given approximately by

$$C_k = \frac{8(1 - R)^2 V_m}{\alpha \bar{c} A} \Lambda \tag{4.8-17}$$

where α is the effective accommodation coefficient (giving the probability that a solute molecule "sticks" to the surface on collision), $\bar{c}$ the mean molecular velocity of solute molecules, V_m/A the ratio of mobile phase volume to the area of the adsorbing surface, and Λ is the *heterogeneity factor* of the surface. For an absolutely uniform surface Λ is unity. As a very rough rule Λ is given by[48] (see equation (6.5-42))

$$\Lambda = 0.1 \exp(\Delta E_{21}/\mathcal{R}T) \tag{4.8-18}$$

where ΔE_{21} is the spread in adsorption energies associated with the heterogeneous surface. It is found that rather severe heterogeneity, corresponding to a ΔE_{21} of about 10 kcal/mole, is needed to make C_k large enough to be significant. In general the exponential dependence of Λ on the energy spread makes this factor rather sensitive to the nature of the adsorbing surface. We should therefore expect rather large variations in this particular C term. The same arguments probably apply in a qualitative way to that form of adsorption chromatography which has a liquid mobile phase. This matter is discussed in Section 6.5.

A Summary of C Terms

In Table 4.8-1* a compilation of C terms obtained from the generalized nonequilibrium theory (or some special form of the general theory) is given. Several of the terms have been derived by other means: the 1-site adsorption term first obtained by the stochastic theory[52] and the terms of the Golay equation for capillary columns.[23] But most of the terms in the table were first obtained through the nonequilibrium approach outlined in the last two chapters. The table contains most of the terms now known (a few complex terms have been omitted). The source of each term is indicated so that the reader may refer back to the appropriate place for a more complete discussion.

4.9 References

1. J. C. Giddings, *J. Chem. Phys.*, **31,** 1462 (1959).
2. J. C. Giddings, *J. Chromatography*, **3,** 443 (1960).
3. J. C. Giddings, *J. Chromatog.*, **5,** 46 (1961).
4. J. C. Giddings, *J. Chem. Phys.*, **31,** 1462 (1959); *J. Chromatog.*, **3,** 443 (1960); **5,** 46 (1961); *J. Phys. Chem.*, **68,** 184 (1964).
5. K. J. Mysels, *J. Chem. Phys.*, **24,** 371 (1956).
6. J. R. Boyack and J. C. Giddings, *J. Biol. Chem.*, **235,** 1970 (1960).
7. D. D. Fitts, *Nonequilibrium Thermodynamics*, McGraw-Hill, New York, 1962, p. 136.
8. R. C. Tolman, *The Principles of Statistical Mechanics*, Oxford, London, 1938, pp. 164–165.
9. P. H. Hanns, *Theory of Determinants*, Ginn, Boston, 1888, pp. 84 ff.
10. C. R. Wylie, Jr., *Advanced Engineering Mathematics*, 2nd ed., McGraw-Hill, New York, 1960, pp. 8–9.
11. C. R. Wylie, Jr., *op. cit.*, pp. 6–7.
12. H. Margenau and G. M. Murphy, *The Mathematics of Physics and Chemistry*, Van Nostrand, Princeton, 1956, pp. 302–303.
13. P. M. Morse and H. Feshbach, *Methods of Theoretical Physics*, McGraw-Hill, New York, 1953, Chap. 10.
14. R. B. Lindsay and H. Margenau, *Foundations of Physics*, Wiley, New York, 1936, pp. 49–50.
15. E. Glueckauf, *Trans. Faraday Soc.*, **51,** 1540 (1955).
16. D. C. Bogue, *Anal. Chem.*, **32,** 1777 (1960).
17. A. Sommerfeld, *Mechanics of Deformable Bodies*, Academic Press, New York, 1950, p. 19.
18. J. C. Giddings, *Anal. Chem.*, **33,** 962 (1961).
19. J. C. Giddings, unpublished results.
20. J. Bohemen and J. H. Purnell, *J. Chem. Soc.*, **1961,** 2630.

---

* Table 4.8-1 begins on page 190.

21. J. C. Giddings, *Anal. Chem.*, **34,** 1186 (1962).
22. J. C. Giddings, *J. Gas Chromatog.*, **1,** No. 4, 38 (1963).
23. M. J. E. Golay in *Gas Chromatography, 1958*, D. H. Desty, ed., Academic Press, New York, 1958, p. 36.
24. M. A. Khan in *Gas Chromatography, 1962*, M. van Swaay, ed., Butterworths, Washington, 1962, p. 3.
25. F. H. Huyten, W. van Beersum, and G. W. A. Rijnders, *Gas Chromatography, 1960*, R. P. W. Scott, ed., Butterworths, London, 1960, p. 224.
26. M. J. E. Golay in *Gas Chromatography*, H. J. Noebels, R. F. Wall, and N. Brenner, ed., Academic Press, New York, 1961, Chap. II.
27. R. E. Collins, *Flow of Fluids Through Porous Materials*, Reinhold, New York, 1961, pp. 4, 5.
28. J. C. Giddings, S. L. Seager, L. R. Stucki, and G. H. Stewart, *Anal. Chem.*, **32,** 867 (1960).
29. R. Kieselbach, *Anal. Chem.*, **33,** 23 (1961).
30. S. Dal Nogare and J. Chiu, *Anal. Chem.*, **34,** 890 (1962).
31. D. D. DeFord, R. J. Loyd, and B. O. Ayers, *Anal. Chem.*, **35,** 426 (1963).
32. R. H. Perrett and J. H. Purnell, *Anal. Chem.*, **35,** 430 (1963).
33. W. L. Jones, *Southwide Chemical Conference, ACS-ISA*, Memphis, Dec. 6, 1956. (See reference 29.)
34. M. R. James, J. C. Giddings, and H. Eyring, *J. Phys. Chem.*, **68,** 1725 (1964).
35. C. J. Krige and J. Pretorius, *Anal. Chem.*, **35,** 2009 (1963).
36. R. L. Martin, *Anal. Chem.*, **33,** 347 (1961).
37. R. L. Pecsok, A. de Yllama, and A. Abdul-Karim, *Anal. Chem.*, **36,** 452 (1964).
38. R. L. Martin, *Anal. Chem.*, **35,** 116 (1963).
39. F. Helfferich, *Ion Exchange*, McGraw-Hill, New York, 1962, Chap. 6.
40. R. Turse and W. Rieman, III, *J. Phys. Chem.*, **65,** 1821 (1961).
41. J. C. Giddings, *Anal. Chem.*, **35,** 439 (1963).
42. J. C. Giddings, *J. Chromatog.*, **2,** 44 (1959).
43. J. C. Giddings, *Nature*, **188,** 847 (1960).
44. J. C. Giddings and H. Y. Chen, unpublished results.
45. D. D. DeFord, R. J. Loyd, and B. O. Ayers, *Anal. Chem.*, **35,** 426 (1963).
46. G. H. Stewart, private communication.
47. K. L. Mallik and J. C. Giddings, *Anal. Chem.*, **34,** 760 (1962).
48. J. C. Giddings, *Anal. Chem.*, **36,** 1170 (1964).
49. J. C. Giddings and R. A. Robison, *Anal. Chem.*, **34,** 885 (1962).
50. J. C. Sternberg and R. E. Poulson, *Anal. Chem.*, **36,** 1492 (1964).
51. A. J. P. Martin and R. L. M. Synge, *Biochem. J.*, **35,** 1358 (1941).
52. J. C. Giddings, *J. Chem. Phys.*, **26,** 1755 (1957).

Table 4.8-1 Principal C Terms Obtained from the Generalized Nonequilibrium Theory. The plate height contribution H is given by Cv. The mean velocity of the mobile phase (expressed as distance traveled per unit time by an inert tracer) is v; the fraction of solute in the mobile phase at equilibrium is R (and thus the mean solute velocity is the fraction R of the inert tracer velocity v); the diffusion coefficients in mobile and bulk stationary phases are D_m and D_s, respectively; the mean particle diameter is d_p. All other terms are defined in the table.

Value of H	Applicability
I. Diffusion in the Stationary Phase	
1. $C_s = \frac{2}{3} R(1 - R) \frac{d^2}{D_s}$ d = depth of stat. phase	to stationary phase existing as a uniform film on a solid surface or to pores of uniform bore and depth d. (Source: equation (3.3-35), (4.4-32).)
2. $C_s = \frac{1}{16} R(1 - R) \frac{d_p^2}{D_s}$ d_p = rod diameter $d_p/2$ = pore diameter	to stationary phase existing in rod-shaped units (length $\gg$ diameter and thus end effects negligible), or in narrow pores with unit taper factor (see no. 5, below). In particular, to paper chromatography. (Source: equation (4.4-38).)
3. $C_s = \frac{1}{30} R(1 - R) \frac{d_p^2}{D_s}$ d_p = sphere diameter $d_p/2$ = pore diameter	to bulk stationary phase in spherical configuration or narrow pores with a taper factor of 2. Applicable to ion exchange chromatography. (Source: equation (4.4-44).)
4. $C_s = \frac{1}{12} R(1 - R) \frac{d^2}{D_s}$ d = contact point to meniscus distance, or pore depth	to small amounts of stationary phase accumulated around the contact points between glass beads, or generally, to any narrow pore with a taper factor of 3. (Source: equation (4.4-56) ff.)
5. $C_s = \frac{2}{(n + 1)(n + 3)} R(1 - R) \frac{d^2}{D_s}$ d = pore depth x = distance from pore bottom	to any narrow pore whose cross-sectional area increases from the bottom as $a = a_0 x^n$, where n is the taper factor. To any wide unit of stat. phase which can be divided into narrow units by symmetry. (Source: equation (4.4-56).)
6. $C_s = \frac{4}{(n + 2)(n + 3)} R(1 - R) \frac{d^2}{D_s}$ d = pore depth x = distance from pore bottom	to wide pores whose cross-sectional area increases as $a = a_0 x^n$. (Source: equation (4.4-59).)

Table 4.8-1 (cont'd.)

Value of H	Applicability
7. $C_s = qR(1 - R)\dfrac{d^2}{D_s}$ q = configuration factor d = depth	to a column in which each unit of stationary phase is the same. The geometry of each unit may be highly complex. This will affect q, obtainable as $2\bar{\bar{\theta}}$, equation (4.4-24). (Source: equation (4.4-25).)
8. $C_s = \Sigma q_j \left(\dfrac{V_j}{V_s}\right) R(1 - R)\dfrac{d_j^2}{D_s}$ q_j = configuration factor for unit j V_j = volume of unit j V_s = total stat. phase volume d_j = depth of unit j	*general combination law* applicable to any complex assemblage of unequal units of stationary phase. The value of q_j is defined mathematically in equation (4.4-24). (Source: equation (4.4-23).)
9. $C_s = \dfrac{R(1 - R)}{30(1 - \phi)}\dfrac{d_p^2}{D_s}$ ϕ = fraction of retained solute which is adsorbed	to ion-exchange beads with fraction ϕ of immobilized solute tied up by adsorption. (Source: equation (4.6-43).)
10. $C_s = \dfrac{2}{3} R(1 - R)\dfrac{d^2}{D_s} \times \left[1 + 2\phi + \dfrac{3\phi^2}{1 - \phi}\right]$ ϕ = (see above)	to a uniform film with an underlying surface of adsorption which combines with the fraction ϕ of immobilized solute. (Source: equation (4.6-33).)
11. $C_s = \dfrac{2}{3} R\dfrac{d^2}{D_s} \times \left(X_s^* + 3X_a^* + 3\dfrac{X_a^{*2}}{X_s^*}\right)$ X_s^* = solute fraction in bulk stat. phase X_a^* = fraction adsorbed on underlying surface	uniform stationary film with adsorption at either interface. (Source: equation (4.6-39) ff.)
II. Diffusion in the Mobile Phase	
12. $C_m = \dfrac{\omega\, d_p^2}{D_m}$ ω = dimensionless constant	to mobile phase nonequilibrium where all diffusion and flow processes locate themselves on a scale proportional to d_p. (Source: equation (4.5-4) ff.)

Table 4.8-1 (cont'd.)

Values of H	Applicability
13. $C_m = \frac{1}{24}(6R^2 - 16R + 11)\frac{r_c^2}{D_m}$ r_c = tube radius	to any open tube (capillary) column with circular cross section and parabolic flow. (Source: equation (4.5-17).)
14. $C_m = \frac{(1 - R)^2 r_c^2}{4D_m}$ r_c = tube radius	to an open tube (capillary) column with uniform rather than parabolic flow. (Source: equation (4.5-18).)
15. $C_m = \left(\frac{\nu - 1}{\nu + 1}\right)^2 \frac{r_c^2}{24\gamma D_m}$ ν = outside/center flow velocity in column	to a symmetric, quadratic velocity variation over the column cross section. Primary use is for preparative columns. (Source: equation (4.5-22).)
16. $C_m = \frac{r_c^2}{\gamma D_m}\sum\sum \frac{G_n G_m}{(n+2)}\left[\frac{2}{(m+4)} - \frac{4}{(n+2)(n+m+4)} - \frac{n(n+6)}{(m+2)(n+2)(n+4)}\right]$ r_c = tube radius γ = obstructive factor for diffusion G_n, G_m = velocity coefficients n, m = summation indices (except in D_m)	to a symmetric velocity variation over column cross section of the general form $v = \bar{v}\Sigma G_n(r/r_c)^n$. Main use is for preparative columns. (Source: equation (4.5-21).)
17. $C_m = \frac{2(1 - \Phi R)^2}{60(1 - \Phi)\gamma_p}\frac{d_p^2}{D_m}$ Φ = fraction of mobile phase in interparticle space γ_p = obstructive factor for diffusion within particle	to diffusion within a support particle containing mobile phase. (Source: equation (4.526).)
III. Adsorption Kinetics	
18. $C_k = 2R(1 - R)/k_d$ k_d = desorption rate constant	to adsorption on 1-site surfaces or to any process described by the kinetic analog $A_1 \rightleftarrows A_2$. (Source: equations (3.2-36) and (3.4-37).)
19. $C_k = 2R(1 - R)\bar{t}_d$ $\bar{t}_d$ = mean desorption time	to adsorption on multisite surfaces. Direct mass transfer between different kinds of sites assumed negligible. (Source: equation (4.2-45).)

Table 4.8-1 (cont'd.)

Values of H	Applicability
20. $C_k = \dfrac{2(1-R)^2}{k_{12}f_2 + k_{14}f_4}$ k_{12},k_{14} = rate constants for initial adsorption step f_2,f_4 = fraction proceeding from initial to complete adsorption	to the adsorption of complex (bifunctional) solute where alternate mechanisms of adsorption exist. (Source: equation (4.2-52).)
21. $C_k = 2R(1-R)/k_{sm}$ k_{sm} = rate constant for transfer from stat. phase to mobile phase	to interfacial resistance between the mobile and the stationary phase. (Source: equation (4.6-36).)
22. $C_k = 2R(1-R)\phi/k_{as}$ ϕ = fraction of retained solute which is adsorbed k_{as} = desorption rate constant	to adsorption on an underlying surface or sites in the case of simultaneous partition-adsorption processes. (Source: equations (4.6-37) and (4.6-44).)
23. $C_k = \dfrac{2R(1-R)^2}{X_\alpha^* k_{\alpha m}}$ X_α^* = equil. solute fraction at interface $k_{\alpha m}$ = rate constant for escape into mobile phase	to adsorption on interface between stationary and mobile phase where escape to the mobile phase is rate limiting. (Source: equation (4.6-41).)
24. $C_k = \dfrac{2R(X_s^* + X_a^*)^2}{X_\alpha^* k_{\alpha s}}$ X_α^* = (above) X_s^* = equil. solute fraction in bulk stat. phase X_a^* = equil. solute fraction on underlying surface $k_{\alpha s}$ = rate constant for escape into stationary phase	to adsorption on interface between stat. and mobile phase where escape to the stationary phase is rate limiting. (Source: equation (4.6-42).)

Chapter Five

Packing Structure and Flow Dynamics

5.1 Introduction

The solid particles in a chromatographic medium restrain, deflect, and disperse the flow and diffusion of solute in the mobile phase. Hence it is imperative to gain insight into the packing structure where these influences originate. In this fashion we can learn something about the nature of flow in chromatographic columns. Although largely ignored in the chromatographic literature, the details of flow have perhaps a greater influence on the outcome of chromatographic separation than any other process except diffusion.

The main function of flow is displacement. It provides the mechanism of transport for the down-column motion of chromatographic zones. The flow process, however, is not limited to the simple and rather passive role of transporting zones. It is an integral component of the dynamics of chromatography, and its proper use requires that the laboratory system be designed largely for its control.

A close examination of flow—any flow—shows the reason for its important role in dynamics. All known flow in confined spaces is differential in nature; different points in the flow stream have different velocities. Thus zone transport is not uniform, a factor of great significance to zone spreading. The differential nature of flow originates because all fluids are viscous (in varying degrees) and because the motion of any fluid layer in contact with a fixed surface is stilled. Flow is thus achieved by pushing outer layers over the quiescent ones beneath. The shear of one layer with respect to another requires the application of a force (in the form of pressure) since viscous forces resist the shearing of layers. Simultaneously a nonuniform flow profile is generated and the necessity for a finite pressure gradient is established. The magnitude of the pressure drop is one of the factors to be considered in column design, and limits the ultimate performance of chromatographic systems (see Section 7.5).

The precise form of the flow pattern is determined by the geometry of the interstitial flow space. The flow space, in turn, is fixed by the packing structure—by the orientation and relative position of the packing particles. There are two extreme difficulties here in developing a quantitative understanding of chromatographic flow. First, the packing structure is so complicated that it has defied all efforts to come to grips with it mathematically. Second, even if the structural geometry were tractable, the flow pattern could only be obtained approximately by extremely difficult numerical analysis. Thus our practical understanding of the flow pattern must be limited to inexact and intuitive concepts supported by experience. We must rely on simple models and empirical parameters in order to obtain the mathematical tools necessary for theoretical development. However inexact this procedure may be, it is made necessary by the very real importance of flow phenomenon and structural factors in chromatographic separations.

While the internal structure of chromatographic beds is of paramount importance in determining flow pattern, it is also important in influencing diffusion rates and in establishing the pressure gradient needed to maintain a given level of flow. These matters and several others will be considered here and in Chapter 6.

In discussing flow and structural phenomena it is not intended to start from first principles. The theoretical basis of much of this subject, unlike that of the dynamics of chromatography, has already been exposed in several monographs.[1–4] This chapter will be largely interpretive, concerned mainly with applying the known elements of the science of porous media to the chromatographic process.

Some of the chromatographic implications of the flow pattern have already been explored in Chapter 2 (Sections 2.8 and 2.9). Here we hope to establish more fully the scope and significance of the flow phenomenon to the extent that it is of interest in chromatography.

5.2 Packing Structure

The space within most chromatographic beds is composed of two parts. There is the solid part, provided by the particles, and the free part between the particles and within their pores. One common and continuous surface (i.e., the contacting surfaces of the adjoining particles) defines both parts so that the description of one also specifies the other. The complementary relationship of these two parts makes it impossible to separate them, although emphasis is frequently directed toward only one.

The geometry of the common surface determines the packing structure and all of its characteristics. Because the detailed geometry is so complex, however, it is necessary to look beyond this directly to the separate structural characteristics. In many cases crude geometric arguments and observations are used to implement structural concepts, but the absolute role of geometry in describing structure is greatly weakened by its own vast complexity. In the following we shall focus on a few of the important structural characteristics.

Interconnection in the Interstitial Network

The apparent "solidity" of chromatographic beds disguises to some extent the thoroughly interpenetrating nature of the interstitial space. As emphasized in Chapter 2, a molecule at any given location between particles can reach any other such location by an immense number of paths. In fact the molecule may even start out in a large number of ways, finding interconnections in almost every direction. The physical models (some capillaric models) in which material is assumed to be trapped in lengths of capillaries, with egress only through the ends, is certainly an unrealistic description of granular materials. Two-dimensional diagrams compound this error because exits above and below are concealed.

The great accessibility between interstitial points means that diffusion can occur in the packing at a rate not much below its value in the free fluids. This will be discussed in Chapter 6. It also means that the mobile fluid may flow freely in almost any direction if there is a simple disturbance such as that caused by an obstacle blocking the channel. This leads to the conclusion that lateral transport occurs readily and that mobile fluid will immediately flow into those regions which provide a relatively unobstructed path for flow. Outflow occurs just as quickly when the path becomes blocked. The latter effect is apparently of considerable significance in creating a velocity bias (of the short-range interchannel type) and leading to the spreading of zones.

Porosity

Porosity f is defined as the fraction of free (nonsolid) space within a certain volume element of porous material. It is a measure of the room available for the mobile phase.† This parameter is basic to most studies of porous materials. In chromatography it is important because it affects the volume V_m of mobile phase and thus the migration parameter R. It

† In partition chromatography mobile phase is excluded from both the solid support and the stationary phase volumes. We must then specify whether porosity refers to the non-solid volume fraction (i.e., to the volume occupied by both mobile and stationary phases) or to the mobile phase volume fraction, V_m. The former is more correct, but the latter has some chromatographic advantages.

has some influence on the flow rate-pressure drop relationship. Variations in the local value of porosity are largely responsible for the velocity biases which lead to zone spreading.

The statistical nature of porosity is sometimes important in chromatographic studies. Applied to extremely small volume elements, the porosity of chromatographic packing ranges from unity in the interstitial space to zero in the solid material. If the volume element is large enough to incorporate several particles, porosity will not reach these extremes, but will still vary perhaps 30–40% from place to place in the column. If the volume element is made extremely large, such that it incorporates thousands of particles, porosity will remain essentially constant throughout the medium. The reason is that the porosity referred to a large volume element is the average porosity of all the small elements within. In common with most statistical parameters, the average falls in a narrower range as more components (small volume elements) are weighted.†

Under ordinary circumstances porosity is applied to large masses of material and thus approaches the status of a fixed bulk property. However, we shall occasionally wish to refer to the more variable *local porosity* of some smaller regions, usually of such a size as to contain several particles.

Porosity, by definition, is the fraction of a total volume element occupied by all free space. Some of this space may occur within the particles themselves if they are porous (as are the diatomaceous earth particles of gas chromatography). The free space which permits the main movement of mobile fluid, however, is that occurring in the interstitial channels and gaps between particles.‡ Hence for some purposes we may be concerned with a porosity which gives the fraction of interparticle free space only.§ This basic parameter, *interparticle porosity*, will be identified by f_0. This quantity is smaller than, or at most equal to, the (total) porosity f. Thus we may write

$$\Phi = f_0/f \tag{5.2-1}$$

† The book by Collins[4] treats the variations in average porosity. (See pp. 14 ff.)

‡ It is impossible to make an absolute distinction between interparticle and intraparticle free space in connection with flow. All interparticle space is not engaged in flow because the velocity approaches zero at all solid surfaces and at certain stagnation points. Conversely, all intraparticle space is not totally impassive to flow. Flow velocity varies roughly as channel size squared, so intraparticle channels 1/3 of interparticle size, for instance, may well carry 10% or so of flow. This may be an important consideration in normal gas liquid chromatography.

§ Still another porosity used commonly in studies of porous materials is the "effective porosity." This denotes the extent of interconnected pore space only. Since most pore space in chromatographic columns is thoroughly interconnected the distinction is rarely necessary.

where Φ is the fraction of total free space existing between the particles. This parameter was used earlier in connection with the plate height related to transparticle diffusion, equation (4.5-26).

The interparticle porosity (as measured for a large bulk of material) of most hard, granular matter lies in the range 0.35–0.50. A fairly well-packed chromatographic column will typically have $f_0 = 0.4 \pm 0.03$. The true porosity f is, of course, more variable, being higher if the particles are porous. The porous supports used in gas liquid chromatography, for instance, may have half of their total pore space tied up within the particles, giving $\Phi \sim 0.5$, and thus $f = f_0/\Phi \sim 0.4/0.5 \sim 0.8$.

Variations in porosity have long been known to occur for a variety of reasons. Agitation and compaction will reduce the porosity of hard-grain materials to only a moderate extent (about 20%). Soft or fragile materials may, on the other hand, suffer enormous compaction under moderate stresses. With the same degree of compaction, angular particles will maintain a higher porosity than rounded particles. This is due to particle bridging. The extreme of bridging occurs in fibrous materials (paper chromatography) where porosity may be as large or larger than 0.9.

It is an observed fact that large particles pack more densely than small particles. This matter has not been studied in any detail for chromatographic packings. From studies outside the field we may surmise that this variation will only be measurable for very fine materials with a particle diameter <0.005 cm.†

Mixtures of large and small particles pack more densely than either alone. This is a consequence of the fitting of the fine particles within the interstices of the coarse material. A related phenomenon is the increase in porosity observed for columns which are reduced in diameter to several particle diameters.[5,6] It is presumably difficult to fit the particles together at maximum density with the nearby column wall restricting particle configuration.

Measurement of Porosity

The many classical methods for measuring porosity are discussed in the standard references and need not be reviewed here. However there is a simple chromatographic method for measuring porosity which deserves mention. When an inert (nonsorbing) zone reaches a point L in the column, then this zone will have swept through the entire free volume preceding point L (this assumes that the stationary phase volume is negligible). This volume must equal the volume passing out of the column in the

† Carman[2] indicates that the increased porosity of fine materials is due to surface forces. He finds the effect significant only as d_p approaches one micron. However, the effect has been observed for larger particles also.

same time—the product of volumetric flow rate $\dot{V}$ and the time t. The ratio of this free volume $\dot{V}t$ to the total column volume in distance L, $\pi r_0^2 L$, is just the porosity f.

$$f = \dot{V}t/\pi r_0^2 L \tag{5.2-2}$$

If the zone is sorbed and has a retention ratio of R, then Rt must replace t. If the mobile fluid is highly compressible, further allowances must be made. For an ideal gas the above expression must be multiplied by $3(P^2 - 1)/2(P^3 - 1)$, where P is the ratio of inlet to outlet pressures.

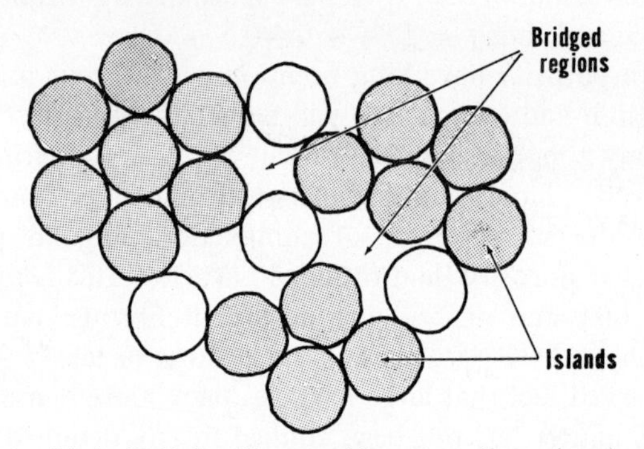

Figure 5.2-1 Bridging illustrated by a two-dimensional model.

Bridging

The structure of granular masses results as a compromise between two opposing forces. The influence of gravity (or pressure), which causes packing in the first place, induces a state of maximum density and thus maximum order. The process of bringing the granules together, on the other hand, is highly random, leading to a low density, disordered configuration. The final structure will consist of small islands where the particles have achieved nearly maximum order and density, and regions of poor fit between the islands, structurally imperfect because the islands do not join one another smoothly.† This is illustrated in Figure 5.2-1. The precise bounds of any given island cannot easily be defined in practice because the closeness of fit can vary for irregular particles, and it is not always clear which particles should be considered as part of the island. In a very approximate way we may think of typical islands having 10 or so particles. Between these islands are found the bridged regions.

The concept and magnitude of bridging can be brought into better focus by considering the packing of perfect spheres. The arrays of maximum

† There is a strong resemblance here with the molecular structure of liquids. The compromise in that case is a balance of the randomness of thermal motion with the ordering effect of intermolecular forces.

possible density, rhombohedral and face-centered cubic, have an interparticle porosity f_0 of only 0.259.[7]† A natural packing of such spheres will have $f_0 \sim 0.40$. The tightly packed islands in the natural aggregate of spheres will likely have a porosity approaching the minimum value of 0.26. However, the average (or bulk) porosity is close to the larger value, 0.40, due to the bridging. This increase shows, if the bridging hypothesis is correct, that the bridged regions account for a considerable volume fraction of the aggregate, and that bridging is by no means a minor effect.

Pore-Size Distribution

The size and relative location of the pores, channels, and cavities of a chromatographic bed determine the rate of fluid flow and the nature of various diffusion and mass transfer phenomena. Although these interrelationships have received little attention in the chromatographic literature,‡ they undoubtedly have an important role in the effectiveness of separation.

Due to geometrical complexities, the size of a pore is more an intuitive than a rigorous concept. We could, for example, take pore volume as an indication of its size, but it is necessary to be arbitrary in deciding where the volume of one pore begins and another ends. We could also take some maximum or minimum dimension, but again the choice would be arbitrary and would rarely provide a satisfactory description of size. Scheidegger[10] has suggested that a pore diameter might be associated with each point in the free space by taking it as the diameter of the largest sphere which can fit into the free space while still enclosing the given point. Each of these approaches is necessarily arbitrary and somewhat limited in quantitative significance. For this reason much can be said for a simple intuitive approach in which the pore size is regarded as some "typical" dimension of a unit of void space. This dimension can often be tied in with experimental data on pore-size distribution. Evidence on pore shape, if obtainable, will further enhance the understanding of the role played by pores.

The "pore" space of a granular material may be divided into that between the particles (related to the interparticle porosity f_0) and, when the particles themselves are porous, that within the particles (contributing a porosity $f - f_0$). The interparticle free space forms a three-dimensional, interconnected network throughout the material. Along any given path

† Scheidegger also points out that the structure with the smallest known density, but still stable, has an interparticle porosity $f_0 = 0.875$. By comparison the simple cubic lattice has $f_0 = 0.476$.

‡ The most extensive work yet reported on a chromatographic material is that by Ruoff, Stewart, Shin, and Giddings on paper chromatography.[8,9]

the channel (or pore) will alternately constrict and expand. The size of the channel will be in direct proportion to that of the particles.

The nature of the intraparticle pore space will be determined more by the composition of the material than by the size of its grains. The pores will generally be smaller, occasionally ranging down to molecular size (as in molecular seives).

The precise distribution of pore "size" in chromatographic packings is rather difficult to obtain. The pores are usually of such a size (>100 angstroms) that mercury injection is a convenient method for pore size analysis.† (This method works by forcing mercury into smaller and smaller pores, meanwhile recording the volume injected and the pressure of injection to determine the volume occupied by pores of each size range.) Although mercury injection is invalid if there are numerous ink-bottle pores (large caverns connected through small openings), it is undoubtedly valuable for characterizing chromatographic materials. In recent years mercury injection has done much to clarify the role of solid supports in gas liquid chromatography.[11,12] It is used primarily for internal pores since the ink-bottle effect is present to some extent in the interparticle network.

The final result of a full pore-size investigation is the pore-size distribution $\alpha(\delta)$. This function gives the amount of free space lying in a given pore-size range. The free volume, for instance, occupying pores with a diameter between δ and $d\delta$ is $\alpha(\delta)\, d\delta$.

Occupancy of Pore Space by Mixed Fluids

In partition chromatography two different fluids generally occupy the available pore space. The retentive medium is usually a liquid and the carrier either a gas or a liquid. The ability of these two (usually immiscible) fluids to function in the normal way depends on their relative interaction with the surface of the solid support. Some of the important factors involved in these interactions are discussed below.

Of basic importance to the partitioning equilibrium between phases is the relative amount of each. In studies of porous media a term known as the *saturation* defines the fraction of total pore volume occupied by a given phase. If this is given by s_i for phase i, then obviously $\Sigma s_i = 1$. In partition chromatography this can usually be written as $s_s + s_m = 1$, where the two saturations refer to the stationary liquid phase and the mobile phase, respectively. The relative quantities of mobile phase compared to stationary liquid is given by s_m/s_s. Dal Nogare and Chiu[13] have emphasized the importance of this usually large ratio, called β, in gas-liquid chromatography.

† The materials of paper chromatography have been treated only by capillary-rise methods.

In addition to the relative amounts, chromatographic performance is strongly influenced by the configuration and stability of these two phases, particularly the stationary phase. If the stationary phase spreads effectively over the solid support, it will expose a large surface area as required for rapid equilibration. However, spreading or "wetting" is governed by surface energy considerations and is not always achieved without forethought. Given two fluids competing for the force field of a solid surface, the fluid with the lowest fluid-solid interfacial energy will wet most successfully. Thus if the stationary phase is to wet the solid satisfactorily, its surface tension with the solid, γ_{ss}, must be lower than that of the mobile phase with the solid, γ_{ms};

$$\gamma_{ss} < \gamma_{ms} \tag{5.2-3}$$

If this condition is achieved, the stationary phase will usually cover the surface and fill the smaller pores of the solid support. Otherwise the stationary phase will form unattached globules and perhaps be carried off by the mobile phase. This is but one aspect of the whole problem of stability, as will be shown.

Even when the stationary liquid wets the solid, much of the initial deposit of this phase can be lost in subsequent operations. First, if the mobile phase is not saturated with this liquid, the latter will be gradually stripped away during development. Second, excessive amounts of stationary liquid will be carried off by flow or gravity. For any system there is an *irreducible saturation*[14] which is a measure of the maximum saturation stable with respect to the flow of mobile phase. When the support particles are porous, the capacity of the system is naturally very large; the particles' interior may be totally saturated with stationary liquid (of course, column efficiency may suffer at such extremes). In any case when stationary liquid must cling to the outside of particles, either because they are nonporous or already filled, the loose, excess liquid (beyond the irreducible saturation) will be quickly carried away. This would be the case with a heavily loaded glass bead support. The smaller the particles, the higher the irreducible saturation. At this time there is, unfortunately, little data on the irreducible saturation of chromatographic supports.

A more detailed treatment of stationary phase structure will be considered in relation to specific systems in Parts II and III.

5.3 The Laminar Flow Process

In this section we shall be concerned with flow pattern and flow rate under laminar flow conditions. The onset of turbulence at high flow rates will be discussed in the following section. Chromatographic flow is nearly always laminar in nature.

Physical Basis of Flow

The science of hydrodynamics treats fluid flow in a highly mechanical fashion.[15] The basic equations are established by assuming the conservation of matter, an appropriate equation of state (pressure-volume-temperature relationship) and Newton's second law. The latter equates the mass of a fluid element times its acceleration to the force exerted on the element—a procedure identical to that used to obtain the equation of motion of solid masses. The difference between this and simple mechanical motion is that the differential (viscous) nature of flow introduces a more complex set of forces. Instead of a solitary exterior force (such as gravity), most fluid elements will be subject to the shear forces of viscous drag, the forces derived from a pressure gradient, and the exterior force as well. Stokes[16] and Navier[17] were the first to formulate the general force-acceleration equations for fluids over a century ago. The nature of the Navier-Stokes equation can be illustrated by writing a one-dimensional form with flow occurring only along the z axis. We have the following balance of forces applying to any small element of fluid

$$\underbrace{\rho \frac{dv'}{dt}}_{\text{accel-eration}} = \underbrace{-\frac{\partial p}{\partial z}}_{\text{pres-sure}} + \underbrace{F}_{\text{ex-ter-nal force}} + \underbrace{\eta \nabla^2 v' + \frac{1}{3}\eta \frac{\partial^2 v'}{\partial z^2}}_{\text{viscous force}} \tag{5.3-1}$$

where ρ and η are fluid density and viscosity, respectively, v' the z-component of local flow velocity, and ∇^2 the Laplacian (differential) operator. The solution of this equation yields a detailed velocity profile. Since the equation shows no basic distinction between gases and liquids, the flow characteristics of these two fluids may be regarded as identical except for differences born of changes in ρ and η (the higher compressibility of gases is negligible in chromatographic flow and is also neglected in the above equation).

In practice the Navier-Stokes equation has been found extremely intractable even for moderately simple geometries. The complex geometry of the flow space in chromatographic packings is therefore far beyond exact treatment.

Although the formal equation given above may not convey much to most readers, the fact that the flow described by the equation originates in a simple mechanical fashion (force = mass × acceleration) should greatly implement our intuitive understanding of the flow process.

Darcy's Law and Permeability

The mathematical difficulties associated with the Navier-Stokes equation have led to various empirical approaches. Foremost among these, and of great value in the classical studies of porous materials, is Darcy's law.[18] This law is essentially Ohm's law of fluid flow. It states that the flux of fluid (current) per unit cross section is proportional to the pressure drop (voltage) per unit length:†

$$q = -K\,dp/dz \tag{5.3-2}$$

The constant of proportionality K is an empirical parameter (the negative sign is used to assure the proper direction of flow—a forward flow when p decreases toward the outlet). Darcy's law is useful in predicting total flow through a known medium under given hydrostatic conditions. It is, however, limited to laminar flow, and its constant must be acquired by independent means for each case. One of its main shortcomings in connection with chromatography is that it indicates nothing about the flow pattern, as would solution of the Navier-Stokes equation. However, total flow is of sufficient interest that Darcy's law and its refinements must be pursued further.

Darcy's empirical law is most valuable when it is used in association with more detailed concepts of flow, including those originating in the Navier-Stokes equation. In the development below we will show how this approach leads to useful flow equations for chromatographic columns.

It has long been established that Darcy's law is valid at low velocities‡ where the resistance to flow is caused by the viscous shear of one fluid layer over another.§ Hence simple intuitive reasoning indicates that the fluid flux, at a given pressure drop, will increase in inverse proportion to the fluid viscosity η. Combining this and the statement of Darcy's law, equation (5.3-2), we may write the following for volumetric fluid flux per unit area

$$q = -\frac{K_0}{\eta}\frac{dp}{dz} \tag{5.3-3}$$

where K_0 is the *specific permeability* with dimensions of length squared. (K_0 is often given in darcy units where 1 darcy is 9.87×10^{-9} cm^2.)

† Pressure gradient is written here in differential form so that the expression is valid for gas as well as liquid chromatography. In the former the pressure gradient varies throughout the column length; equation (5.3-2) is valid for each point along this length.

‡ Nearly all present chromatographic work is carried out in the range of velocities covered by Darcy's law, i.e., with slow laminar flow.

§ This corresponds to the viscous-force term of the Navier-Stokes equation, (5.3-1).

Specific permeability is a much-used parameter in the study of porous materials, and has been applied to gas chromatography.[19,20]

In chromatography we are more often interested in the downstream fluid-flow velocity v then in fluid flux per unit area q. These two are simply related through porosity f by $v = q/f$.† Using (5.3-2), we have

$$v = -\frac{K_0}{\eta f}\frac{dp}{dz} \tag{5.3-4}$$

which relates flow velocity to the pressure gradient, viscosity, porosity and the parameter K_0.

Perhaps the most important factor controlling mobile phase flow in chromatography is the particle size. The superior resolution of finely divided materials is often impossible to utilize because of the difficulty in getting adequate flow throughout. Thus it is important to relate flow quantitatively to particle size (in the last two equations a particle-size change is reflected by large but as yet undiscussed variations in K_0). This is best done by considering the so-called capillaric models of fluid flow.

Capillaric Models

The capillary analog is based on the principle that flow through the interstices is much like flow through fine capillary tubes. Accordingly, numerous capillaric models have been devised. In the simplest case a granular material is pictured as a bundle of uniform, equal capillaries directed along the flow axis. Other models assume capillaries of unequal diameter, capillaries with a nonuniform bore, interconnected capillaries, nonparallel capillaries, etc. These models will not be discussed in any detail here since our main purpose lies only in developing a general understanding of flow in chromatographic columns and in showing the equations which are applicable thereto.

Quite naturally the multiple capillary approach to flow is based on Poiseuille's law for flow in single capillaries. In a sense this is a return to the basic flow dynamics of the Navier-Stokes equation since flow in a capillary (Poiseuille's equation) is simple enough for an exact treatment (i.e., in the form of Poiseuille's law). This law shows that the fluid flux Q (volume of fluid per unit time) through a single capillary is given by

$$Q = -\frac{\pi d_c^4}{128\eta}\frac{dp}{dz} \tag{5.3-5}$$

† In a column devoid of packing ($f = 1$) q would equal v since it is the number of cm^3 of fluid moving into a unit area in unit time. If q is held constant as packing is introduced into the column, v will increase because the same flow must be carried in a reduced free space. The value of v is, in fact, inversely proportional to the fractional free space f available for flow.

where d_c is the capillary diameter and dp/dz is the pressure gradient along the net flow or z axis.

The simplest application of capillary flow, leading to a useful equation which incorporates particle size, can be made as follows. First it is clear that the diameter of the interstitial "capillary" d_c increases in direct proportion to particle diameter d_p, providing there is no gross change in packing structure. Hence through a given interstitial passage the flow rate increases with d_p^4. However, in a unit cross-sectional area the number N of passages available for flow decreases as their size increases; $N = \text{const.}/d_p^2$. The flux q through a unit cross section equal to NQ thus increases only as d_p^2. Hence the mean fluid velocity through a cross section, $v = q/f$, increases as d_p^2. It also increases, as can be deduced from equation (5.3-5), in proportion to the pressure gradient and inversely with viscosity. Thus for structurally similar materials we can write a general expression

$$v = -\text{const.}\,(d_p^2/\eta)(dp/dz) \tag{5.3-6}$$

If the constant factor is written as $1/2\phi'$, the value of ϕ' is found empirically to have a value of about 300 for most aggregates of nonporous particles. When each particle has its own microporous network, the (mean) velocity v is reduced because much of the fluid is held essentially stagnant within the small internal pores. The velocity will, in fact, be essentially proportional to Φ, the fraction of void space (fluid) in the interparticle channels where the bulk of flow is occurring. Hence the last equation can be written

$$v = -\frac{\Phi}{2\phi'}\frac{d_p^2}{\eta}\frac{dp}{dz} \tag{5.3-7}$$

or, in terms of the pressure gradient in the column,

$$-\frac{dp}{dz} = \frac{2\phi'\eta v}{\Phi d_p^2} = \frac{2\phi\eta v}{d_p^2} \tag{5.3-8}$$

where ϕ is the overall *flow resistance parameter*. For nonporous solids (e.g., glass and ion exchange beads) $\Phi = 1$, while for porous solids (e.g., diatomaceous earth) Φ may vary from unity to slightly less than 0.5. Thus the flow resistance parameter ϕ ranges from about 300 to 600. Fibrous materials (paper chromatography) have much higher porosities and consequently much smaller ϕ' and ϕ values (the magnitude of these have not been specifically investigated for paper chromatography). It is well known that the flow velocity varies strongly with interparticle porosity, and hence the utility of the above two equations with $\phi' \sim 300$

depends on porosity and how sensitive flow properties are with respect to its variations. This will be considered shortly in connection with the Kozeny-Carman equation.

A direct comparison of equations (5.3-4) and (5.3-7) yields a relationship between *specific permeability* K_0, and the dimensionless parameter ϕ'. We get

$$K_0 = \frac{f\Phi\, d_p^{\,2}}{2\phi'} = \frac{f_0\, d_p^{\,2}}{2\phi'} \tag{5.3-9}$$

The second equality comes from writing the interparticle porosity f_0 ($\sim$0.4) as $f\Phi$ (see equation (5.2-1)).

Most of the following material will deal specifically with the constant ϕ' rather than ϕ because the former is almost a hydrodynamic constant in many chromatographic columns. In practical chromatographic studies (see Chapter 7) ϕ, although of similar magnitude, is a more direct and useful parameter.

Dependence on Porosity: The Kozeny-Carman Equation

Many investigators, starting with Slichter,[21] have obtained expressions for the dependence of permeability on porosity. By far the best-known expression is the Kozeny-Carman[22,23] equation, although its content was largely anticipated by Blake.[24] These approximate equations can be derived from capillaric models using an empirical parameter. The derivation will not be repeated here since it is discussed in the standard references (e.g., Carman and Scheidegger). The Kozeny-Carmen equation gives the specific permeability as

$$K_0 = \frac{d_p^{\,2}}{180}\,\frac{f_0^{\,3}}{(1-f_0)^2} \tag{5.3-10}$$

The parameter ϕ', related to K_0 in equation (5.3-9), is thus found to be

$$\phi' = 90(1-f_0)^2/f_0^{\,2} \tag{5.3-11}$$

and is thus a function only of the interparticle porosity f_0. If it is assumed that $f_0 = 0.4$, this equation yields $\phi' = 202$. The empirical value, as mentioned earlier, is closer to 300. The same magnitude of discrepancy has been noted by Bohemen and Purnell[25] and by Dal Nogare and Juvet[26] for gas chromatographic supports. Hence the factor 300 would appear to be quite reasonable for most chromatographic materials with $f_0 \sim 0.4$.

The validity of assuming a near-constant ϕ' value (about 300) would appear to hinge on the constancy of interparticle porosity f_0 for column packings. It is certainly true that a large variation in f_0 could be obtained on comparing a loosely filled column and one in which the granular material is agitated and packed to maximum density. However, the latter course of action is standard practice with chromatographic columns so that a comparison need be made only for well-packed columns. Dal Nogare and Juvet[26] reported an f_0 range of 0.40–0.41 for particles varying from 163–545 μ. Bohemen and Purnell[25] found $f_0 = 0.42 \pm 0.03$ for a wide range of particle sizes. Bernard and Wilhelm[27] found an f_0 range of 0.36–0.43 for well-shaken spherical packings. From these results it is safe to conclude that f_0 will only occasionally vary by more than 0.03 from a normal value of 0.40 for well-packed granular materials in chromatography. The significance of this magnitude of variation on the flow parameter ϕ' can best be ascertained by finding the rate of change of ϕ' with respect to f_0. From equation (5.3-11) we find that

$$d\phi'/df_0 = -180(1 - f_0)/f_0^3 \qquad (5.3\text{-}12)$$

With $f_0 = 0.40$, this gives $d\phi'/df_0 = -1700$. A variation of 0.03 in f_0 gives a variation in ϕ' of $\Delta\phi' = (d\phi'/df_0)\,\Delta f_0 = -1700 \times 0.03 \sim -50$. Thus ϕ' would normally be expected to fall in the range 250–350. This is confirmed by results obtained on various materials in the author's laboratory and summarized in Table 5.3-1. It was necessary to approximate Φ (interparticle fraction of void) as 0.5 to calculate ϕ' for alumina and Chromosorb. The results show that ϕ' does generally lie in the range from 250–350.

Table 5.3-1 Range of ϕ' Values

Material	ϕ' value
50/60 mesh glass beads	250–280
30/40 mesh alumina	300
50/60 mesh alumina	260
60/80 mesh Chromosorb W (5% DNP)	350
60/80 mesh Chromosorb W (20% DNP)	330

Since ϕ' does not normally vary from 300 by more then 20%, the most practical approach to the study of pressure drop—particle size—flow relationships in granular-bed chromatography is through the use of equations (5.3-7) and (5.3-8). The parameters of this equation (except for Φ, which is normally unity and rarely <0.5) are readily available. Use of Kozeny-Carman equation requires either an estimation of f_0 (as well as Φ), in which case the ϕ' equations can be used directly, or a direct measure of f_0. For fibrous materials, of course, f_0 varies so widely that the Kozeny-Carman equation must be used. Although this equation has been criticized[28] for fibrous materials, it is probably valid to a fair approximation except at very high porosities.[29]

Flow Pattern

The classical studies of flow in porous materials have almost ignored the flow pattern as compared to the gross flow effects of the Kozeny-Carman type. It is true that some theoretical treatments have been devised for miscible displacement (equivalent to zone spreading in chromatography) based on certain flow pattern models. However, our understanding of this phenomenon is probably derived more from experimental and theoretical work in chromatography than from the classical studies. Even then it is a difficult subject not very well understood.

The flow pattern (i.e., the detailed distribution in relative flow velocity and its direction) is essentially the sum of all the local velocity biases or differentials which are so important to zone spreading (see Chapter 2). In theory the flow pattern can be obtained as a solution to the rigorous Navier-Stokes equation. In practice this is too complex, and it is necessary to make rather crude estimates of the various flow differences. The zig-zag lateral flow which results as each streampath skirts the solid obstacles is ignored for the most part in the approximate treatments.†

The essence of the various components of flow pattern has already been discussed in Sections 2.8 and 2.9. There is no need to repeat those deductions here. However, there are certain features of the flow pattern which deserve further elaboration in the light of the more basic approach of this chapter.

As indicated earlier, the simplest capillaric model assumes that flow in porous materials occurs along a bundle of parallel, equal capillaries, uniform throughout their length. Thus each interstitial flow "channel" is in a sense equivalent to one capillary of the model. However, this simple model has numerous shortcomings. It is clear, first, that each capillary is not uniform along its length. The numerous constrictions will constitute the main obstacle to flow‡ and will often force the fluid into neighboring capillaries (this requires numerous interconnections, a feature not possessed by the simple model). Second, the capillaries are not equal

† Lateral transport plays a critical role in transcolumn nonequilibrium as was explained in Section 2.9.

‡ The constriction and tortuosity of the channels reduce the flow velocity in paper chromatography (and presumably in other media) by a factor of 10^2 below the value predicted for uniform capillaries using pore-size distribution data.[8,9]

to one another. Where the particles fit together poorly (bridged regions) a much larger flow will occur. Finally, the capillaries are not parallel, but zig-zag randomly. Thus for a detailed study of flow pattern the simplest capillaric model is a failure. Each type of velocity bias must be treated independently using the best theoretical estimates available. This approach has been followed by the author in calculating the C_m (mobile phase mass transfer) terms and their origin in the flow differences of gas chromatography.[30] Some of the general conclusions will be discussed below.

The most localized of all the flow pattern effects is connected with flow in single interstitial channels, giving the transchannel effect (see Section 2.8). In a practical sense there is but little recourse to regarding this flow as if it were occurring in a round capillary of uniform bore (this gives a parabolic flow pattern in which the velocity is greatest at the center and falls off to zero at the wall). This gross simplification is justified mainly by the fact that the transchannel velocity bias has only a minor chromatographic effect.

When the individual particles are porous, the mobile phase frequently occupies the internal pore space made available. This part of the mobile phase is generally regarded as stagnant because the internal pore structure is so fine (recall that mean velocity increases with the square of channel diameter). Thus a strong velocity bias (leading to the transparticle effect) will generally exist between intraparticle and interparticle fluid. However, when large internal pores exist the internal fluid may not be stagnant. Its movement will be enhanced by the fact that the normal channels between particles are so constricted and tortuous that some fluid may choose any reasonable alternative pathway. The largest "round" channel penetrating a close packed structure is 15% of d_p. Combine this with the fact that flow between particles is restricted to less than half (about 40%) of the total volume and is even then very tortuous, and it is easy to see why large, internal pores might be competitive in allowing the passage of fluids. As an estimate we may assume that intraparticle flow is comparable to interparticle flow if there are many large internal channels of diameter $d_c > 0.1\, d_p$. This condition may be approached in some cases by the "white" forms of diatomaceous earth (e.g., Chromosorb W) used in gas-liquid chromatography.

The foregoing conclusions are verified to some extent by Carman. This author indicates that the permeability of charcoal (with fine pores) may be calculated ignoring the internal surface, but that the permeability of Kieselguhr (coarse pores) is related to the total surface area.[31]

Next on the increasing scale is the important short-range interchannel velocity bias. In Section 2.8 this was identified with the loosely filled space between tightly packed regions. We now recognize this as the bridged space between close-packed islands. The increased porosity in the bridged regions permits a much higher flow rate than is found in the surrounding islands. If we assume that the packing of the bridged regions resembles a simple cubic structure ($f_0 = 0.48$) and the islands a close-packed structure ($f_0 = 0.26$), the Kozeny-Carman equation (to the extent it is valid for such local flow disturbances) tells us that the interparticle fluid velocity is seven times greater in the bridged regions than in the islands. This is confirmed by noting[30] that the interstitial aperture in the simple cubic structure is larger than that in the close-packed structure by a factor of 2.7, giving a $(2.7)^2$- or approximately 7.5-fold velocity increase.

The main unsolved problem in connection with this effect is whether the bridged "channels" are too isolated to permit such increased flow. If, for instance, each large capillary in a capillaric model is dead end, it contributes nothing to flow. There is no doubt that flow is hindered to some extent within the bridged regions since ingress and egress are not entirely free at each end. In spite of this it is believed that a large velocity bias still exists because of the extensive interconnections between flow channels (Section 5.2) and perhaps between bridged regions themselves.

The long-range interchannel effect refers to flow inequalities over a distance of $\sim 10d_p$. This effect is due to long-range random variations in permeability. Some studies have been made on random variations in permeability and porosity,[4] but these have not yet been translated accurately into the corresponding chromatographic effect. The latter would require the use of the generalized nonequilibrium theory of the last chapter.

The transcolumn effect is a manifestation of column-wide flow differences. Such differences can arise from variations in particle diameter (or porosity) over the column cross section or from the bending or coiling of columns. Giddings and Fuller[32] observed that the larger particles of a granular material would move preferentially to the outside as a stream of particles impacted at the center of a large tube. The reason for this is somehow tied up in the complex, undeveloped mechanics of the flow and migration of granular materials. It is possible that the larger particles roll to the outside while the smaller particles are trapped more easily in surface "pockets." The author has also observed that large particles rebound more on impact with the bulk granular material than do small particles, thus giving a differential distribution. This may be due to their larger terminal velocity. More work is needed to identify the precise mechanism at work. In any case the flow velocity increases roughly with the square of mean particle size (see equation (5.3-7)) any may thus vary widely over the column cross section.

Another source of transcolumn variations in flow is related to the nature of the flow channels where wall and packing meet. These are undoubtedly different than the normal interstitial channels. Golay has postulated that this "wall effect" leads to a velocity about 5 times higher than average within $\frac{1}{2}$ particle diameter from the wall.[33] Reduced tortuosity would certainly lead to some increase in velocity. However, the wall channels would still be constricted at frequent intervals. Also irregular particles might flatten themselves against the wall and reduce flow even further. It is questionable whether the wall velocity is as great as that found in the bridged network of the packing (in a sense the structure near the wall is a bridged structure). Experimental studies of packed-column flow are not conclusive on this point either, indicating that a maximum velocity occurs 1 particle diameter from the wall.[34] The work of Benenati and Brosilow[6] shows that the average porosity of a wall layer $d_p/2$ thick is $\sim$0.44; application of the Kozeny-Carman equation suggests that the velocity increase would be less than 50% in such a layer.

The bending or coiling of a chromatographic column will introduce a velocity gain at the inside of the bend and a velocity lag at the outside. The reason is that these paths have different lengths with the same pressure drop, and therefore the pressure gradient is greatest for the shorter, inside path. The velocity profile is given by the ratio of the regional velocity v and the mean cross-sectional velocity $\tilde{v}$.[35]

$$\frac{v}{\tilde{v}} = \frac{R_0}{R_0 + r \sin \theta} \tag{5.3-13}$$

where R_0 is the radius of curvature, r is the distance from the tube center, and θ is the angle with respect to a line extending through the tube center parallel to the coil axis. The chromatographic implications of this flow pattern have been investigated mathematically.[35,36]

The transcolumn flow pattern is of great importance in preparative-scale chromatography. Huyten, van Beersum, and Rijnders have measured the velocity profile in large gas chromatographic columns and have demonstrated the increased velocity toward the outside.[37] In recent work Giddings and Jensen have studied the chromatographic implications of the flow pattern and have obtained experimental results which corroborate the size-sorting hypothesis.[38] The detailed study of preparative-scale processes in gas chromatography will be reserved for Part II of this treatise.

Gas Flow: The Effect of Compressibility

With a given interstitial geometry and slow laminar flow the detailed flow pattern of gases and liquids is identical. The basic mechanical equation of Navier-Stokes makes no distinction between gases and liquids—differences arise only as a result of variations in density and viscosity (a full comparison of gas and liquid chromatography will be given in Section 7.5). At sufficiently low velocities (including those of usual chromatographic interest) even these variations are unimportant as far as the flow pattern itself is concerned.

The identity of flow pattern does not carry over to properties related to compressibility. Liquids may be regarded as imcompressible under the moderate pressures applied to chromatographic columns. However, these same pressures may compress a gas to a fraction of its atmospheric volume (compressibility is still negligible insofar as local flow mechanics is concerned). Thus as a gas stream moves down a column from regions of high pressure to low pressure, it expands considerably. Since the same mass of gas must pass through each cross section in unit time (when the flow is in steady state), a greater volume flux will be observed where the gas is considerably expanded. This means that the flow velocity (proportional to volume flux) is largest where the pressure is least—at the column outlet. In view of the proportionality of velocity and volume, Boyle's law can be written as $pv =$ constant, where pv replaces pV.

In liquid chromatography both flow velocity and pressure gradient are constant throughout any column of uniform bore. Such a column can be regarded as ideally uniform in the lengthwise direction. The velocity and pressure distributions in a gas chromatographic column depart from ideal as shown below.

Equation (5.3-8) shows that the column pressure gradient is given by

$$-dp/dz = 2\phi\eta v/d_p^2 \tag{5.3-14}$$

where we recall that the flow resistance parameter ϕ replaces the constant ϕ'/Φ. Since flow velocity v varies throughout the column, this equation shows immediately that the pressure gradient dp/dz is also nonuniform (the viscosity η is, of course, constant with changes in pressure). Since pv is constant throughout the column, we may write $pv = p_i v_i$ where subscript i stands for inlet values. Thus $v = p_i v_i/p$. With this substituted back into equation (5.3-14) and the latter rearranged, we may write

$$-p\,dp = (2\phi\eta p_i v_i/d_p^2)\,dz = I\,dz \tag{5.3-15}$$

where I is a column constant given by the terms in parenthesis. Integrating from some point z with pressure p to the column outlet ($p = p_0$, $z = L$), we have†

$$-\int_p^{p_0} p\,dp = I\int_z^L dz \tag{5.3-16}$$

† In liquid chromatography, with v constant, we have $-dp/dz =$ a constant; thus on integration p is linear in distance.

or

$$p^2 - p_0^2 = I(L - z) \qquad (5.3\text{-}17)$$

which gives the pressure distribution as a function of the distance $L - z$ from the outlet. The velocity distribution can easily be obtained in terms of p since pv is constant. The implications of these pressure and velocity nonuniformities will be discussed in Part II.

Capillary Flow: Paper and Thin Layer Chromatography

A unique kind of flow process occurs in paper and thin layer chromatography. The main driving force for such flow is capillary in nature; it originates, as in "capillary rise," by the tendency of wetting liquids to flow into empty capillary or pore space.

Simple capillaric models for flow in paper were established by Lucas,[39] Washburn,[40] and Bosanquet.[41] Of the subsequent refinements, the one most applicable to chromatographic flow is the interconnected capillary model of Ruoff et al.[8,9]

The simplest model can be used to illustrate the basic nature of flow in paper and thin layers. Assuming that these materials are equivalent to bundles of uniform, parallel capillaries, the problem is similar in many ways to the classical studies of capillary-rise equilibrium. In each case surface tension, γ, is the driving force for capillarity. For dynamic flow liquid transport is hindered mainly by viscosity, while in the static rise experiments surface tension is balanced against gravity. In mathematical form the flow velocity (rate of advance of the front located at z_f) dz_f/dt is proportional to the driving force γ and inversely proportional to the restraining force (viscosity) η multiplied by the length z_f of capillary through which liquid is being drawn. Thus

$$\frac{dz_f}{dt} = \text{const.} \times \frac{\gamma}{\eta z_f} \qquad (5.3\text{-}18)$$

or rearranged

$$z_f\, dz_f = \text{const.} \times (\gamma/\eta)\, dt \qquad (5.3\text{-}19)$$

When this is integrated from time 0 to t and over the flow path from 0 to z_f, we have

$$z_f^2 = 2\ \text{const.} \times (\gamma/\eta)t \qquad (5.3\text{-}20)$$

or simply

$$z_f^2 = \kappa t \qquad (5.3\text{-}21)$$

The latter is the *parabolic flow law*, indicating that the liquid front at z_f advances only with the square root of time. The constants in these equations are found to be proportional to capillary radius.

The parabolic flow law is valid for uniform, rectangular paper strips and thin layers in the absence of net gravity forces (i.e., in horizontal flow). Other geometries follow modified flow laws.[42]

The main shortcoming of the simple models is the failure to come to grips with the strong gradients in the concentration of the penetrating liquid. Visual inspection shows that a paper strip is "wetter" near the liquid reservoir than at the liquid front. A velocity gradient also exists along the flow path. To explain these phenomena we must invoke an interconnected capillary model.[8] The interconnections permit the flow of liquid from large capillaries to small ones, thus permitting a condition of partial liquid saturation at each point. The interconnected model is consistent with (and provides a generalization of) the parabolic flow law.

Since zone migration rates depend on the amount of mobile phase available for the partitioning equilibrium, and on the mobile phase velocity, the gradients in mobile phase concentration and velocity can seriously influence chromatographic migration.[9] These gradients explain why R_f values are not reproducible if we change the starting position of the zone, and why there is no simple, exact relationship between the R_f's in rectangular, radial, and other geometrical modifications. A complete exposition of this subject is beyond the scope of the present volume. A detailed treatment will be given in Part III.

Other Flow Processes and Concepts

This brief section on flow has barely touched on a number of interesting phenomena and theories because they have only a marginal chromatographic interest. The literature must be consulted for a complete treatment.

Aside from turbulent flow (Section 5.4) we have ignored flow processes which do not conform to normal laminar flow. In so doing, we have passed over slip flow and Knudsen (molecular) flow in gases. Slip flow occurs as the mean free path becomes a considerable fraction of the channel diameter, and is marked by an apparent slippage (increased permeability) of the gas over the solid surfaces. Knudsen flow occurs when the mean free path exceeds the channel diameter and thus only collisions with the wall are important. At normal pressures the mean free path will range from 0.01–0.1 μ. The interparticle channels in gas chromatography will range from 10–100 μ. Thus either kind of flow is normally unimportant. However, when reduced pressures are used, or when flow (and diffusion) in the fine internal pore space is considered, these processes should be taken into account.

Much of the literature of porous materials revolves around concepts such as specific surface and hydraulic radius. The specific surface A is simply the surface area per unit volume of porous material. The hydraulic radius is a measurable parameter given by $r_h = f/S$. For a bundle of capillaries r_h is $\frac{1}{4}$ of the capillary diameter d_c. For complex materials, where a "true" channel radius does not exist, the hydraulic radius often serves as an "effective" radius for capillary flow. A whole class of hydraulic-radius theories have sprung up around this concept. This includes the Kozeny theory and its various modifications (e.g., the Kozeny-Carman equation).

5.4 Turbulent Flow

Until now we have dealt almost exclusively with laminar flow (also known as viscous or streamline flow). With randomly arranged channels (e.g., as in a granular chromatographic bed) a small element of fluid in laminar flow will zig-zag erratically as it strives to avoid the solid particles in its path. The zig-zag pattern is unvarying, however, in the sense that any subsequent fluid element, started in the same position, will follow its predecessor's path exactly, neglecting diffusion. This is a consequence of the fact that the two fluid elements are subject to the same relative forces at each point, and thus move off on identical trajectories despite the time interval between them. Thus the laminar flow of chromatographic interest is spatially erratic but temporally constant—the flow pattern, although complex, remains fixed with the passage of time.

Nature of Turbulent Flow

An increase in flow velocity to a certain level will produce turbulence. Turbulent flow is characterized by rapid fluctuations in velocity, direction of flow and pressure at any given point. Thus turbulent flow is random in both space and time. A series of fluid elements initiated at the same point will fail to follow the same streampath through the medium. The strong fluctuations in velocity observed in turbulent flow will increase the lateral transport of momentum (thus making the fluid appear more viscous) and the effective rate of diffusion.

Turbulent flow occurs when the inertial or acceleration term of the Navier-Stokes equation becomes large. The inertial term increases more rapidly than the viscous terms as velocity increases,† and inertial effects

† The inertial term $\rho\, dv/dt$ increases rapidly with velocity for two reasons; velocity differences (dv) are magnified and, the time scale (dt) contracts when flow past two successive points is considered. By way of a mechanical analog, a centrifugal force, which has an inertial origin, increases with the square rather than the first power of angular velocity.

eventually become dominant, leading to turbulence. The detailed reasons for the transition to turbulent flow are tied up in the complexities of the Navier-Stokes equation.† However, a simplified picture will be given here. From a physical viewpoint viscous forces may be regarded as restraining forces with a tendency to dampen out relative motion and velocity fluctuations. Not so with inertial forces. A fluid element in motion will tend to remain in motion without a change in direction no matter how vigorous are its collisions with other fluid elements.

The degree of turbulence can be gaged by the well-known Reynold's number, *Re*. In granular materials this is given by‡

$$Re = \rho\, v d_p / \eta \tag{5.4-1}$$

This is a dimensionless quantity proportional to velocity. Turbulence develops gradually from a minor to a dominant role as *Re* increases through the range§ from 1 to 100. This is in strong contrast to straight tubes where turbulent flow occurs suddenly as the Reynold's number exceeds 2100. The gradual onset of turbulence in granular materials is undoubtedly due to the fact that many flow channels are involved, each with a different size and configuration.[45] At $Re \sim 1$ we can envision turbulence in only a few of the largest channels. As *Re* increases, turbulence gradually expands into the remaining channels.

The Reynold's number, or Reynold's criterion as it is sometimes called, logically reflects the relative strength of inertial and viscous forces. Inertial forces increase with density and with velocity squared—ρv^2. Viscous forces increase with viscosity and velocity, and inversely with the channel (or particle) diameter††—$\eta v/d_p$. The ratio of these two terms is simply the Reynold's number, *Re*. It may appear strange that turbulence would begin when this number is so small (1–100) in granular materials as compared to a straight tube (2100).‡‡ However, Muskat has pointed

† Some question has been raised as to whether the Navier-Stokes equation is entirely valid in turbulent flow. In any case there is no exact theory of turbulence, even in flow through straight tubes, as the Navier-Stokes equation becomes intractable.

‡ Strictly speaking v should usually be replaced by the interparticle velocity $v_0 = v/\Phi$. However, the uncertainty in the precise happenings at a given *Re* value make this distinction temporarily unnecessary.

§ Collins[43] speaks of a transition range from $Re = 1$–10. However, Perkins and Johnston[44] indicate that a full transition is obtained only when $Re > 10^3$. For chromatographic purposes the range 1–10^2 is probably valid as an approximate description.

†† At a given mean velocity, viscous shear is much greater in small tubes than large ones.

‡‡ It must be pointed out that there is not unanimous agreement that tubulence begins at $Re \sim 1$–100. Scheidegger[46] argues that inertial effects can become important without turbulence, and that the Reynold's number is not always an adequate criterion of turbulence. Muskat[47] however, has pointed out that turbulence can actually be observed with a stream of fluorescein in a packed column. While we do not intend to resolve the argument here it would seem that our present picture of turbulence, essentially in agreement with Muskat, is based on the best evidence.

out that flow is taking place in channels with very irregular features.[47] The frequent changes needed in velocity and direction would certainly increase the importance of inertial effects. This can be observed by noting that turbulence occurs around a bend in a tube much before it does in a straight portion of the same tube.

Failure of Darcy's Law

It is found empirically that at high flow rates fluid velocity and pressure gradient are no longer proportional. In order to account for growing inertial effects, a quadratic velocity term must be added to the latter:

$$-dp/dz = av + bv^2 \tag{5.4-2}$$

Darcy's law is valid only when v is so small that the bv^2 term is negligible by comparison to av. This condition is found when $Re < 1$. The departure from Darcy's law when $Re > 1$ constitutes the principal evidence that inertial effects are becoming dominant and that turbulence is occurring. As mentioned in the previous footnote, however, Scheidegger disagrees that turbulence is necessarily connected with the bv^2 term. In contrast to most authors he suggests that equation (5.4-2) may be valid for laminar flow.

The fact that a pressure drop must be used to force fluid through a column in steady flow indicates that there is some form of energy dissipation. The av term, interpreted energy-wise, is a measure of energy dissipated as heat in viscous shear (similar to the heat generated in frictional shear between solid bodies). The bv^2 term is actually proportional to the kinetic energy of the fluid. This energy is dissipated as heat in the numerous eddies and cross-currents of turbulent flow. Even here the ultimate mechanism of energy dissipation is viscous in nature, and is related to the decay of eddies due to viscous resistance.

The heat generated during flow will ordinarily remain with the fluid, thus increasing its temperature. In an ideal gas, however, this heat gain is equaled by the heat loss due to gas expansion.

Turbulence in Chromatographic Flow

As we shall see shortly, most chromatographic runs are made safely within the range of laminar flow. Recent theoretical developments (particularly related to coupling, Section 2.10) indicate that there is much to be gained by operating a chromatographic column at very high flow rates. This section is written, therefore, in anticipation of a growing role for turbulent flow in chromatography. Turbulence itself can be a useful force in chromatographic analysis, as pointed out recently by Sternberg and Poulson.[48] This matter will be investigated shortly.

In order to put turbulence in proper perspective, it is useful to estimate the Reynold's number for current chromatographic practice. High Reynold's numbers are most readily achieved in gas chromatography, mainly because of the high flow rates. For He and N_2 flowing at 50°C through a 60–80 mesh nonporous support ($d_p \sim 0.02$ cm)

$$\left.\begin{array}{ll} Re \sim 0.015v & \text{He} \\ Re \sim 0.12v & \text{N}_2 \end{array}\right\} \text{nonporous particles} \qquad (5.4\text{-}3)$$

In a porous (e.g., diatomaceous earth) support the interchannel velocity v_0 of main concern to turbulence is roughly twice the cross-sectional mean velocity v. Thus

$$\left.\begin{array}{ll} Re \sim 0.03v & \text{He} \\ Re \sim 0.25v & \text{N}_2 \end{array}\right\} \text{porous particles} \qquad (5.4\text{-}4)$$

The higher value for N_2 is due mainly to its high density. We see for porous supports and N_2 carrier that turbulence may begin at $v \sim 4$ cm/sec and be well developed at $v \sim 400$ cm/sec. Other heavy carrier gases (e.g., CO_2, A) would develop turbulence in this same approximate range. For He (and roughly for H_2) turbulence in porous supports would develop over the range $v \sim 30$–3000 cm/sec.

The Reynold's number of liquid chromatography can be estimated by reference to benzene and *n*-butyl alcohol mobile phases flowing through a 270–325 mesh ($d_p \sim 0.005$ cm) nonporous material at 50°C

$$\left.\begin{array}{ll} Re \sim v & \text{C}_6\text{H}_6 \\ Re \sim 3v & \text{C}_4\text{H}_9\text{OH} \end{array}\right\} \text{nonporous particles} \qquad (5.4\text{-}5)$$

The values would be somewhat higher for porous particles.

The foregoing Reynold's values indicate that some degree of turbulence may be expected occasionally in gas chromatography, particularly with the heavier carrier gases. At a typical velocity of 10 cm/sec a moderate turbulence would be developed with N_2 but not with He. For liquid chromatography, with its lower velocities, turbulence would rarely be encountered. For instance, Hamilton et al.,[49] using higher than average velocities (up to 0.18 cm/sec), have noted *Re* values in the range 0.003–0.2. These conclusions in no way alter the fact that flow in the turbulent range may offer significant advantages as will be shown below.

Chromatographic Implications of Turbulence

The most obvious effect of turbulence in chromatography is the increase in mass transfer rates. But this is not the only major effect. With the low

Reynold's numbers encountered in chromatography an outgrowth of turbulence, described here for the first time, which we shall term the *velocity equalization effect*, may be more significant.

It will be recalled from Chapter 2 that a large fraction of zone spreading is caused by velocity biases in the mobile phase. It will be recalled further that turbulence begins in the large, fast flowing channels and gradually spreads to other channels as velocity increases. As shown by the bv^2 term of equation (5.4-2), turbulence causes an increase in flow resistance† and thus a reduction in flow rate. Consequently, the effect of the high flow channels (mostly caused by bridging), which contribute so much to the velocity bias in laminar flow, will be considerably suppressed by the onset of turbulence. This is illustrated in Figure 5.4-1 where large and small channels are represented by two sizes of capillary. Laminar flow is

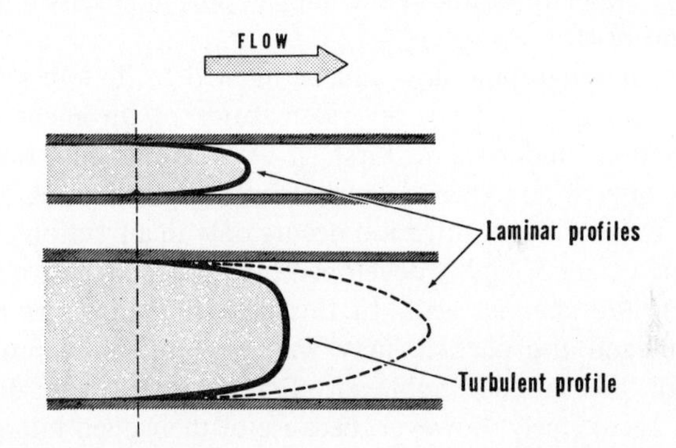

Figure 5.4-1 An illustration of velocity equalization in two hypothetical channels.

occurring in the smaller capillary, and the flow profile is of the typical "bullet" shape for such flow. The turbulent flow of the larger capillary is characterized by a flattened velocity profile. Its mean velocity has been decreased as shown by comparison to the laminar profile (dashed line) in the large channel.

The velocity equalization effects work for other forms of velocity bias as well, because in nearly all imaginable cases turbulence will develop first where velocity is greatest. Even in a single tube (the transchannel effect) the very process of flattening the profile will equalize velocities.

Although currently of less practical concern, the velocity equalization effect will apply also to the very high velocities of fully developed turbulence. This situation has been discussed in connection with flow through porous rocks.[44] It was shown that the ratio of velocity extremes under

† This applies to separate flow channels as well as the medium as a whole.

laminar conditions ν_L would reduce to ν_T with fully developed turbulence, where

$$\nu_T = \nu_L^{1/4} \tag{5.4-6}$$

Thus with the important short-range interchannel effect where normally $\nu_L \sim 7$, the turbulent velocity ratio would only be $\nu_T \sim 1.6$. This would institute a significant reduction in zone spreading.

The other main consequence of turbulence in chromatography is the increase in mass transfer due to the wild excursions of the flow path. Where before much mass transfer was accomplished by molecular diffusion, with turbulence the same may often be achieved by erratic shifts in the entire flow stream—by *turbulent diffusion*. The basic mathematical treatment of turbulent diffusion is far from complete, as seen in a number of thorough discussions of the subject.[50–52] In porous materials very little at all is understood about the subject so our discussion will be qualitative and brief.

If chromatographic flow can be likened to flow in straight tubes (i.e., using capillaric models), several features of turbulent diffusion can be immediately understood. First, it is obvious that turbulent diffusion occurs only where there is turbulence. In chromatography this usually means that turbulent diffusion occurs only in a fraction of the flow space. Even in a state of highly developed turbulence there is a *laminar boundary layer* at the channel wall. In this case the transport rate between the channel and the particle may well be limited by molecular diffusion through the laminar sublayer. Transport between different channels would occur freely, however, because of their open interconnections.

Both longitudinal and lateral transport are enhanced by turbulent diffusion. In the center of a flow channel the enhancement is about equal (i.e., isotropic). Near the walls, however, longitudinal displacements are somewhat larger than lateral.

Turbulent diffusivity increases roughly in proportion to mean flow velocity[51] as suggested by Sternberg and Poulson[48] (this applies, of course, only in those regions where turbulence has developed). Hence in a state of fairly well-developed turbulence we may replace mobile phase diffusivity D_m by a term proportional to v. Consequently, the simple plate height equation for mobile phase effects $H_m = B/v + C_m v$ becomes $H_m = H_t$, a constant independent of velocity, in view of the inverse dependence of C_m on D_m and the direct dependence of B on D_m. This is the same result as that obtained for the coupling theory of eddy diffusion at high velocities (Section 2.10). The relationship between these two phenomena will be discussed presently.

Several efforts have been made to estimate the magnitude of the constant H_t.[53–55] It has been postulated that each void space between flow constrictions becomes a "mixing vessel" as a consequence of turbulence—each trace of solute entering the void cell is immediately mixed throughout its volume. As indicated in Section 2.3, a mixing cell is equivalent to a theoretical plate. Thus the length of the void cell is equal to the limiting plate height H_t. This has been assumed as about one particle diameter, $H_t \sim d_p$. However, it may well be that a more "realistic" mixing unit would be the closely packed islands between the bridged regions, or perhaps even some parcel of larger dimensions. The author feels that the assumption $H_t = d_p$ does not adequately account for long-range non-uniformities, and that H_t should be $>d_p$. The answer to this question must await experimental investigation. It should be noted that the usual interpretation of experimental evidence obtained in connection with mixing-cell models is suspect because no account is taken of coupling.

While velocity equalization and turbulent diffusion have the most direct effects on column performance, the increased pressure drop of turbulent flow is far from negligible. This will be made clear in Section 7.5 where the maximum pressure drop imposed by the equipment is shown to be one of the basic limitations on further advances in resolution and speed.

Turbulence and Coupling

Turbulent transfer and coupling (Section 2.10) have much the same effect on plate height at high flow rates; they restrain the increase in the mobile phase contribution to some constant value. The question of which phenomenon first becomes effective with increased flow is important from the standpoint of designing chromatographic systems to take advantage of the beneficial effects.

Since the degree of turbulence is measured by Reynold's number, $Re = \rho v d_p/\eta$, and the degree of coupling by reduced velocity, $\nu = d_p v/D_m$, the ratio of one to another can be used to estimate relative importance

$$\frac{\nu}{Re} = \frac{\eta}{D_m \rho} \tag{5.4-7}$$

This ratio is about 10^3 for liquids and 1 for gases. In Section 2.10 it was shown that coupling reached half of its full effectiveness at some particular value, $\nu_{1/2}$, of the reduced velocity. The most significant effect (excluding transcolumn terms), related to short-range interchannel velocities, is shown by Table 2.10-1 to have a reduced transition velocity of $\nu_{1/2} \sim 2$. The gradual transition to turbulence will be characterized similarly by a certain value of Re—$Re_{1/2}$. This will be some intermediate value pertaining to partially developed turbulence, probably $Re_{1/2} \sim 10$. Equation (5.4-7) shows that a given value of ν is obtained $\eta/D_m\rho$ times more readily than a given value of *Re*. Since *Re* must reach about 10 (about 5 times higher than $\nu_{1/2}$) for effective turbulent transport, coupling is achieved $\sim 5\eta/D_m\rho$ times more readily than its turbulent counterpart. This factor changes from about 5 to 5000 from gas chromatography to liquid chromatography. In view of the approximations made the former is enough of a borderline case that it should be verified experimentally. (A distinction could be obtained by studying gas flow and liquid flow in the same column.) Unfortunately, there is no data of this kind yet available.

For capillary (open-tube) chromatography there are, unfortunately, no coupling processes. Turbulence is also difficult to achieve because of the high Reynold's number (about 2000) required in "pipe" flow. A gas velocity of $>10^4$ cm/sec would be needed in a typical capillary for turbulence. Once achieved, turbulence would show both the velocity equalization and the enhanced mass transfer effects. Although out of practical reach now, turbulence in capillaries might have an enormous potential if the stationary phase mass transfer terms can be made small enough.

Even if coupling overrides turbulent transport in most chromatographic columns, the velocity equalization effect still affixes great potential value to the use of turbulence in chromatography.

5.5 References

1. M. Muskat, *Flow of Homogeneous Fluids*, McGraw-Hill, New York, 1937.
2. P. C. Carman, *Flow of Gases through Porous Media*, Academic Press, New York, 1956.
3. A. E. Scheidegger, *Physics of Flow through Porous Media*, Macmillan, New York, 1957.
4. R. E. Collins, *Flow of Fluids through Porous Materials*, Reinhold, New York, 1961.
5. J. C. Sternberg and R. E. Poulson, *Anal. Chem.*, **36**, 1492 (1964).
6. R. F. Benenati and C. B. Brosilow, *Am. Inst. Chem. Engrs. J.*, **8**, 359 (1962).
7. A. E. Scheidegger, *op. cit.*, p. 19.
8. A. L. Ruoff, G. H. Stewart, H. K. Shin, and J. C. Giddings, *Kolloid-Z.*, **173**, 14 (1960).
9. J. C. Giddings, G. H. Stewart, and A. L. Ruoff, *J. Chromatog.*, **3**, 239 (1960).
10. A. E. Scheidegger, *op. cit.*, p. 7.
11. W. J. Baker, E. H. Lee, and R. F. Wall, in *Gas Chromatography*, H. J. Noebels, ed. et al., Academic Press, New York, 1961, Chap. 3.
12. D. M. Ottenstein, *J. Gas Chromatog.*, **1**, No. 4, 11 (1963).
13. S. Dal Nogare and J. Chiu, *Anal. Chem.*, **34**, 890 (1962).
14. R. E. Collins, *op. cit.*, p. 30.
15. See M. Muskat, *op. cit.*, Chap. 3.
16. C. L. M. H. Stokes, *Ann. Chim. Phys.*, **19**, 234 (1821).
17. G. G. Stokes, *Trans. Cambridge Phil. Soc.*, **8**, 287 (1845).
18. H. Darcy, *Les Fontaines Publiques de la ville de Dijon*, Dalmont, Paris, 1856.
19. A. I. M. Keulemans, *Gas Chromatography*, 2nd ed., Reinhold, New York, 1959, Chap. 5.

20. J. C. Reisch, C. H. Robison, and T. D. Wheelock, *Gas Chromatography*, N. Brenner et al., ed. Academic Press, New York, 1962, Chap. VII.
21. C. Slichter, *19th Ann. Rep. U.S. Geol. Survey*, **2,** 305 (1897–8).
22. J. S. B. Kozeny, *Akad. Wiss. Wien., Abt., IIa*, **136,** 271 (1927).
23. P. C. Carman, *Trans. Inst. Chem. Eng. London*, **15,** 150 (1937).
24. F. C. Blake, *Trans. Amer. Inst. Chem. Engrs.*, **14,** 415 (1922).
25. J. Bohemen and J. H. Purnell, *J. Chem. Soc.*, **1961,** 360.
26. S. Dal Nogare and R. S. Juvet, Jr., *Gas-Liquid Chromatography*, Interscience, New York, 1962, p. 135.
27. R. A. Bernard and R. H. Wilhelm, *Chem. Eng. Progr.* **46,** 223 (1950).
28. J. L. Shearer, *Textile Res. J.*, **29,** 467 (1959).
29. P. C. Carman, *Flow of Gases through Porous Media*, Academic Press, New York, 1956, Chap. 1.
30. J. C. Giddings, *Anal. Chem.*, **34,** 1186 (1962).
31. P. C. Carman, *op. cit.*, p. 39.
32. J. C. Giddings and E. N. Fuller, *J. Chromatog.*, **7,** 225 (1962).
33. M. J. E. Golay in *Gas Chromatography*, H. J. Noebels et al., ed. Academic Press, New York, 1961, p. 11.
34. C. E. Schwartz and J. M. Smith, *Ind. Eng. Chem.*, **45,** 1209 (1953).
35. J. C. Giddings, *J. Chromatog.*, **3,** 520 (1960).
36. J. C. Giddings, *J. Gas Chromatog.*, **1,** No. 4, 38 (1963).
37. F. H. Huyten, W. van Beersum, and G. W. A. Rijnders, "*Gas Chromatography, 1960*, R. P. W. Scott, ed. Butterworths, London, 1960, p. 224.
38. J. C. Giddings and G. E. Jensen, *J. Gas Chromatog.*, **2,** 290 (1964).
39. R. Lucas, *Kolloid-Z.*, **23,** 15 (1918).
40. E. W. Washburn, *Phys. Rev.*, **17,** 276, (1921).
41. C. H. Bosanquet, *Phil. Mag.*, *Series 6*, **45,** 525 (1923).
42. A. L. Ruoff and J. C. Giddings, *J. Chromatog.*, **3,** 438 (1960).
43. R. E. Collins, *Flow of Fluids through Porous Materials*, Reinhold, New York, 1961, p. 51.
44. T. K. Perkins and O. C. Johnston, *Soc. Pet. Engs. J.*, March, 1963, p. 70.
45. M. Muskat, *op. cit.*, pp. 63–64.
46. A. E. Scheidegger, *op. cit.*, p. 32.
47. M. Muskat, *op. cit.*, p. 62.
48. J. C. Sternberg and R. E. Poulson, *Anal. Chem.*, **36,** 1442 (1964).
49. P. B. Hamilton, D. C. Bogue, and R. A. Anderson, *Anal. Chem.*, **32,** 1782, (1960).
50. A. Sommerfeld, *Mechanics of Deformable Bodies*, Academic Press, New York, 1950, Chap. VII.
51. R. B. Bird, W. E. Stewart, and E. N. Lightfoot, *Transport Phenomena*, Wiley, New York, 1960, Chap. 5, 20, etc.
52. R. C. L. Bosworth, *Transport Processes in Applied Chemistry*, Wiley, New York, 1956, Chap. IX.
53. R. Aris and N. R. Amundson, *Am. Inst. Chem. Engrs. J.*, **3,** 280 (1957).
54. J. J. Carberry, *Am. Inst. Chem. Engrs. J.*, **4,** 1, 13M (1958).
55. J. M. Prausnitz, *Am. Inst. Chem. Engrs. J.* **4,** No. 1, 14M (1958).

Chapter Six

Diffusion and Kinetics in Chromatography

6.1 Molecular Displacement and Chromatography

The dynamic process of central importance in chromatography is the migration of solute molecules. Chromatography, after all, is a process of differential molecular movement, and the various elements of molecular displacement are therefore the building blocks of the whole technique.

The displacement steps of chromatography generally result from flow, diffusion, and sorption kinetics. The nature of flow was studied in Chapter 5. Diffusion and kinetic processes will be studied here. All these processes are fundamental to many branches of science and much is known about them. The present treatment is far from comprehensive, being designed mainly to acquaint the reader with those elementary concepts which frequently have a bearing on chromatographic separation.

6.2 Nature of Diffusion

Diffusion is one of the most commonly observed phenomena of nature. It is characterized by the dilution and spreading out of concentrated units of matter, and the concomitant intermixing of components initially separated. Diffusion is a natural consequence of one of science's most respected laws—the second law of thermodynamics. The dilution of matter is part of the unceasing creation of entropy, or randomness, in the natural world. Any parcel of matter with unbound molecules† will tend to expand its volume and, in so doing, increase its randomness. Diffusion is the physical mechanism for achieving this gain in entropy and randomness.

At the molecular level diffusion is found to be a consequence of the unceasing thermal agitation of molecules. Each molecule will set out on

† When all molecules are bonded together, as in a crystal lattice, diffusional expansion is prevented by energy considerations. This, too, can be reduced to an entropy criterion, and in the end one finds that a solid will not break apart because the entropy of the "universe" would thereby be decreased.

an erratic course which will, after a very long time, pass through each small region of available space. Thus an initial concentration pulse will be leveled as each constituent molecule explores neighboring regions along its own zig-zag path.

The leveling of a concentration pulse (such as a chromatographic zone) logically means that there must be a net flux of solute from the concentrated center to the dilute edges. In fact, as we shall see shortly, there is always net diffusional transport along a concentration gradient. This transport is quite frequently pictured, erroneously, in terms of each molecule faithfully migrating toward the low concentrations. It is essential to the basic understanding of diffusion to realize that each molecule's path (except in certain nonlinear cases) is independent of its companion molecules. Thus a given molecule is not even "aware" of a concentration gradient as it pursues its random course, and cannot possibly direct itself to the dilute regions. Hence another explanation must be sought for the observed transport to the dilute regions. Such an explanation follows.

Due to the random nature of diffusional migration, any small region of space will send molecules out equally in all directions. If a concentrated region and a dilute region exist side by side, each will receive a certain fraction of the others original store in a given time interval. However, since the concentrated region started with more molecules, it yields more to the dilute region than vice versa. There is no preferential direction at all in the molecular motion involved in this process. Nonetheless the dilute region gains at the expense of the concentrated regions, in accord with observation.

The exact form of a molecule's path is of some interest in calculations regarding the magnitude of diffusion coefficients. This will be discussed in Section 6.3.

The theory of diffusion is composed mainly of two parts. One involves obtaining concentration profiles for a given diffusivity as discussed next. The other is concerned with the molecular processes which determine the magnitude of diffusivity (Section 6.3).

The principal rules governing the precise nature of diffusional spreading are Fick's laws.[1,2] Fick's first law states that the flux J of a given component, say solute, across a unit area fixed in the medium is

$$J = -D\, \partial c/\partial z \tag{6.2-1}$$

where D is the diffusion coefficient and $\partial c/\partial z$ is the concentration gradient. The quantity J will equal the number of moles passing through a unit area (normal to the concentration gradient) in unit time. This equation is usually applied to true molecular diffusion, but it is found to be applicable to a great and varied number of natural transport phenomena (e.g., the horizontal flow of water in unsaturated soils, the mixing of air masses in the atmosphere, the penetration of light into snow or clouds, the flux of neutrons in a pile, etc.). These processes may be termed effective diffusion processes, and each has an effective diffusion coefficient, simply because they obey Fick's first law of diffusion. One such process has already been described in Chapters 3 and 4 (see equation (3.2-31) and following) where it was seen that nonequilibrium leads to solute zone spreading in accord with equation (6.2-1).

The mass transport processes which obey Fick's first law also obey his second law†

$$\frac{\partial c}{\partial t} = \frac{\partial}{\partial z} D \frac{\partial c}{\partial z} \tag{6.2-2}$$

Usually D can be considered as a constant and the equation written in its more usual form

$$\frac{\partial c}{\partial t} = D \frac{\partial^2 c}{\partial z^2} \tag{6.2-3}$$

This is a partial differential equation whose solution, consistent with the appropriate initial conditions, describes the concentration profile as it evolves with time. If, for instance, the initial profile is a sharp spike (corresponding to a narrow injected zone), the solution to this equation is in a Gaussian form

$$c = \frac{M}{t^{1/2}} \exp(-z^2/4Dt) \tag{6.2-4}$$

where M is a constant proportional to the total mass of diffusing solute. This can be easily verified as a solution by substituting it back into the differential equation. The maximum in concentration, at $z = 0$, can be seen to vary as $1/t^{1/2}$. This shows how the flattening of a zone occurs as diffusion proceeds (e.g., during column migration). In chromatography, at a given flow rate, time t is proportional to the distance migrated so that the maximum concentration diminishes with L according to $1/L^{1/2}$.

† Fick's second law can be derived in a way similar to the analogous flow accumulation expression, equation (3.2-17). We consider a thin slab of unit area normal to the concentration gradient. In unit time the amount of solute diffusing into the slab through the face located at z is $J_z = -D(\partial c/\partial z)_z$. The amount flowing out the other face located at $z + dz$ is $J_{z+dz} = -D(\partial c/\partial z)_{z+dz}$. Thus the rate of accumulation of solute in the slab, influx minus outflux, is $J_z - J_{z+dz} = -(\partial J/\partial z)\, dz$. The rate of concentration increase, $\partial c/\partial t$, is the accumulation rate per unit volume, $(J_z - J_{z+dz})/dz$, i.e., $\partial c/\partial t = -\partial J/\partial z$. With J substituted into this from equation (6.2-1) this gives Fick's second law directly. The generalization of Fick's laws to three dimensions can be achieved with little additional difficulty.

No simple rule like this is applicable if the flow rate changes since the various effective diffusion coefficients (plate height terms) are sensitive to flow velocity.

Gaussian profiles are usually written in a form involving the standard deviation σ

$$c \propto \exp(-z^2/2\sigma^2) \tag{6.2-5}$$

A direct comparison of the last two equations shows that

$$\sigma^2 = 2Dt \tag{6.2-6}$$

This is Einstein's relationship as discussed in Section 2.5. This equation indicates the relationship between the zone quarter-width σ and the molecular (or effective) diffusion process which is causing the spreading. And, of course, since $H = \sigma^2/L$, the plate height is related to the effective diffusion coefficient by $H = 2Dt/L = 2D/Rv$, where the zone displacement velocity Rv replaces L/t.

Fick's laws describe diffusion in regions which are fixed in the medium. If the medium is a flowing fluid, as is often the case in chromatography, then we can simply move our coordinates along with the flow displacement. However, if we choose to fix our coordinates to a stationary point, then we must add the effects due to flow transport through each cross section. The latter phenomenon has been described thoroughly in equations (3.2-14) to (3.2-17). The combined effects have already been presented in equation (2.3-9). Fick's second law must in this case be replaced by

$$\frac{\partial c}{\partial t} = -Rv\frac{\partial c}{\partial z} + D\frac{\partial^2 c}{\partial z^2} \tag{6.2-7}$$

where Rv is the velocity of the zone in question. If we are concerned with a lateral diffusion phenomenon, then of course we need not add the term for longitudinal flow.

Occasionally the diffusion process is complicated by a nonconstant diffusion coefficient. The diffusion of solute in mobile and stationary phases usually occurs at such dilutions that the diffusion problem is linear—D is constant. Various effective diffusion processes in chromatography are nonlinear. The flow of liquid in paper and thin layers obeys a diffusion equation in which D is strongly dependent on liquid concentration.[3] The effective D value for a zone with nonlinear sorption is also concentration dependent. These cases require special approaches, some of which will be discussed in the subsequent volumes.

6.3 Diffusion Rates in Liquids and Gases and on Surfaces

Diffusion rates in gases are four or five orders of magnitude greater than in liquids. The molecular basis of this enormous difference is readily apparent. Diffusion occurs as a random walk (or flight), the solute molecules stepping this way and that in a random fashion. Due to the wide distance between gaseous molecules, the random steps in a gas are of considerable length, being equal roughly to the distance covered between collisions. This distance, the mean free path, is the order of 100 molecular diameters. In liquids the random steps are the order of one molecular diameter, and can be pictured as occurring when the molecule slips by its neighbors to an adjacent equilibrium position. The much longer random steps of gaseous diffusion are mainly responsible for the relatively large diffusion rates as compared to liquids. In fact the diffusion coefficient increases roughly in proportion to step length squared,† so we expect a variation in the order of $(100)^2 = 10^4$, much as observed. Diffusion along surfaces is usually comparable in rate to liquid diffusion.

The detailed theory of gas, liquid, and surface diffusion is beyond the scope of this book. We shall proceed with separate discussions of gases and liquids, amplifying some of the final theoretical and empirical equations with these broad concepts rather than with concise theory.

Liquid Diffusion

A wide spectrum of liquid diffusion processes occur in chromatography. In most forms of liquid chromatography we are concerned with the diffusion of small or medium size solute molecules in solvents (stationary or mobile phase) of a similar molecular size. Occasionally we are concerned with the diffusion of very large solute species (e.g., proteins) in smaller solvent species. Conversely, in gas liquid chromatography we are usually interested in the diffusion of medium size solute molecules in the high molecular weight species of the stationary phase. Most concepts of liquid diffusion are applicable in part to the entire range of solute/solvent size. For concreteness we will mainly discuss the interdiffusion of two medium size species with reference where needed to the size extremes mentioned above. All discussions of numerical diffusion values in liquids, including this one, are hampered to a certain extent by the lack of adequate experimental values. In view of the fact that liquid molecules are more or less bound into place by rather strong intermolecular attractions, and must

† By combining the random walk expression, $\sigma^2 = l^2 n$ (see equation (2.5-2)), and the Einstein relationship, $\sigma^2 = 2Dt$, we get $D = l^2\dot{n}/2$, where $\dot{n}$ is the number of steps per second, n/t, and l is the step length along a given axis. The assumption that D increases with l^2 is valid only when $\dot{n}$ is nearly constant. By accident this is nearly true in comparing gases and liquids. The step of a gas molecule is slow because the molecule must travel a relatively long way between collisions. The step of a liquid molecule is slow because it must accumulate sufficient energy to escape the confines of its molecular cage. Although exceptions can easily be found, these effects are nearly equal in many gases and liquids.

gain considerable thermal energy to break these bonds and achieve displacement, diffusional transport is logically an activation process (see Section 6.5) with the usual exponential dependence on temperature,†

$$D_l = D_0 \exp(-W_l/\mathscr{R}T) \qquad (6.3\text{-}1)$$

where W_l is the activation energy required for molecular displacement, $\mathscr{R}$ the gas constant, T the absolute temperature, and D_0 a term with only slight temperature variation. This equation can be used to calculate the temperature dependence of liquid diffusivity in chromatographic systems. We must, of course, search out a value for W_l, a quest that by itself leads to several interesting conclusions. There are two methods of estimation.

It is well known that the viscous shear of liquids is an activation process and that $1/\eta$ shows an exponential temperature dependence much like D_l in the previous equation. The interesting thing is that the activation energy W_l is found experimentally to be nearly equal for D_l and $1/\eta$. Since viscosity data are much more prevalent (being easier to acquire) than diffusivities, we can often obtain W_l in this way. If η is known at a series of temperatures, a plot of $\log(1/\eta)$ versus $1/T$ will yield a line of slope $-W_l/2.3\mathscr{R}$.

The fact that diffusion in a binary mixture (solute and solvent) shows the same activation energy as a process in the pure solvent (viscous shear) indicates that molecular displacement in diffusion is governed somehow by the slip of solvent molecules. In the hole theories of liquids[4] it has been surmised that the rate controlling process for diffusion is the creation of a void space next to the solute which the solute can then fall into with little effort. The energy W_l is approximately equal to that needed to create the hole.

Another method for estimating W_l is related to the heat of vaporization. Eyring and co-workers have determined that the energy needed to create a hole in a liquid is one third of that needed to vaporize one of its molecules. Hence W_l should be roughly $\frac{1}{3}$ of the vaporization heat of the solvent. For convenience in gas chromatography it has been assumed[5] that W_l could be approximated by 0.35 times the heat of vaporization of solute from solvent, a parameter measured readily from gas chromatographic data. The inadequacy of this assumption is shown by the fact that W_l may even exceed solute vaporization heats. The vaporization of isobutylene from DNP, for instance, requires only 4560 cal/mole, whereas W_l is 8000

† In the random walk expression of the previous footnote, $D_l = l^2\dot{n}/2$, the term $\dot{n}$ is responsible for the exponential dependence. Since the displacement process is akin to a unimolecular reaction, the reaction rates (steps per second) is of the order (Section 6.5) of $\dot{n} = 10^{13} \exp(-W_l/RT)$. If l is one molecular diameter, this gives about the right magnitude for diffusivity, i.e., $D_0 \sim 10^{-2}$ or 10^{-3} cm²/sec.

cal/mole.[6] Further correlation should be sought between these two parameters because of the ease with which the former is acquired.

If it is desired to obtain numerical values for D_l itself, and not just of W_l, a large number of equations can be invoked with varying degrees of success. Gambill[7] has given an excellent summary of the matter. Nearly all equations relate D_l to viscosity. The first such equation with any success was due to Einstein[8] (the Stokes-Einstein equation).

$$D_l = \mathscr{R}T/6\pi r_1 \eta N \qquad \text{(Stokes-Einstein)} \qquad (6.3\text{-}2)$$

where r_1 is the "radius" of the solute molecule and N is Avogadro's number. (Hereafter subscript 1 will apply to solute and 2 to solvent.) Assuming a spherical molecule with molar volume V_1 this equation reduces to[7]

$$D_l = 10^{-7}T/\eta V_1^{0.33} \qquad (6.3\text{-}3)$$

in which D_l is inversely proportional to the cube root of solute volume and thus approximately of molecular weight. In reality there should be a shape factor in this equation; D_l will be reduced somewhat for rod shaped molecules as opposed to spheres.

The Stokes-Einstein equation was derived for large Brownian (colloid) particles, and is not applicable to solute molecules less than about 1000 in molecular weight. This situation is sometimes remedied by using a "slip" form of the equation in which the constant 6 is replaced by 4. However, the Eyring theory and its attendant equation are more directly designed for the diffusion of small molecules, particularly in cases of self diffusion. The equation is

$$D_l = \mathscr{R}T/(\lambda_b \lambda_c/\lambda_a)\eta N \qquad \text{(Eyring)} \qquad (6.3\text{-}4)$$

where the λ's are molecular dimensions as defined in the reference.[9] The term $\lambda_b \lambda_c/\lambda_a$ replaces $6\pi r_1$ of the Stokes-Einstein equation, and is smaller (D_l is larger) by factors of about 10. The Eyring equation does not improve the accuracy of D_l calculations beyond the Stokes-Einstein equation, but it does give clear insight into the nature of liquid diffusion.

Along with the Stokes-Einstein and Eyring equations, numerous empirical expressions have been developed. The Wilke-Chang[10] equation is perhaps best

$$D_l = \frac{7.4 \times 10^{-8}(\Psi_2 M_2)^{0.5}T}{\eta V_1^{0.6}} \qquad \text{(Wilke-Chang)} \qquad (6.3\text{-}5)$$

where M_2 is the molecular weight of solvent and Ψ_2 is an association factor—unity for nonpolar solvent, 2.6 for water, 1.9 for methanol, and 1.5 for ethanol. This equation is applicable to small and medium size molecules with an accuracy of about 10%. Note that D_l is much more sensitive to the molecular size of the solute than indicated by the Stokes-Einstein equation, changing with $V_1^{-0.6}$ instead of $V_1^{-0.3}$.

The numerous other equations for D_l would require too much space for a full discussion. Gambill[7] has listed several others—the Arnold equation, in which D_l varies† with $\eta^{-1/2}$ as opposed to $1/\eta$, the Olson-Watson relationship with surface tension, etc.

In order to establish a frame of reference for D_l values, some typical cases with small to medium size solute and solvent molecules around room temperature are shown in Table 6.3-1.

Table 6.3-1 D_l for Medium and Small Molecules

Solute-solvent pair	Temperature, °C	D_l, cm²/sec
glycine—H_2O	25	1.06×10^{-5}
glucose—H_2O	15	0.52
H_2—H_2O	25	3.36
CCl_4—CH_3OH	15	1.7
$C_2H_5NO_3$—CH_3OH	15	2.2
TNT—C_6H_6	15	1.39
acetic acid—C_6H_6	14	1.92
phenol—CS_2	19	3.4
phenol—$CHCl_3$	19	1.6

These values group roughly around 10^{-5} cm²/sec. The latter value may be assumed as a norm for the following variations.

Solvent Dependence. The Wilke-Chang equation along with most others show that D_l depends strongly on the solvent. The main variation is in reciprocal viscosity, $1/\eta$, but the effect of this is tempered somewhat by the fact that D_l increases with the square root of solvent molecular weight. Viscosity increases with polarity, association and molecular weight. This is seen by the ascending viscosity in centipoises, at 25°C, of the following series: ether, 0.23; acetone, 0.33; benzene, 0.65; water, 1.0; ethyl alcohol, 1.2; acetic acid, 1.3; *n*-butyl alcohol, 2.9; ethylene gylcol, 19; *m*-cresol, 21; glycerol, 1500. This shows that solvent viscosity is by far the most significant of the two factors affecting D_l; in this series $1/\eta$ covers a range of ~6000 while $M_2^{1/2}$ varies only by ~2.5. We may thus presume that D_l varies by ~2000 in this series of solvents, a large variation indeed.

† A comparison of the two viscosity relationships in gas liquid chromatography has been made by Hawkes and Mooney.[11]

Gas chromatographic solvents usually contain large molecules (to avoid solvent vaporization) and are highly viscous. However, the temperature is frequently high enough to partially remove the high viscosity. Hawkes[11,12] has accumulated much data on this matter, and has indicated the viscosity level which, in different circumstances, will seriously hinder liquid mass transfer and thus resolution. As taken from his work, viscosities in centipoises at 25°C can be represented by the following typical group: DNP, 21; squalane, 35; TOTP, 64; polypropylene sebacate, 4730. Thus viscosities are 20–4000 times higher, and presumably D_l 20–4000 times lower (ignoring $M_2^{1/2}$), than the values ($\eta \sim 1$ centipoise) for ordinary solvents. The situation is much remedied at 100°C for which the range of viscosities has been reduced to 3–150.

It can be seriously questioned whether the inverse viscosity relationship holds for some of the long snake-like molecules used as solvents in gas liquid chromatography. Eyring et al.[9] point out that the activation energy for viscosity approaches a constant, ~8000 cal/mole, for increasing chain length. They interpret this as meaning that viscous (and thus presumably diffusional) displacement occurs through segments of the molecule containing 20–25 carbon atoms.† Viscosity as a whole still increases with the chain length Z of long molecules as follows

$$\eta = \text{const.} \exp(\alpha Z^{1/2}) \tag{6.3-6}$$

because the coordination of segments necessary for viscous shear becomes more difficult with increasing Z. We may assume, however, that such coordination is not necessary to achieve diffusion with small solute molecules. Thus it is likely that D_l will not decrease indefinitely with Z as does $1/\eta$. For cross-linked chains (e.g., ion-exchange materials, rubber) viscosity may go essentially to infinity because there can be no gross displacement of chains. There can, however, be a local movement, sufficient to allow the passage of solute molecules, and D_l will therefore remain finite.

There is no doubt that D_l does fall off rapidly with increasing molecular size up to the presumed segment length. Reamer, Opfell, and Sage,[14] for instance, find the $D_l \times 10^5$ for methane at 4.4°C decreases in cm²/sec as follows for a series of hydrocarbon solvents: pentane, 9.2; heptane, 6.7; decane, 4.0; white oil (about 24 carbons), 0.46. Conclusions regarding longer solvent chains must await further experimental work.

Solute Dependence. The value of D_l does not depend nearly as much on solute as solvent. The Wilke-Chang equation shows a dependence on $V_1^{-0.6}$, where V_1 is the molar volume of the solute. The Stokes-Einstein equation shows an even weaker dependence—$V_1^{-0.33}$.

† Shorter diffusional segments are suggested by Van Geet and Adamson.[13]

The insensitivity of D_l to solute structure has been emphasized by Frenkel.[15] He gives the following examples for $D_l \times 10^5$ in water solvent at room temperature: radon, 2.3; oxygen, 1.6; HCl, 1.39; sugar, 0.34 (all in cm^2/sec). He points out that the numerical difference is quite insignificant when compared with the difference in chemical constitution and size. The 7-fold difference shown here is indeed quite small compared to the 2000-fold difference predicted for a series of solvents.

A more direct comparison between solute and solvent factors can be illustrated with the following data taken from Volume 5 of the *International Critical Tables*. As solutes (in benzene at 15°C) ethyl ether and isoamyl alcohol have $D_l \times 10^5$ values (cm^2/sec) of 2.2 and 1.5, respectively. As solvents (using phenol at 19°C as solute) these two exhibit $D_l \times 10^5$ values of 3.6 and 0.2. As solutes the ratio in D_l values is 1.5, while as solvents the ratio is a much larger 18 (the latter is in rough agreement with the viscosity ratio, 24, at 10°C).

The relatively mild dependence of D_l on solute does not mean that widely different values cannot be be obtained if one seeks the extremes. Gosting[16] reports D_l values for proteins and other solutes of very high molecular weight that range from 0.003×10^{-5} to 0.1×10^{-5} cm^2/sec, a variation of up to 300 from common D_l values.

Gosting also presents some results on the effect of chain length and chain branching. The α-amino acids with butyric, valeric, and caproic skeletons, in water at 25°C, show $D_l \times 10^5$ decreasing in the order 0.83, 0.77, and 0.72. Chain branching has very little effect; in the same series the butyric and isobutyric skeletons have values of 0.83 and 0.81, respectively.

Temperature Dependence. It was indicated earlier that liquid diffusion is an activation process with an exponential temperature dependence, $D_l = D_0 \exp(-W_l/\mathscr{R}T)$, equation (6.3-1). The strong dependence on temperature can also be seen by noting that nearly all equations predict $D_l\eta/T$ to be constant, and η is known to vary rapidly with temperature. Due to the exponential dependence of D_l its value will ordinarily double for temperature increases of 15–70°, the precise value depending on the activation energy and the temperature range of concern. For instance, the diffusion of methane in pentane[14] shows a relatively mild dependence, $D_l \times 10^5$ values changing only from 9.2 to 18.1 between 4 and 71°C. Over a comparable temperature range, 0–75°C, the value[15] for isobutylene in DNP changes from 0.026 to 0.61, a 20-fold change compared to the 2-fold change of the methane-pentane system. The latter values are of particular interest because they are the only wide range measurements applicable to a common gas liquid chromatographic solvent. Because of this uniqueness a digression will be made to this subject.

Houghton et al. have compared experimental values for the isobutylene-DNP system in the range 0–75°C with various theoretical expressions. The Eyring expression yields values too small by a factor of about 100. The Wilke-Chang expression is also in serious error with values too small by ratios of 11 up to 44 (the latter at 0°C). Furthermore, values of $D_l\eta/T$, constant in nearly all equations, varied from 4.7 to 1.2×10^{-7} cm^{-1}. (The variation in this term is not nearly so great in most systems; see, e.g., Gosting.[16]) Thus until further evidence is available, calculations regarding small solute molecules in the large solvent molecules of gas liquid chromatography are suspect.

Concentration Dependence. The diffusion of solutes in chromatography usually occurs at very low concentration levels corresponding to the linear region. It is inevitable, however, that higher concentrations will occasionally be encountered, in which case D_l will vary with concentration. The classical example of this in chromatography is the study by Helfferich and Plesset[17] on nonlinear diffusion, influenced by electrical forces, in ion exchange beads. In nonionic liquids we may presume that in covering the extreme range between pure solvent and pure solute D_l will change roughly in proportion to the variation in $1/\eta$, usually a very large variation. Small concentration changes will be proportionately significant. The data presented by Gosting[16] indicates that various solutes at unit molarity in water will change D_l by about 10%. This high sensitivity makes it necessary to keep solute concentrations minimal to establish linearity.

Diffusion in Gases

Although gaseous diffusion is mainly of interest in the single area of gas chromatography, we will discuss the elementary concepts and equations in this volume for the sake of continuity.

The process of main interest in gas chromatography is the diffusion of organic (occasionally inorganic) molecules, sometimes quite large, through the small inert molecules of the gas phase. Most theoretical and experimental studies of diffusion in gases have been concerned with the interdiffusion of one small molecular species in another, say, He in N_2. Very little data is available concerning the diffusion of large solute molecules as encountered in gas chromatography. Hence once again we are hampered by the paucity of experimental work. Much of the data now available has been summarized in a recent review,[18] to which the present discussion owes much.

Since molecular displacement in gases is not an activation process (requiring a minimum energy for its execution), the diffusion coefficient, D_g, is not a strong, exponential function of temperature. It is obvious, however, that the increased molecular agitation of high temperatures will cause some increase in D_g with temperature. It is found experimentally that D_g increases approximately as $T^{1.7}$ at constant pressure.[19] In addition, D_g is inversely proportional to pressure p except at high pressures ($p >$ 20 atm).[20,21] These variations are all described, with different degrees of precision, by the equations of the kinetic theory of gases, below. This theory is in much better shape than liquid theory because gas molecules have fewer and simpler encounters with neighboring molecules. The theories themselves are no simpler, however, and in their most rigorous form require extensive mathematical development. Our main object is to present the chromatographic uses of these theories rather than the theories themselves. For those who wish a physical picture of gas phase diffusion, a simple random walk approach can be used.†

The early Stefan-Maxwell equation describes diffusion in a gas composed of hard sphere molecules

$$D_g = \frac{\alpha}{n{\sigma_{12}}^2}\left[\frac{8\mathscr{R}T}{\pi}\left(\frac{1}{M_1} + \frac{1}{M_2}\right)\right]^{1/2} \qquad \text{(Stefan-Maxwell)} \quad (6.3\text{-}7)$$

where n is the number of gas phase molecules per cm^3, σ_{12} the collision diameter (separation between molecular centers on collision of species 1 and 2), $\mathscr{R}$ the gas constant, and M_1, M_2 the molecular weight of solute and carrier gas, respectively. The constant α is variously given as $1/3\pi$,[22] $\frac{1}{8}$,[23] $1/2\pi$,[24] and $\frac{3}{32}$.[25] Inasmuch as the latter constant is obtained from the rigorous Chapman-Enskog theory of gases, it may be regarded as the correct value. On substituting numerical values for this and the other constants in the above equation we get the useful numerical form[26]

$$D_g = \frac{0.00263T^{3/2}}{p{\sigma_{12}}^2}\left[\frac{1/M_1 + 1/M_2}{2}\right]^{1/2} \qquad (6.3\text{-}8)$$

† Section 2.5 and the beginning of this section describe the random walk model. A footnote at the start of this section shows that $D = l^2\dot{n}/2$, where $\dot{n}$ is steps (collisions) per second and l the component step length (mean free path). Consider an increase in pressure. The crowding together of molecules reduces the path length between them so that l falls off as $1/p$, but increases the collision frequency giving $\dot{n}$ proportional to p. Substituted above this gives $D_g \alpha 1/p$, as observed.

Consider an increase in T at constant p. First, molecules move further apart as the gas expands; reversing the pressure arguments, we get $l \alpha T$, $\dot{n} \alpha 1/T$ and thus $D_g \alpha T$. However, molecular speed (and thus collision frequency) increases with $T^{0.5}$ giving in total, $\dot{n} \alpha T^{-1.5}$ and thus $D_g \alpha T^{1.5}$. This agrees with the simpler kinetic theories. The rigorous theories account for the fact that, due to the softness of molecules, a grazing collision at high temperatures is essentially no collision at all. Consequently, l is increased and n decreased by a certain small fraction at high temperatures. The l value is more influential than the $\dot{n}$ since D_g depends on the square of the former. Thus D_g increases slightly faster than $T^{1.5}$, i.e., roughly $T^{1.7}$.

where p is the gas pressure in atmospheres and σ_{12} the collision diameter in angstroms. This equation shows proportionality to $T^{3/2}$ (typical of hard sphere models) and $1/p$ as suggested earlier. It also shows that D_g falls off slowly with increasing molecular weight and rapidly with molecular size.

In view of the fact that only limited data are available for σ_{12}, Gilliland[27] and Arnold[28] modified the above equation, replacing σ_{12} values by more accessible molar volume terms. The constant in front was obtained by a best fit to experimental data. Gilliland gives

$$D_g = \frac{0.0043T^{3/2}(1/M_1 + 1/M_2)^{1/2}}{p(V_1^{1/3} + V_2^{1/3})^2} \quad \text{(Gilliland)} \qquad (6.3\text{-}9)$$

where V_1 and V_2 are molar volumes in cm^3 at the boiling point and p is again in atmospheres. Molar volumes can be obtained directly, or better, they can be estimated as an additive sum of the volume of molecular constituents as proposed by Kopp, Le Bas, and others.[21,29]

Arnold attempted to improve on the $T^{3/2}$ dependence of the above equations by introducing a second temperature dependent term in the denominator to account for molecular "softness"

$$D_g = \frac{0.00837T^{3/2}(1/M_1 + 1/M_2)^{1/2}}{p(V_1^{1/3} + V_2^{1/3})^2(1 + c_{12}/T)} \quad \text{(Arnold)} \qquad (6.3\text{-}10)$$

where c_{12} is Sutherland's constant. The latter can be estimated in various ways.[18] Depending on the temperature, the above equation shows a dependence varying from $T^{3/2}$ to $T^{5/2}$.

The first rigorous diffusion theory of "soft" gas molecules was presented by Hirschfelder, Bird, and Spotz following the Chapman-Enskog kinetic approach combined with the Lennard-Jones intermolecular (6-12) potential function.[26] Their equation reads

$$D_g = \frac{0.00186T^{3/2}(1/M_1 + 1/M_2)^{1/2}}{p\sigma_{12}{}^2\Omega_{12}} \quad \text{(Hirschfelder-Bird-Spotz)} \qquad (6.3\text{-}11)$$

with p in atmospheres and σ_{12} in angstroms. The term Ω_{12} is a collision integral depending in a complicated way on temperature, and the interaction energy of colliding molecules. Values of Ω_{12} have been tabulated.[20] The σ_{12} value, exactly interpreted, is the minimum approach distance of two molecular centers when the molecules collide with zero (or nearly zero) initial kinetic energy.

The main disadvantage of the Hirschfelder-Bird-Spotz equation is the difficulty encountered in evaluating σ_{12} and Ω_{12} values. These depend on molecular dimensions and interaction energies, values that are not readily available for the majority of molecules. Most values have been obtained from viscosity parameters. For chromatographic purposes various approximations are probably most useful in obtaining the desired parameters. Hirschfelder, Bird, and Spotz relate these parameters to the critical temperature and volume of the component species (boiling temperatures and volumes can also be used). Details can be found in the review.[18]

Perhaps the most explicit approximation to the Hirschfelder-Bird-Spotz equation is one given by Chen and Othmer[30]

$$D_g = \frac{0.43(T/100)^{1.81}(1/M_1 + 1/M_2)^{1/2}}{p(T_{c1}T_{c2}/10^4)^{0.1405}[(V_{c1}/100)^{0.4} + (V_{c2}/100)^{0.4}]^2} \quad \text{(Chen-Othmer)} \tag{6.3-12}$$

where critical temperatures and volumes are given by T_c and V_c, with subscripts 1 and 2 added for solute and carrier gas, respectively.† Both T_c and V_c values can be estimated in various ways.[21]

The difficulty in estimating parameters can be avoided, without sacrificing accuracy, using the recent method of Fuller, Schettler and Giddings.[31] These authors developed a successful equation in which atomic and structural volume increments and other parameters have been obtained by a least squares fit to over 300 measurements. The equation is

$$D_g = \frac{0.00125T^{1.75}(1/M_1 + 1/M_2)^{1/2}}{p[(\sum v_i)_1^{1/3} + (\sum v_i)_2^{1/3}]^2} \quad \text{(Fuller-Schettler-Giddings)} \tag{6.3-12a}$$

with p in atmospheres and the volume increments v_i in cm^3. A short table of volume increments is an accessory to this equation; no further parameters are needed. Some of the more important increments are C = 25.4, H = 0.8, O = 6.3, N = 8.6, and an aromatic ring = −50.2. For carrier gases, He = 5.4, H_2 = 11.1, N_2 = 22.5, and CO_2 = 32.8.

Table 6.3-2 is a condensation of the data presented in the Fuller-Giddings review. The percentage error in estimating experimental D_g values is listed for five equations. This brief summary does not cover all the methods for estimating D_g. For instance, Wilke and Lee proposed a mass modification to the Hirschfelder-Bird-Spotz equation. A corresponding state method proposed by Slattery and Bird gives an average error of 14%.

The results of these calculations show that the Gilliland equation (the simplest form of the original Stefan-Maxwell expression) incurs rather large errors, the order of −35% for He carrier gas and −15% for the

† Since interdiffusion in gases, unlike liquids, is nearly independent of composition, subscripts 1 and 2 can be exchanged in all these equations without affecting D.

Table 6.3-2 Accuracy of Diffusion Equations. (a = Gilliland equation, b = Arnold equation, c = Hirschfelder-Bird-Spotz equation, d = Chen-Othmer equation, e = Fuller-Schettler-Giddings.)

System	T, °K	D_g, exp cm²/sec	% Error a	b	c	d	e
$He—C_6H_6$	298	0.384	−30	−10	−4	−4	−16
$He—n—C_6H_{14}$	417	0.574	−36	−6	−2	−1	−18
$He—n—C_8H_{18}$	303	0.248	−20	+6	−4	+8	−7
$He—CH_3OH$	423	1.032	−31	−5	+7	+3	0
$He—C_2H_5OH$	298	0.494	−31	−14	−4	−5	−14
$He—C_2H_5OH$	523	1.173	−33	−1	+4	+11	−3
$N_2—C_2H_4$	298	0.163	−19	0	−4	+1	+1
$N_2—n—C_8H_{18}$	303	0.073	−16	−7	−18	−13	+2
$A—n—C_8H_{18}$	303	0.059	−7	−5	−13	−4	+17
47 systems, average absolute error			20	9	7	8	6

heavier gases. It has been pointed out by Reid and Sherwood[21] that the error found with helium carrier can be greatly reduced by replacing its molar volume by unity. With this change the overall error of the Gilliland method is only 10%, and thus compares favorably with the more complex methods. However we note that the other light carrier gas, H_2, is not represented here, and that a large error might also be found for this gas. Furthermore, it is known that the Gilliland temperature dependence is wrong ($T^{1.5}$ instead of $\sim T^{1.7}$) and serious errors may thus be found in the high or low temperature extremes of gas chromatography.

For high precision (and confidence) in estimating D_g values, the more complicated methods derived from the Hirschfelder-Bird-Spotz equation should be used. These methods, based on detailed gas dynamics, are undoubtedly valid for unusual systems (e.g., very large molecules, very high temperatures) for which the Gilliland and Arnold equations are untested. The Fuller-Schettler-Giddings equation, however, provides the best practical combination of accuracy and simplicity.

Surface Diffusion

The phenomenon of surface diffusion has been almost totally ignored in the chromatographic literature, in spite of the fact that nearly all interfaces provide a two dimensional path for molecular transport. In this brief summary we will attempt to indicate the nature and possible role of surface diffusion in chromatography.

Diffusion on solid surfaces is ordinarily an activation process much like liquid diffusion—molecules must acquire considerable activation energy to move from site to site on the surface. (Each step between sites is subject to the rate theories of Section 6.5.) In common, then, with equation (6.3-1) the surface (or interfacial) diffusion coefficient depends on temperature in an exponential fashion

$$D_I = D_1 \exp(-W_I/\mathscr{R}T) \tag{6.3-13}$$

The magnitude of this parameter is more comparable to liquid than gas D values because of the exponential term and also because each molecular displacement covers a distance of only one or so molecular diameters (instead of a mean free gas path). For instance Carman and Raal[32,33] report D_I values at $-50°C$ ranging from 0.1–2×10^{-5} cm²/sec for solute molecules ranging in size from CO_2 to *n*-heptane on carbon black and silica gel. At higher temperatures, near the boiling point of the solute, D_I ranged from 3–7×10^{-5} cm²/sec. The activation energy W_I ranged from about 30 to 80% of the heat of adsorption; it is typically about 50% of the latter. Most D_I values are very sensitive to surface concentration, increasing to a temporary maximum at one monolayer coverage.

In view of the lack of data, we can only speculate on the role of surface diffusion in chromatography. The major effect of surface diffusion, if there is one, would be to provide a competing mechanism for the transport of solute in and out of the pores of the solid particle. For gas-solid chromatography surface diffusion would have to compete with the gas phase diffusion which normally accounts for intraparticle transport. Ordinarily the great speed of gaseous diffusion would cause it to swamp out surface effects, but there are several cases for which this might not be true. First, in very narrow pores gas phase diffusion is hindered because the mean free path is reduced by wall collisions (Knudsen transport). At the same time the smallness of the pore (large area to volume ratio) will lead to an increased equilibrium ratio of adsorbed solute as compared to gas phase solute within the pore;† hence more surface than gas transport may be occurring despite the higher rate of the latter. These two factors may cause surface diffusion to dominate in certain instances.

In gas-liquid chromatography most surface diffusion would likely occur at the gas-liquid interface, particularly for those components which tend to adsorb at the interface. In some cases this diffusion would be nonlocalized[34] (solute molecules not tied to individual sites) and thus quite rapid. In the right combination of circumstances (high adsorption, rapid diffusion) this mechanism might become competitive or even dominant.

† The rate of diffusional transport by any mechanism is proportional to the amount of solute subject to that mechanism as well as to the diffusion coefficient. Hence a high surface concentration may more than offset the small diffusion coefficient.

In liquid chromatography surface diffusion would occur mainly at solid-liquid interfaces. While this diffusion would be slow, perhaps even slower than gas-solid surface diffusion, the competitive liquid diffusion is also slow. The importance of surface diffusion, like the previous cases, would depend on the amount of solute on the surface and thus available for surface diffusion, and the precise ratio of diffusion coefficients.

The phenomenon of surface diffusion in chromatography obviously deserves sufficient additional study to determine its role as a transport mechanism.

6.4 Diffusion Through Porous Materials

While migrating through a granular medium, a chromatographic zone will expand—in part due directly to molecular diffusion and in part due to various nonequilibrium and differential flow phenomena (which, of course, may involve molecular diffusion). The latter effects have been treated extensively in this volume. With the exception of the very short Section 2.6, longitudinal molecular diffusion has been ignored. It is true that there is less variation here than elsewhere, but, as Knox has pointed out,[35] the precise identification of this effect can be very useful in isolating the other spreading terms experimentally. Hence we shall discuss this phenomenon in more detail.

Diffusion in the Mobile Phase

As shown in Section 2.6, longitudinal diffusion in the mobile phase contributes a plate height term

$$H = 2\gamma D_m/v \tag{6.4-1}$$

where D_m is the diffusion coefficient for solute in the mobile phase and γ is an obstructive factor indicating the degree to which diffusion is hindered by the granular material

$$\gamma = D/D_m \tag{6.4-2}$$

the ratio of the effective to the molecular diffusion coefficient. The value of γ probably ranges from 0.4 to 0.9 when all types of chromatographic materials are taken into account.

The only attempts made to measure γ in connection with chromatography are apparently related to gas chromatography. Of these, only the work of Knox and McLaren[36] is sufficiently detailed in both its theoretical and experimental aspects to lead to firm conclusions. The deductions of these authors are in no sense limited to gas chromatography. In fact the theoretical basis of γ was largely established in connection with a treatment of ion migration in packed materials (electrophoresis).[37] Thus there is every reason to believe, given a certain packing structure, that γ will be essentially identical for all liquids and all gases—it is a structural constant. For this reason we will discuss the origin of γ, following Knox and McLaren, assuming that the same concepts are valid for all forms of chromatography.

The longitudinal diffusion of solute is best envisioned by assuming that all flow has ceased. Diffusion will then occur along the stilled fluid in each of the interstitial flow channels. However, we may deduce with the help of Chapter 5 that these channels will have two main characteristics which hinder diffusion. First, the channels zig-zag through the material in a tortuous manner, thus preventing the diffusion of molecules along the shortest, direct path. Second, the channels alternately vary from wide to narrow. The narrow constrictions slow diffusion considerably. Hence there are two factors—tortuosity and constriction—which hinder diffusion through chromatographic materials. At one time the entire obstructive effect was commonly blamed on tortuosity. But comparison of theory and experiment, in both gas chromatography[36] and electrophoresis,[37,38] shows that the constriction effect is also needed to explain the full obstruction to diffusion.

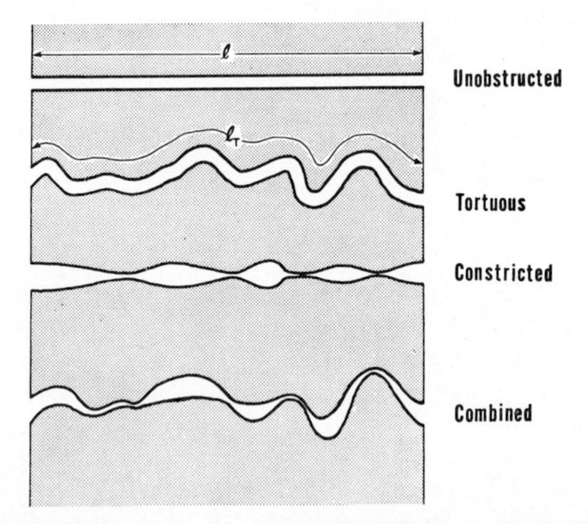

Figure 6.4-1 The nature of tortuous and constricted paths in a chromatographic packing.

The nature of tortuosity and constriction is shown in Figure 6.4-1. The unobstructed path provides no hindrance to diffusion. The tortuous path slows diffusion by virtue of the increased path length. The constricted path creates "traffic bottlenecks" at random points. The combined path provides both kinds of hindrances.

Theory of Tortuosity

The magnitude of the tortuosity effect can be established as follows. If a solute spike is allowed to diffuse outward, the root mean square displacement (or standard deviation) σ is given by Einstein's relationship

$$\sigma^2 = 2Dt \tag{6.4-3}$$

where D is the effective diffusion coefficient. However, the actual process of molecular diffusion is occurring along a path that is $l_T/l = \mathscr{T}$ times longer, so that from a molecular viewpoint the displacement is $\mathscr{T}\sigma$ instead of the apparent displacement σ. Hence

$$(\mathscr{T}\sigma)^2 = 2D_m t \tag{6.4-4}$$

where D_m is the molecular diffusion coefficient in the mobile phase, in contrast to the apparent diffusion coefficient D. On dividing equation (6.4-3) by (6.4-4), we have

$$D/D_m = \mathscr{T}^{-2} \tag{6.4-5}$$

As shown by equation (6.4-2) this ratio equals γ. Thus

$$\gamma = \mathscr{T}^{-2} \tag{6.4-6}$$

where $\mathscr{T}$, termed the *tortuosity*, is simply the average ratio of the real to the shortest path length through a porous medium. The real path length is, of course, very difficult to come to grips with because of the complications of interstitial geometry.

Theory of Constriction

The constriction effect can also be derived in a relatively simple way. We start by writing the solute flux J_c through a cross section of a single channel beset with constrictions but not tortuosity. By Fick's first law this is

$$J_c = -D_m(\partial c/\partial z)A \tag{6.4-7}$$

where A is the cross section of the channel, a quantity that varies from point to point as a result of the constrictions. This equation may be arranged to $\partial c/\partial z = -J_c/D_m A$ and then integrated over a short length l of the channel

$$\Delta c = -(J_c/D_m)\int_0^l dz/A \tag{6.4-8}$$

where Δc is the concentration difference over the length l. The integral in this equation is simply l times the mean value of $1/A$, or $\overline{(1/A)}l$. Hence

$$\Delta c = -(J_c/D_m)\overline{(1/A)}l \tag{6.4-9}$$

The quantity J_c has been assumed constant throughout the segment because in the steady state (nearly always valid for short segments) the same flux must pass each cross section along the segment. Now we may write this flux in terms of the entire segment length l using the effective diffusion coefficient D and the mean cross-sectional area $\bar{A}$

$$J_c = -D(\Delta c/l)\bar{A} \tag{6.4-10}$$

where, of course, $\Delta c/l$ is the concentration gradient. On solving for Δc

$$\Delta c = -(J_c/D)(1/\bar{A})l \tag{6.4-11}$$

we have another equation in Δc comparable to (6.4-9). When the two expressions are equated, we find $\gamma = D/D_m$ to be

$$\gamma = D/D_m = 1/\bar{A}(\overline{1/A}) \tag{6.4-12}$$

in which γ is one divided by the product of the mean area and the mean reciprocal area. For a uniform channel this is unity. For a highly constricted channel, where at certain points A goes almost to zero and thus $1/A$ goes almost to infinity, the mean value of $1/A$ is disproportionately large and γ thus small. The tendency of $1/A$ to become very large in restrictions is a mathematical reflection of the fact that such restrictions unduly hinder diffusion.

Magnitude of γ

To a good approximation the total obstructive factor γ is the product of the two independent terms for tortuosity and constriction. Thus

$$\gamma = [\mathscr{T}^2\bar{A}(\overline{1/A})]^{-1} \tag{6.4-13}$$

Various cell models have been used to get numerical values for the factors in this equation.[36,37] Although the relative contribution of tortuosity and constriction varies with the type of model chosen (particularly on whether there are some straight paths or only staggered paths through the medium), the calculations are in remarkable close agreement with measurement. As an example, Knox and McLaren found a γ value of 0.60 ± 0.02 for glass beads as compared to the theoretical value of 0.63. Other results were even closer. This value, 0.60, is probably a good approximation for most nonporous, densely packed granular materials in chromatography.

Experimental results on porous solids (the white and pink forms of diatomaceous earth) showed some anomalies. The very "open" white form has $\gamma = 0.74$, while the denser pink form (firebrick) showed $\gamma = 0.46$. As shown by Knox and McLaren, these results must account for the diffusive flux through, as well as around, the particles. Giddings and Boyack[38] have derived a "retardation factor" which is meant to account for a similar phenomenon in paper electrophoresis, but no attempt has yet been made to transfer this to chromatographic diffusion.

Low density materials such as paper can be expected to show γ values from 0.7 to 0.9, corresponding to the electrophoretic value.

The theory of tortuosity and constriction, given earlier in this section, shows that γ is essentially a structural constant and thus independent of the particular mobile phase or solute being considered. However, there is very likely a slight velocity dependence to γ. When the reduced velocity ν exceeds unity (such that flow displacement exceeds diffusive displacement over a distance d_p), the constriction effect is probably reduced in importance by virtue of the fact that solute is carried directly through the constrictions by the flow stream. Tortuosity is likely to change some also, but it is difficult to say how much. The theory of γ's velocity dependence has not been worked out; as things stand now, we must assume that the velocity effect is negligible.

Lateral Diffusion

Lateral diffusion through a chromatographic bed is important in establishing equilibrium between unequal flow paths. There is, of course, an effective lateral diffusivity based on a flow mechanism as discussed in Section 2.9. As shown in Table 2.10-1, however, the flow mechanism does not become dominant until the reduced velocity exceeds about 40, much higher than the practical operating range, at least in gas chromatography. Below this velocity molecular diffusion is dominant. The effective diffusion coefficient may be assumed to equal γD_m, the same as for longitudinal diffusion. The similarity of the two processes was stressed in Section 4.5 in connection with transcolumn effects. The medium must be isotropic if lateral and longitudinal γ's are to be the same. Most chromatographic materials fall in this category. An exception is paper where a preferred fiber orientation is introduced during manufacture. The orientation is not complete, and it is questionable whether γ in the flow ("machine") direction is ever more than 0.1 in excess of the lateral γ.

The desirability of increasing the lateral γ and decreasing the longitudinal γ has been stressed for preparative-scale gas chromatography.[39] We are unfortunately limited by the fact that γ cannot exceed unity in any direction. By the same token it is doubtful if γ can ever be reduced much in the flow direction without seriously hindering flow (as with a plate structure).

Diffusion Within Particles

Diffusion into and out of porous particles affords the mechanism for transporting solute from the flow stream to the stationary phase. Diffusion within particles, however, may be presumed to be hindered by the internal structure of the particle much as external diffusion is hindered by the particles themselves. Hence there will be an obstruction factor γ_p for intraparticle diffusion (see Section 4.5 for the application of this parameter). If the particle is reasonably open and not too dense, γ_p is probably in the same general range as other γ values—0.4 to 0.9.

Diffusion in the Stationary Phase

Equilibration between phases in partition chromatography depends to a large extent on stationary phase diffusion. The diffusion of molecules in and out, leading to equilibration, is accompanied by diffusion up and down, leading to longitudinal molecular diffusion. In Section 2.6 it was shown that the plate height contribution of this effect is

$$H = \frac{2\gamma_s D_s}{v} \frac{1 - R}{R} \tag{6.4-14}$$

where γ_s is the obstructive factor and D_s the diffusion coefficient for the stationary phase. The factors controlling γ_s are somewhat different than those discussed earlier in connection with γ. The stationary phase is not nearly as continuous as the mobile phase. It tends to be broken up into small units with very thin bridges between. While in theory transport can occur through these choked regions, as described earlier in the theory of constriction, it will more often occur by means of a desorption step followed by migration through the mobile phase over an equivalent path. Examples of such narrow constrictions are the contact points between ion exchange beads and the thin film between the filled pores in gas liquid chromatography.

The theory of obstruction for the above diffusion has not been worked out. In many cases γ_s may be comparable to other γ values, say, $\gamma_s \sim 0.5$. However, in the case of strong retention we can imagine a molecule, trapped in a given unit by the excessive energy needed for escape, diffusing back and forth from edge to edge without achieving any real displacement. In this case the effective γ_s might be very small.

Surface Diffusion

Diffusion along surfaces will also be hindered by the geometrical complexities of a chromatographic packing. This case will be analogous to bulk diffusion. Thus surface diffusion from point to point is obstructed by the tortuosity of the surface, and by the constrictions in the diffusion path along the surface (e.g., as will occur at contact points).

6.5 Kinetics of Single-Step Processes: Adsorption-Desorption Phenomena

Diffusion occurs as a cumulative result of an enormous number of molecular displacements, each single step being negligible. Adsorption, desorption, or chemical reaction, on the other hand, are complete in just one or several steps. We shall be mainly concerned with the latter here although the basic concepts of chemical kinetics are applicable to the constituent steps of both kinds of processes (with the exception of gas phase diffusion).

It was pointed out in connection with liquid diffusion that each displacement requires a minimum (activation) energy, and thus the rate of displacement is an exponential function of temperature. It is well known that the exponential dependence applies to all chemical and physical processes requiring energy, a fact expressed by the classical Arrhenius equation[40]

$$k = \nu \exp(-W/\mathscr{R}T) \tag{6.5-1}$$

where k is the rate constant (see below), ν the frequency factor, and W the activation energy needed for the process. Although W is largely an empirical parameter (since attempts to establish its value by quantum mechanics are not sufficiently accurate), ν has been predicted closely, using the kinetic theory of gases, for simple bimolecular reactions in the gas phase. For more complicated processes, including those often found in chromatography, the absolute reaction rate theory of Eyring's should be used. The gas-kinetic theories are a special case of the latter. This theory gives the reaction rate constant as[9]

$$k = \kappa \frac{\mathscr{R}T}{Nh} \exp(\Delta S^{\ddagger}/\mathscr{R}T) \exp(-\Delta H^{\ddagger}/\mathscr{R}T) \tag{6.5-2}$$

where κ is the transmission coefficient (usually near unity), N is Avogadro's number and h is Planck's constant (6.62×10^{-27} erg-sec). The enthalpy of activation ($\sim W$) is given by $\Delta H^{\ddagger}$ and the entropy of activation is $\Delta S^{\ddagger}$. One of the most useful aspects of this equation is the relationship of k to $\Delta S^{\ddagger}$. It shows that the reaction path must be clear and straightforward such that the reacting species can easily acquire a configuration favorable to reaction (only after this can the reaction itself occur with the involvement of the $\Delta H^{\ddagger}$ or W terms). A simple example of this illustrated by the fact that a bimolecular reaction at normal pressures in the gas phase, requiring that two molecules be at the same place at the same time, is $\sim 10^3$ slower than a unimolecular reaction, involving only one molecule, with an identical activation energy. We will refer back to this concept of a probability or entropy factor later in connection with chromatographic adsorption.

While values of the frequency factor can vary widely for different kinds of processes, unimolecular reactions (such as a simple desorption step[41]) invariably have $\nu \sim 10^{13}$ sec^{-1}, the same order of magnitude as the frequency of vibration in a molecule. In these cases $\nu \sim \mathscr{R}T/Nh$, where the latter is near 10^{13} sec^{-1}.

The Rate Laws

Most reaction processes of chromatography (adsorption and desorption) follow a rate law that is first order in solute concentration c

$$-dc/dt = kc \tag{6.5-3}$$

The rate of loss of solute (say, by adsorption), $-dc/dt$, is proportional to its concentration. The constant of proportionality is the rate constant k. The first-order law fails in nonlinear chromatography (the linearity of the adsorption isotherm and of reaction kinetics will generally cover the same range of concentration).

Some diffusion processes, although multistep, follow the above equation approximately. In those cases we can assume an "effective" rate constant for diffusion-controlled reactions. This was done in Section 2.8 in connection with diffusion in bulk stationary phases.

Equation (6.5-3) holds for each transfer process in chromatography. While solute in the mobile phase is being lost through sorption, for example, it is being gained through desorption. Thus the rate of increase in mobile phase concentration c_m is composed of the desorption step rate constant times stationary phase concentration—$k_d c_s$—minus the adsorption rate constant times mobile phase concentration—$k_a c_m$

$$dc_m/dt = s_m = k_d c_s - k_a c_m \tag{6.5-4}$$

where s_m is the net rate of reaction as discussed in Section 3.2. The above is identical to equation (3.2-8). This type of rate equation has been used as the starting point for nearly all kinetic theories of chromatography. As indicated in Chapter 4, this two-step mechanism is probably a gross oversimplification for real chromatographic systems. We may find a whole series of reactions, each involving a given kind of site or perhaps even chemical change. The rate law must then sum over all possible reaction paths as shown in equation (4.2-7). Since each of the many rate constants will have different ν and W values, the relative importance of different processes may shift rapidly with temperature.

It can be seen from the rate equations above that the rate constant k has the dimensions of reciprocal time. One over k thus has the dimensions of time. It was shown intuitively in Section 2.7 that $1/k$ is the mean reaction time for the single process involved. This is shown rigorously as follows.

The integration of equation (6.5-1) gives

$$c = c_0 \exp(-kt) \tag{6.5-5}$$

where c_0 is the concentration at the beginning of the reaction. The mean lifetime is given by

$$\bar{t} = \int_{c_0}^{0} t(-dc/c_0) \tag{6.5-6}$$

This integral expression arises in considering the fraction of reactant, $-dc/c_0$, disappearing in the short interval dt at time t. This fraction is obtained from (6.5-5) as $k \exp(-kt)\,dt$. The substitution of this for $-dc/c_0$ in equation (6.5-6) leads quite naturally to an integration over the variable t, and yields, after integration by parts, the desired result, $\bar{t} = 1/k$.

Adsorption Kinetics on Uniform Surfaces

While no chromatographic surface has identical adsorption sites everywhere, we may take this ideal case to establish the basic concepts of adsorption kinetics.†

The study of adsorption is concerned with two different kinds of sorptive process: physical (or van der Waals) adsorption and chemisorption (or activated adsorption).[42] The former involves a weak (<15 kcal/mole) interaction with the surface utilizing the type of forces that hold liquid molecules together. The latter involves the formation of a chemical bond with the surface and thus an interaction of ~20–100 kcal/mole. Most normal chromatographic adsorption is undoubtedly of the physical type. However, chemisorption may be presumed to occur frequently, involving perhaps only a fraction of the solute molecules in any one run, but in so doing leading to such anomalies as irreversible adsorption, tailing, streaking, and ghost spots.‡

Physical adsorption generally occurs with little or no activation energy W. Chemisorption, on the other hand, usually requires an activation energy ranging up to that of the chemical bond itself. Consequently, chemisorption is usually very slow by comparison to physical adsorption. In fact the former may be considered as a two-step process, first involving physical adsorption and then the final step with bond rearrangement. If the latter is slow enough, we may see little trace of it during the residence

† For an elementary discussion of adsorption dynamics see de Boer.[41]

‡ A discussion of various anomalies or "artifacts" in paper chromatography is given by Zweig.[43]

of solute in a column. However, once a molecule is chemisorbed it is also slow in desorbing, a fact which may account for the anomalies mentioned above (see Section 2.13). While the tendency to chemisorb can sometimes be predicted by the chemical nature of the surface and the solute (e.g., an acid surface and a basic solute), its real existence can be confirmed only by observation, mainly using such clues as provided by the chromatographic anomalies themselves. Since these anomalies, particularly tailing, may have several sources (e.g., nonuniform surfaces, nonlinearity), it is important to understand the effects of each to determine the culprit in any particular case. This matter was discussed to some extent in Section 2.13.

Physical adsorption in chromatography is often nonlinear. On a uniform surface nonlinearity could generally be considered as an indication that the number of adsorbed molecules is of the same order of magnitude as the total number of adsorption sites. Assuming the sites to be spaced 3×10^{-8} cm apart, 10^{15} sites would be available on each cm^2. An adsorbent with 10 m^2/gm would have $\sim 10^{20}$ sites/gm, or approximately 10^{20} sites per cm^3 of column. If uniform, these sites could easily accommodate 10% coverage or 10^{19} molecules, $\sim 10^{-5}$ moles, without nonlinearity. A chromatographic zone spread over the length 4σ in a column of radius r_c will occupy an effective volume of $4\sigma\pi r_c^2$. We should, therefore, be safely in the linear range if the number Y of moles of solute is restricted by

$$Y < 10^{-5}(4\sigma\pi r_c^2) \text{ moles} \tag{6.5-7}$$

or for an arbitrary surface area A (in cm^2) per unit of column volume

$$Y < 10^{-10}A(4\sigma\pi r_c^2) \text{ moles} \tag{6.5-8}$$

Since $\sigma = L/\sqrt{N}$ and $\pi r_c^2 L$ equals the column volume V_{col}

$$Y < 10^{-10}A(4V_{col}/N^{1/2}) \text{ moles} \tag{6.5-9}$$

an equation valid also for geometries other than circular columns (e.g., thin layers). Assuming a column with radius 0.25 cm and length 100 cm, generating 1000 plates, this inequality reduces to

$$Y < 2.5 \times 10^{-10}A \text{ moles} \tag{6.5-10}$$

for normal adsorbents ($A \sim 10^5$ cm^{-1}) this reduces to $Y < 10^{-5}$ (about 1 mg of average solute). The above derivation assumes that nonlinearity at the head of the column where zones are concentrated is unimportant providing linearity is achieved for the final zone width. Since a margin of 10 or so is built into the derivation, this approximation should be safe.

The above treatment indicates the limits which must be imposed on sample size to maintain linearity on uniform surfaces. Tailing or other zone distortion that occurs within this limit can usually be attributed to the excessive nonlinearity of nonuniform surfaces, a kinetic effect on nonuniform surfaces, or chemisorption.

Assuming a linear range for adsorption on a uniform surface, kinetic arguments can be used to fix the plate height contributions of adsorption-desorption within a reasonable range.[44] We start with the plate height expression for 1-site adsorption, equivalent to equation (4.2-36)

$$H = 2(1 - R)^2 v / k_a \tag{6.5-11}$$

where k_a is the rate constant for adsorption (see equation (6.5-4)). This constant can be written as follows

$$k_a = \frac{\text{adsorption rate}}{\text{no. mobile phase molecules}} \tag{6.5-12}$$

Within a unit volume of the column the adsorption rate (molecules adsorbing per second) may be written as αZA—as the product of the number Z of molecules striking a unit surface area per second times the area A in unit volume, all multiplied by the fraction α of molecules which stick on collision. The number of mobile phase molecules in the unit volume is equal to nV_m where n is the number of molecules per unit mobile phase volume and V_m is the fractional volume of the mobile phase. Thus[44]

$$k_a = \frac{Z}{n} \frac{\alpha A}{V_m} \tag{6.5-13}$$

When substituted into equation (6.5-11), this yields the plate height for any kind of adsorption chromatography with an ideal, uniform surface

$$H = \frac{2(1 - R)^2 v}{\alpha (Z/n)(A/V_m)} \tag{6.5-14}$$

The parameters of this equation may be estimated as follows. Since adsorption may involve an activation energy W, the sticking coefficient α will depend exponentially on temperature

$$\alpha = \alpha_0 \exp(-W/\mathscr{R}T) \tag{6.5-15}$$

Most physical adsorption from gases occurs with negligible activation energy and the exponential part may be ignored: $\alpha = \alpha_0$. For such systems α (or α_0) values normally range from 0.1 to 1; values less than 0.01 may be considered as exceedingly rare.[44] It is reasonable to suppose that similar

α_0 values will be found for adsorption from liquids. The α value for liquids, however, may be considerably smaller; say, $\sim 10^{-4}$, because W is finite.†

The parameter Z/n is given simply by $\bar{c}/4$ in gases, where $\bar{c}$ is the mean molecular velocity. The value of this ratio for liquids in uncertain; we will assume it is comparable to the gas phase value.‡

The ratio A/V_m is the surface area per unit volume of mobile phase. This is equivalent to $\rho A_w/f = \rho_0 A_w(1-f)/f$, where A_w is the surface area per unit mass (gram), f the porosity, ρ_0 the density of solid involved, and ρ the bulk density of the granular solid. For most solids $A/V_m \sim A_w$.

If the foregoing expressions are substituted back into equation (6.5-14), we have

$$H = \frac{8f(1-R)^2 v}{\alpha_0 \bar{c} \rho A_w} \exp(W/\mathscr{R}T) \tag{6.5-16}$$

For gas chromatography, assuming $f = \frac{5}{8}$, $R = 0.5$, $\alpha_0 = 0.25$, $\bar{c}$ (roughly the velocity of sound) $= 5 \times 10^4$ cm/sec, $A_w = 10^5$ cm²/gm, $\rho = 1$ gm/cm³, and $W = 0$, we have[44] in cgs units

$$H \sim 10^{-9} v \tag{6.5-17}$$

that is, the nonequilibrium or C_k term is $\sim 10^{-9}$ sec. This is negligibly small, as will be seen shortly.

For liquid chromatography, using the above parameters except that $W = 6$ kcal/mole ($T = 300$°K), we have

$$H \sim 10^{-5} v \tag{6.5-18}$$

giving $C_k \sim 10^{-5}$ sec.

The significance of the plate height values calculated above for adsorption kinetics can be inferred from Section 2.11. It was shown there that in a satisfactory chromatographic system the reduced nonequilibrium term Ω should be less than unity

$$\Omega_k = C_k D_m / d_p^2 < 1 \tag{6.5-19}$$

In gas chromatographic systems $D_m \sim 0.4$ cm²/sec and $d_p \sim 0.02$ cm; since $C_k \sim 10^{-9}$ sec, $\Omega_k \sim 10^{-6}$, a millionfold below the value which would make a significant contribution to column plate height. In liquid chromatographic systems $D_m \sim 10^{-5}$ cm²/sec and $d_p \sim 0.01$ cm; with

† It is probable that a solute molecule adsorbing from a liquid will have to displace an adsorbed solvent molecule. In this case an activation energy of 2 to 10 Kcal/mole will be involved, giving $\exp(-W/\mathscr{R}T) \sim 10^{-4}$.

‡ Collision with the surface from a liquid would be of a vibrational type, i.e., a molecule next to the surface would, in effect, collide with the surface at each vibration. Since the average kinetic energy involved in vibration is essentially the same as that of translation, a similar number of collisions should be involved. Transport up to the surface, of course, occurs by diffusion, and is accounted for in a separate plate height term.

$C_k \sim 10^{-5}$, again $\Omega_k \sim 10^{-6}$, a negligible contribution. The small contribution in each case is due to the very high speed of ordinary adsorption and desorption.[41]

The above results show quite conclusively that adsorption-desorption kinetics on uniform surface will have a negligible effect on plate height. We will see that this situation changes drastically for surfaces with a high degree of nonuniformity.

Nonuniform Surfaces

On real surfaces adsorption and desorption are complicated by variable adsorption sites, surface migration, etc. Somewhere between the almost trivial case of an ideal (uniform) surface and the unknowable details of a real surface one must seek compromise models which are reasonably close to the truth without being unbearably complicated. The two-site and multisite models of Section 4.2 are attempts in this direction.

Physical adsorption on nonuniform surfaces will duplicate some of the features of partial chemisorption on uniform surfaces—some molecules will be bound more tightly than others and their slow desorption can lead to tailing and other anomalies. With nonuniform surfaces, of course, unequal adsorption can result from both mechanisms.

It is probably wishful thinking to imagine that total adsorption linearity is achieved for many chromatographic zones in adsorption chromatography. Nonlinearity, however, will be of a mixed nature, rather severe on some kinds of sites and negligible on other kinds. The most tenacious sites of real surfaces (perhaps even on relatively homogeneous carbon blacks) will adsorb so strongly that most such sites will become essentially saturated within the most dilute solute zones. The low energy sites may stay linear, however, for much larger concentrations.

The nature of mixed nonlinearity can be shown by means of the following model. Suppose there is a normal (type 1) low energy adsorption site covering most of the surface. If θ_1 is the fraction of such sites occupied then $\theta_1/(1 - \theta_1)$ will be the ratio of occupied to unoccupied sites (this is the same ratio appearing in the Langmuir type isotherm). A particular kind (type i) of high energy sites will correspondingly have a ratio $\theta_i/(1 - \theta_i)$. If ΔE_i is the energy difference between the normal sites and those of type i, these ratios are related approximately by

$$\frac{\theta_i/(1 - \theta_i)}{\theta_1/(1 - \theta_1)} = \exp(\Delta E_i/\mathscr{R}T) \qquad (6.5\text{-}20)$$

where generally $\Delta E_i > 0$. If we are striving for linearity and thus keeping sample size small in accordance with equation (6.5-9), θ_1 will be small and

$1 - \theta_1 \sim 1$. Thus

$$\frac{\theta_i}{1 - \theta_i} = \theta_1 \exp(\Delta E_i / \mathscr{R}T) \tag{6.5-21}$$

A plot of θ_i versus θ_1 is shown in Figure 6.5-1, assuming $\Delta E_i = 3$ kcal/mole and $T = 300°$K. It is seen that the type i sites exhibit nonlinearity

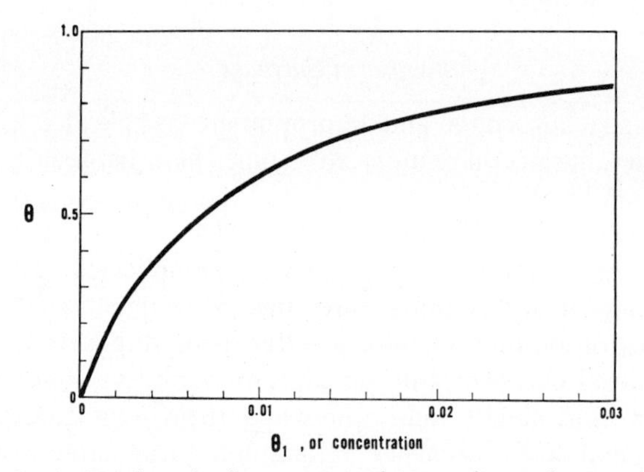

Figure 6.5-1 Nonlinearity in coverage of i-type sites at low concentrations.

($\theta_i \sim 0.5$) before the bulk of the surface has reached 1% coverage. (If ΔE_i were 5 kcal/mole, nonlinearity would occur when coverage of the normal sites was only 1 part in 4000, i.e., with a solute sample of only $\sim 10^{-5}$ g.) Thus we find mixed nonlinearity depending on the adsorptive strength of the site under consideration.

Some unexpected chromatographic results can be expected in view of mixed nonlinearity on nonuniform surfaces. The usual picture relates to uniform surfaces and the chromatographic zone is envisioned as tailing more and more as sample size is increased. However, consider mixed adsorption with the two types of sites considered above. At extremely low concentrations linearity would be essentially complete. With slight increases the type i sites would show nonlinear adsorption and thus, depending on the number of such sites, show some tailing. With further increases these sites would become essentially saturated and the additional adsorbate would be forced onto the normal sites. This would serve to increase the fraction of solute on the normal surface which, if still linear, would swamp out the tailing effect of the high energy sites. Finally, however, when the sample size exceeds the minimum value of equation (6.5-9), all kinds of sites would show nonlinearity and tailing would return with a vengeance. This idealized sequence of events is shown in Figure 6.5-2, with the increasing peak area indicating the change in

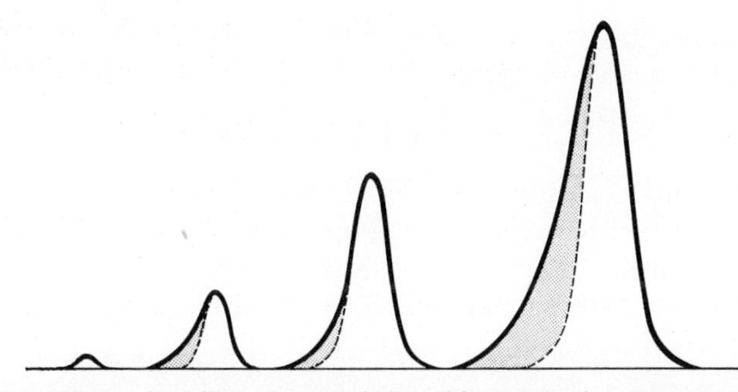

Figure 6.5-2 Change in tailing characteristics with sample size. Due to surface nonuniformity the relative amount of tailing may increase, decrease, then increase again with increasing sample size.

amount of sample involved. In detail things might be quite different; a wide range of adsorption sites might blunt the cyclical trend while kinetic effects (Section 2.13) might enhance it. (The slow desorption from type i sites may be the major source of tailing; this will also be swamped out as that particular site reaches and then passes saturation.) That such a phenomenon exists has been seen in the author's laboratory; tailing has been observed to decrease with increasing sample size.

Nonuniform surfaces also exhibit a different and more complicated form of adsorption-desorption kinetics by comparison to the uniform surface. These changes can have serious chromatographic implications, leading to tailing and increased zone spreading. These will be discussed next.

It was shown in Section 2.13† that tailing could be expected when a few sites with high adsorption energy ("tail-producing sites") captured molecules occasionally and released them back into the mobile stream so slowly that they would form a tail. If the holding (desorption) time for the sites exceeds the time τ needed for a quarter of the eluting zone to pass by, tailing may be expected. Thus

$$1/k_{d2} > \tau \tag{6.5-22}$$

where k_{d2} is the desorption rate constant for the tail producing sites, identified here as type 2 sites. Since $\nu \sim 10^{13}$ for a desorption step, we may write

$$k_{d2} = 10^{13} \exp\left(-W_2/\mathscr{R}T\right) \tag{6.5-23}$$

where W_2 is the activation energy characterizing these sites. Writing

$$\tau = L/Rv\sqrt{N} \tag{6.5-24}$$

† See also reference 45.

where L is the column length, Rv the zone velocity, and N the number of plates, the inequality of equation (6.5-22) reduces to the logarithmic form

$$W_2 > 2.3\mathscr{R}T[13 + \log (L/Rv\sqrt{N})] \tag{6.5-25}$$

where W_2 must exceed the minimum indicated if tailing is to occur. The minimum, of course, varies with temperature, flow velocity, and other column parameters. Assuming a gas chromatographic system with $T = 300°\text{K}$, $L = 100$ cm, $R = 0.1$, $v = 10$ cm/sec, and $N = 10^3$ we find

$$W_2 > 19 \text{ kcal/mole} \qquad \text{(gas chromatography)} \tag{6.5-26}$$

The same parameters in liquid chromatography, except with $v = 10^{-2}$ cm/sec, gives

$$W_2 > 23 \text{ kcal/mole} \tag{6.5-27}$$

These desorption energies are rather high, but they conceivably exist for a few sites on some surfaces. They are certainly not unreasonable for chemisorption, which is also subject to the above theory. In view of the fact that the mechanism for desorption may first involve a step to an adjacent site of the normal kind (type 1), thence perhaps surface diffusion away from the tail-producing site, W_2 may involve only that energy necessary to take the first step, a quantity less than that required for total dissociation from the surface. This conclusion depends on probability considerations and requires that the molecule as an average desorb before being once again trapped with the tenacious site.

Aside from the onset of tailing, surface nonuniformity may have a damaging effect by simply increasing zone width. The theory of this will follow shortly; first, however, we should like to present a new physical explanation for the effect.

We note first that the adsorption constant for uniform surfaces is proportional to surface area A, and thus the plate height is inversely proportional to A. These two facts are shown in equations (6.5-13) and (6.5-14). The proportionality of k_a and A is related to the accessibility of surface; if the adsorbing surface is small, less solute is in a position (right above the surface) for immediate adsorption and the adsorption rate is thus slower. This is related to the entropy concept of Eyring's rate theory—rapid adsorption is contingent on the ease with which a solute molecule acquires a configuration (next to the surface) favorable to reaction.

On nonuniform surfaces the high energy sites adsorb more solute per site than the normal surface. This enhanced adsorption must occur over a very restricted surface (i.e., that covered by these infrequent sites). The number of molecules finding themselves precisely above such sites is limited and thus the rate of adsorption cannot easily keep up with the heavy demand for solute. This tends to create a substantial nonequilibrium and, as shown in Chapters 3 and 4, an increased plate height. Even if molecules reach the high energy sites from adjacent sites (surface diffusion), the scarcity of such sites makes them difficult for solute molecules to find (a consequence, again, of entropy considerations) and thus slows the adsorption rate.

A theory for the kinetic effects of nonuniform surfaces in gas chromatography has been developed.[44] We will present this theory below, modifying it to some extent to account for liquid chromatography.

It was shown in Chapter 4, equation (4.2-45), that the plate height contribution of adsorption-desorption kinetics is

$$H = 2R(1 - R)v\bar{t}_d \qquad (6.5\text{-}28)$$

where $\bar{t}_d$ is the mean desorption time for molecules which have reached equilibrium on the surface. This equation, of necessity, depends on a simplified mechanism of adsorption in which each site acquires and loses solute by direct exchange with the adjacent mobile phase. Surface migration probably does not alter our conclusions much, but it is not explicitly accounted for.

If we assume that a fraction α_i of molecules colliding with the ith kind of site stick to that site, then in exact analogy to the treatment of uniform surfaces, equation (6.5-13), we obtain

$$k_{ai} = \frac{Z}{n}\frac{\alpha_i A_i}{V_m} \qquad (6.5\text{-}29)$$

where the area occupied by type i sites, A_i, replaces the total area A. As before Z is the number of collisions per unit time and area, V_m the fractional mobile phase volume, and n the number of molecules per unit of this volume. The desorption rate constant k_{di}, necessary for the evaluation of $\bar{t}_d$, is obtained from the equilibrium expression

$$k_{di}X_i^* = k_{ai}R \qquad (6.5\text{-}30)$$

The desorption constant times the equilibrium fraction X_i^* of solute on type i sites equals the adsorption constant times the fraction R of solute in the mobile phase. The last two equations yield

$$k_{di} = \frac{R}{X_i^*}\frac{Z}{n}\frac{\alpha_i A_i}{V_m} \qquad (6.5\text{-}31)$$

Now the mean desorption time of surface molecules is the sum of the mean desorption time for each type of site, $1/k_{di}$, weighted by the fraction $X_i^*/(1 - R)$ of surface molecules on that site (X_i^* is the fraction of all solute on type i sites; the factor $1 - R$ reduces this to the fraction of solute found only on the surface). Thus we have (also in accord with equation (4.2-44))

$$\bar{t}_d = \sum [X_i^*/(1 - R)]/k_{di} \tag{6.5-32}$$

On substituting the k_{di} from the previous equation into this, we find

$$\bar{t}_d = \frac{nV_m}{ZR(1 - R)} \sum \frac{X_i^{*2}}{\alpha_i A_i} \tag{6.5-33}$$

The plate height expression, equation (6.5-28), thus becomes

$$H = \frac{2vV_m}{(Z/n)} \sum \frac{X_i^{*2}}{\alpha_i A_i} \tag{6.5-34}$$

This equation can be simplified by employing the dimensionless parameters $a_i = A_i/A$ and $x_i^* = X_i^*/(1 - R)$, where a_i represents the fraction of surface area occupied by type i sites and x_i^* the fraction of surface solute attached to such sites at equilibrium. With these changes

$$H = \frac{2(1 - R)^2 v}{(Z/n)(A/V_m)} \sum x_i^{*2}/a_i\alpha_i \tag{6.5-35}$$

If we now assume that some average sticking coefficient, α, represents the adsorption process, the α_i may be taken outside the summation sign

$$H = \frac{2(1 - R)^2 v}{\alpha(Z/n)(A/V_m)} \Lambda \tag{6.5-36}$$

where

$$\Lambda = \sum x_i^{*2}/a_i \tag{6.5-37}$$

a quantity known as the *heterogeneity factor*.[44] It will be noted that the first part of equation (6.5-36) is identical to the expression for uniform surfaces, equation (6.5-14). Thus the effect of nonuniformity resides in the heterogeneity factor Λ, and we shall naturally wish to inquire about its magnitude.

First, it should be noted that the use of an average sticking coefficient in place of the individual α_i's is probably justified for physical adsorption where only a slight variation from each other and from unity is expected. If nonuniformity is caused in part by chemisorption, the α_i's will differ greatly and should be left in the summation.

Since the first part of equation (6.5-36) is the expression for uniform surfaces, the various factors of this equation can be made more explicit in the same way. Thus by reference to the changes leading to equation (6.5-16) we have

$$H = \frac{8f(1 - R)^2 v}{\alpha_0 \bar{c} \rho A_w} \exp(W/\mathscr{R}T)\Lambda \tag{6.5-38}$$

an equation approximating the adsorption effects in both gas and liquid chromatography.

We shall now investigate the extremely important term Λ. Since x_i^*/a_i is proportional to the surface concentration, we shall term this quantity the relative adsorption density ρ_i. The summation in equation (6.5-37) thus becomes

$$\Lambda = \sum x_i^* \rho_i \tag{6.5-39}$$

The heterogeneity factor is an average value of ρ_i, weighted by the fraction of molecules adsorbed on each kind of site at equilibrium. When the sites are all equivalent (uniform surfaces), we have $\Lambda = \rho \Sigma x_i^* = \rho = 1$. Consequently, Λ goes to the proper limit as we approach uniformity. For illustration purposes it is most convenient to use a 2-site model, the simplest example of a nonuniform surface. Type 1 sites will be normal, low energy sites covering most of the surface, and type 2 sites will be those with considerable adsorption energy. The summation which represents Λ, equation (6.5-39), reduces to

$$\Lambda = x_1^* \rho_1 + x_2^* \rho_2 \tag{6.5-40}$$

The relatively high adsorption energy of type 2 sites will lead to $\rho_2 \gg \rho_1$. Except in the rare cases when x_2^* is negligibly small, Λ can thus be approximated by $x_2^* \rho_2 = x_2^*(\rho_2/\rho_1)\rho_1$. In view of the definition of the ρ's we have $\rho_1 = x_1^*/a_1$. Since, however, we stipulated that type 1 sites cover most of the surface, the relative area term is $a_1 \sim 1$. Hence $\rho_1 \sim x_1^*$. On replacing the final ρ_1 in the term $x_2^*(\rho_2/\rho_1)\rho_1$ by x_1^*, we get

$$\Lambda \sim x_1^* x_2^* (\rho_2/\rho_1) \tag{6.5-41}$$

This form (one of many) is particularly useful in making numerical estimates as we see below.

Since $x_1^* + x_2^* = 1$, the product $x_1^* x_2^*$ goes to 0.25 when an equal number of molecules are adsorbed on each type of site, and approaches zero when all molecules become attached to either site alone. It is likely that the order of magnitude of this term can be taken as 0.1 for a wide range of cases. If this is the case, Λ is simply

$$\Lambda \sim 0.1 \rho_2/\rho_1 \sim 0.1 \exp(\Delta E_{21}/\mathscr{R}T) \tag{6.5-42}$$

where the ratio of adsorption densities has been replaced by its exponential dependence on temperature and the energy, ΔE_{21}, necessary to transfer a mole of solute from type 2 sites to type 1 sites. Although derived for a 2-site surface, ΔE_{21} may be regarded as some kind of characteristic energy spread on a heterogeneous surface. This spread is very large, of course, for extremely heterogeneous surfaces, and Λ may be orders of magnitude greater than unity. For instance if $\Delta E_{21} = 10$ kcal/mole and $T = 300°K$, $\Lambda \sim 10^6$, representing a millionfold increase in plate height relative to uniform surfaces. This presumes, of course, that linearity can be maintained on all sites.

It is clear that the heterogeneity factor can be of sufficient magnitude to make the adsorption-desorption process highly significant to zone spreading. As a rough guide, obtained using equations (6.5-19, 38, 41) and the specific expression $H = C_k v$, we may adjudge the effect of heterogeneity as not serious as long as the energy spread stays within the following bounds

$$\Delta E_{21} < 2.3\mathscr{R}T \log\left(\frac{5d_p{}^2\alpha_0\bar{c}\rho A_w}{4D_m f(1-R)^2}\right) - W \qquad (6.5\text{-}43)$$

A gas chromatographic column at 300°K, with typical parameters used to get equation (6.5-17) and subsequent parameters, gives $\Delta E_{21} < 10$ kcal/mole. A previous estimate of this effect,[44] using different parameters, yields an expression in rough agreement: $\Delta E_{21} < 0.023T$ in kcal/mole. As a rough compromise, then, we may assume that ΔE_{21} in kcal/mole should be restricted by

$$\Delta E_{21} < 0.03T \qquad (6.5\text{-}44)$$

A typical system in liquid chromatography, with the same parameters used for equations (6.5-18) and following, yields about the same answer: $\Delta E_{21} < 0.05T - 6$, or about 9 kcal/mole at room temperature.

The values of ΔE_{21} given above express the energy of one kind of site relative to the other. The energy needed to desorb solute completely from the type 2 sites is that needed to reach type 1 sites, ΔE_{21}, plus the energy of dissociation from these sites, normally 5–10 kcal/mole. The two added together, assuming an upper limit of 10 kcal/mole for ΔE_{21}, gives 15–20 kcal/mole. It is interesting to note that this is the maximum allowable desorption (= activation) energy to avoid the tailing caused by tenacious sites (equations (6.5-26, 27)). We must conclude that type 2 sites and tail-producing sites are much the same thing, and that when these high energy sites exceed a certain maximum "stickiness," a general deterioration of column performance is to be expected.

6.6 References

1. A. Fick, *Ann. Phys. Lpz.*, **170**, 59 (1855).
2. J. Crank, *The Mathematics of Diffusion*, Oxford, London, 1956.
3. A. L. Ruoff, G. H. Stewart, H. K. Shin, and J. C. Giddings, *Kolloid-Z.*, **173**, 14 (1960).
4. H. Eyring, *J. Chem. Phys.*, **4**, 283 (1936); **5**, 896 (1937).
5. J. C. Giddings in *Gas Chromatography*, ed. N. Brenner et al., Academic Press, New York, 1962, Chap. V.

6. G. Houghton, A. S. Kesten, J. E. Funk, and J. Coull, *J. Phys. Chem.*, **65,** 649 (1961).
7. W. R. Gambill, *Chem. Engr.*, June 30, 1958, p. 113.
8. A. Einstein, *Z. Electrochem.*, **14,** 1908 (1908).
9. S. Glasstone, K. J. Laidler, and H. Eyring, *The Theory of Rate Processes*, McGraw-Hill, New York, 1941, Chap. IX.
10. C. R. Wilke and P. Chang, *Am. Inst. Chem. Engr. J.*, **1,** 264 (1955).
11. S. J. Hawkes and E. F. Mooney, *Anal. Chem.*, **36,** 1473 (1964).
12. S. J. Hawkes, Ph.D. thesis, Univ. of London, 1963.
13. A. L. Van Geet and A. W. Adamson, *J. Phys. Chem.*, **68,** 238 (1964).
14. H. H. Reamer, J. B. Opfell, and B. H. Sage, *Ind. Engr. Chem.*, **48,** 275 (1956).
15. J. Frenkel, *Kinetic Theory of Liquids*, Oxford, 1946, Chap. IV.
16. L. J. Gosting in *Advances in Protein Chemistry*, Vol. XI, M. L. Anson et al., ed., Academic Press, New York, 1956.
17. F. Helfferich and M. S. Plesset, *J. Chem. Phys.*, **28,** 418 (1958).
18. E. N. Fuller and J. C. Giddings, *J. Gas Chromatog.*, submitted.
19. S. L. Seager, L. R. Geertson, J. C. Giddings, *J. Chem. Eng. Data*, **8,** 168 (1963).
20. R. B. Bird, W. E. Stewart, and E. N. Lightfoot, *Transport Phenomena*, Wiley, New York, 1960.
21. R. C. Reid and T. K. Sherwood, *The Properties of Gases and Liquids*, McGraw-Hill, New York, 1958.
22. J. H. Jeans, *The Kinetic Theory of Gases*, Cambridge, London, 1940, p. 209.
23. O. E. Meyer, *Kinetic Theory of Gases*, English trans., R. E. Baynes, Longmans, London, 1899, p. 263.
24. E. A. Moelwyn-Hughes, *Physical Chemistry*, 2nd ed., Pergamon, London, 1961, p. 61.
25. S. Chapman and T. G. Cowling, *The Mathematical Theory of Nonuniform Gases* 2nd ed., Cambridge, London, 1952, p. 245.
26. J. O. Hirschfelder, C. F. Curtiss, and R. B. Bird, *Molecular Theory of Gases and Liquids*, Wiley, New York, 1954.
27. E. R. Gilliland, *Ind. Eng. Chem.*, **26,** 681 (1934).
28. J. H. Arnold, *Ind. Eng. Chem.*, **22,** 1091 (1930).
29. S. Glasstone, *Textbook of Physical Chemistry*, Van Nostrand, New York, 1946. pp. 524–525.
30. N. H. Chen and D. H. Othmer, *J. Chem. Eng. Data*, **7,** 37 (1962).
31. E. N. Fuller, P. D. Schettler, and J. C. Giddings, unpublished results.
32. P. C. Carman and F. A. Raal, *Proc. Roy. Soc. London.*, **209A**, 38 (1951).
33. P. C. Carman, *Flow of Gases Through Porous Media*, Academic Press, New York, 1956.
34. H. Eyring, D. Henderson, B. J. Stover, and E. M. Eyring, *Statistical Mechanics and Dynamics*, Wiley, New York, 1964, p. 438.
35. J. H. Knox, *2nd International Symposium on Advances in Gas Chromatography*, Houston, March 23–26, 1964.
36. J. H. Knox and L. McLaren, *Anal. Chem.*, **36,** 1477 (1964).
37. J. R. Boyack and J. C. Giddings, *Arch. Biochem. Biophys.*, **100,** 16 (1963).
38. J. C. Giddings and J. R. Boyack, *Anal. Chem.*, **36,** 1229 (1964).
39. F. H. Huyten, W. van Beersum, and G. W. A. Rijnders in *Gas Chromatography, 1960*, R. P. W. Scott, ed., Butterworths, London, 1960, p. 224.
40. S. W. Benson *The Foundations of Chemical Kinetics*, McGraw-Hill, New York, 1960, p. 66.

41. J. H. de Boer, *The Dynamical Character of Adsorption*, Oxford, London 1953, Chap. III.
42. D. M. Young and A. D. Crowell, *Physical Adsorption of Gases*, Butterworths, Washington, 1962, Chap. 1.
43. G. Zweig, *Anal. Chem.*, **31,** 821 (1959).
44. J. C. Giddings, *Anal. Chem.*, **36,** 1170 (1964).
45. J. C. Giddings, *Anal. Chem.*, **35,** 1999 (1963).

Chapter Seven

The Achievement of Separation

7.1 Theory and Practice

Chromatography is primarily a tool for the separation of compounds. It is a failure when a separation is not realized. It is, furthermore, a partial failure (losing much of its unique value by comparison to other methods) when resolution is poor and when excessive time is required for securing adequate separation. The failures and partial failures of chromatography are as numerous as its successes. Even the successes are rarely achieved under optimum conditions. Thus it is necessary, in this final chapter, to direct an inquiry into the basic dynamics of chromatographic separation.

The present chapter is not intended to provide a complete survey of the selection of optimum parameters for chromatographic systems. This task, because of the diversity of chromatographic methods and experimental requirements, properly belongs with the more specific coverage of Volumes II (gas chromatography) and III (liquid chromatography). Here we wish to discuss the problem of choosing optima in a broad sweeping manner; our main attention will be directed at the general phenomenon of separation and its changing role in different circumstances, under different requirements, and using different techniques.

The efficient laboratory use of chromatography would certainly be strengthened if the technique were simpler and thus subject to elementary operating rules of universal validity. Such is not the actual case. The basis of chromatography, as may have been gathered from the preceding chapters, is a kaleidoscopic blend of interrupted geometry, ubiquitous diffusion, and erratic flow. The practical tasks required of it are equally varied and complicated. The collection of unique rules to cover the nearly unlimited diversity of chromatography will be out of reach for a long time to come.

The best alternative to this unachievable goal is the combined application of principles and intuition. The mastery of each requires a good deal of personal effort. Despite many partisan views on this matter, neither of these complementary approaches will by itself secure the utmost performance from chromatography. Experience, the wellspring of intuition, is very naturally an integral part of productive laboratory work. The use of theory and principle is equally a part of modern chromatography, for the underlying theory is the one common element running through this enormous field. Its mastery makes possible the free movement between the multitude of specialized developments and suggests new approaches.

7.2 Elements of Resolution

A preliminary discussion of resolution appeared in Section 2.5. In mathematical terms, resolution is a measure of the degree of separation of zones

$$Rs = \Delta z/4\sigma \tag{7.2-1}$$

where Δz is the gap created between the centers of gravity of neighboring zones and σ is the mean quarter-width (or standard deviation) of the zones, its magnitude indicating the degree to which the gap is filled and cross-contaminated by zone spreading. Both Δz and σ are given in units of length, but because they appear in ratio they may also be considered as the corresponding quantities in time units or may be measured along the chart paper of a recording system.

Resolution can act as one bridge between theory and practice, for experimentally it becomes a pure number partially characterizing the success of the separation and theoretically it becomes an equation involving the parameters of the chromatographic system. We must emphasize that resolution is not a complete measure of practical success. It provides no inkling of the time which may be consumed in achieving separation nor does it guarantee that the zones are readily detectable, etc. Nonetheless it is a practical measure of some worth if its limitations are kept in mind.

The resolution concept can be applied to neighboring zones from any type of column, although care must be taken in using the proper average for σ when the zones are badly distorted (non-Gaussian). It may be applied with equal validity to uniform columns, columns with gradients of various kinds along their length, and columns with conditions changing as a function of time. However, the nonuniform columns are more difficult to treat theoretically than the uniform columns. Judging by several special cases of nonuniform columns treated in gas chromatography (pressure-velocity gradients[1] and programmed temperature gas chromatography[2]), nonuniformity by itself does not seem to have a major effect on resolution. In this rather sweeping treatment we shall dwell only with uniform columns; the important exceptions to these will be treated in Volumes II and III. First, however, we should like to discuss one aspect of nonuniform columns qualitatively, clearing up a common misconception which holds that gradients can strongly alter resolution.

Gradients and Resolution

Several methods in common use find the component zones gaining speed as they proceed along the column. This occurs, for instance, in gas chromatography where gas decompression is responsible for a velocity increase toward the outlet and in coupled-column systems where the mobile phase is forced into a narrow column from a wide one above. Any initial gap between zone centers will be enlarged because the forward zone will progress more rapidly. Hence it may appear that resolution is enhanced by virtue of the increasing relative displacement of the zones. However, the same gradients which magnify the initial gap are also magnifying the zone width to the same extent; the front regions of a zone will proceed more rapidly than the back due to the same gradient in velocity.

Another way of looking at this is to consider the region between zone centers where separation is assumed to be still incomplete. This means that any given point within the region contains a mixture of both components. An increase in velocity will in no way help disentangle these components because as long as they remain intermixed they are subject to the same velocity change. The only real means for unmixing the pair is to proceed with the migration while maintaining conditions favorable to separation.

Many chromatographic systems operate with an effect opposite to the above; zone migration rates slacken with distance. This occurs in gradient elution[3] (because fluid with stronger displacing properties is approaching from above), in circular paper chromatography (because of the increasing area to be swept out by flow away from the center) and in chromathermography[4] (because the moving temperature gradient is hottest at the rear). In these cases the zones are forced more closely together and, to compensate, they become narrower. Gradient elution and chromathermography are both cases of moving gradients where an effect on resolution may be encountered. The unique characteristics of such systems are best indicated by an excellent theoretical paper on chromathermography.[4]

Relationship of Resolution to Column Parameters

If we hope to establish criteria for improving resolution by manipulating column parameters, we must first relate resolution to these parameters.

We note that the distance migrated by a zone along a uniform column is simply the product of its velocity Rv and its migration time t

$$L = Rvt \tag{7.2-2}$$

This distance is different for each zone because the R factor is a unique characteristic of the component. The gap between two zones may thus be expressed as

$$\Delta z = vt(R_{II} - R_I) = vt\,\Delta R \tag{7.2-3}$$

since $vt = L/R$ as seen by equation (7.2-2), this equality may be written as

$$\Delta z = L\,\Delta R/R \tag{7.2-4}$$

This can be used for the numerator of the resolution expression, equation (7.2-1). It makes little practical difference whether L and R are identified with the mean or with the leading or trailing zone since our discussion is directed at zones in the same vicinity where separability is a problem.

The σ term in the denominator of resolution is best written in terms of plate height. Recalling that the basic definition of plate height in uniform columns is $H = \sigma^2/L$ (see equation (2.3-1)), we have

$$\sigma = \sqrt{LH} \tag{7.2-5}$$

The resolution expression $Rs = \Delta z/4\sigma$ may be written more explicitly with the help of the last two equations, as

$$Rs = \sqrt{\frac{L}{16H}}\,\frac{\Delta R}{R} \tag{7.2-6}$$

Thus resolution is a function of the migration length L, the R and ΔR values, and the plate height H. Since the latter, especially, is a function of nearly every column parameter (e.g., temperature, particle size, flow velocity), the resolution expression is in a form which can be related to all column parameters only by choosing a satisfactory equation for H and for $\Delta R/R$. We shall investigate the latter dependency first.

The quantity $\Delta R/R$ may be termed the *relative velocity difference*, a term whose meaning becomes clear when it is noted that the velocity of a zone is directly proportional to its R value. We may write this quantity as

$$\frac{\Delta R}{R} = \frac{R_{II} - R_I}{R} = \frac{R_I R_{II}}{R}\left(\frac{1}{R_I} - \frac{1}{R_{II}}\right) \tag{7.2-7}$$

The product $R_I R_{II}$ can be closely approximated by R^2 since the zones are close neighbors. Thus

$$\Delta R/R = R\,\Delta(1/R) \tag{7.2-8}$$

The expression on the right is easier to work with than its precursor on the left, an advantage that will now be put to use.

The classical expression for R (the basis of which is discussed in Chapter 1) is

$$R = V_m/(V_m + KV_s) \qquad (7.2\text{-}9)$$

where V_m and V_s are the mobile and stationary phase volumes per unit column volume and K is the apparent distribution coefficient (concentration ratio in stationary and mobile phases). Adsorption chromatography is also covered by this partition-oriented expression if V_s is simply replaced by the area of the adsorbing surface.

The reciprocal of R is easily obtained from the last equation as

$$1/R = 1 + KV_s/V_m \qquad (7.2\text{-}10)$$

Its increment, from one zone to the next, is

$$\Delta(1/R) = \Delta K(V_s/V_m) \qquad (7.2\text{-}11)$$

This along with equation (7.2-9) leads to

$$R\,\Delta(1/R) = V_s\,\Delta K/(V_m + KV_s) \qquad (7.2\text{-}12)$$

Since this has been identified by equation (7.2-8) with $\Delta R/R$, the resolution expression, equation (7.2-6), becomes

$$Rs = \sqrt{\frac{L}{16H}}\,\frac{V_s\,\Delta K}{(V_m + KV_s)} \qquad (7.2\text{-}13)$$

which gives resolution as a basic function of the phase volumes V_m and V_s, and the equilibrium characteristics K and ΔK. We can eliminate the former by multiplying K into numerator and denominator. Since $KV_s/(V_m + KV_s)$ is equal to $1 - R$, we get from this procedure

$$Rs = \sqrt{\frac{L}{16H}}\,\frac{\Delta K}{K}(1 - R) \qquad (7.2\text{-}14)$$

which has the advantage of still relating to the fundamental distribution terms ΔK and K, without ties to column volume parameters. The quantity $\Delta K/K$ is termed the *relative selectivity*.

We may now proceed to relate resolution to the parameters which affect the plate height. The detailed attention given to this matter earlier in this volume shows that plate height, if its components are cataloged fully, is a rather formidable collection of terms. For the rather general approach used here we shall use only those terms which can rarely be ignored. We should recognize that this procedure will exclude some systems; an exclusion to be remedied in the more detailed considerations of the subsequent volumes. For most analytical systems the plate height can be approximated by (see Section 2.11 to assess the validity of this)

$$H = \frac{2\gamma D_m}{v} + \omega \frac{d_p{}^2 v}{D_m} + qR(1 - R)\frac{d^2 v}{D_s} \tag{7.2-15}$$

providing the retention mechanism is primarily one of partition, and

$$H = \frac{2\gamma D_m}{v} + \omega \frac{d_p{}^2 v}{D_m} + 2R(1 - R)\bar{t}_d \tag{7.2-16}$$

if adsorption is dominant, By way of review, γ, ω, and q are all structural factors of order unity related to particle orientation and pore geometry; D_m and D_s are diffusion coefficients for the mobile fluid and the stationary partitioning phase, respectively; d_p and d are particle diameter and the depth of the partitioning liquid; and $\bar{t}_d$ is the mean desorption time.

An explicit relationship between resolution and column parameters can be obtained by substituting the two equations above into equation (7.2-14) (the R terms which remain may be further decomposed through the use of equation (7.2-9)). For partition chromatography we have

$$Rs = \frac{\sqrt{L/16}(\Delta K/K)(1 - R)}{(2\gamma D_m/v + \omega d_p{}^2 v/D_m + qR(1 - R)d^2 v/D_s)^{1/2}} \tag{7.2-17}$$

A similar expression is obtained by using equation (7.2-16) for adsorption chromatography. These equations spell out the dependence of resolution on many of the factors which can be controlled by the operator. In most cases it is easy to find values for these parameters which will yield maximum resolution. This subject will be explored in Section 7.3.

The Separation Function

Resolution is occasionally awkward to work with in the study of optimal separation because the denominator, by virtue of its square root character, is no longer a convenient sum of individual terms. A possible alternative is the *separation function*,[5] defined by

$$F = (\Delta z)^2/8(\sigma_{\mathrm{II}}{}^2 + \sigma_{\mathrm{I}})^2 \cong (\Delta z)^2/16\sigma^2 \tag{7.2-18}$$

The last form on the right shows that F is approximately equal to the resolution squared. As criteria of separability F and Rs are essentially equal; a good separation is indicated when either one equals or exceeds unity. However, F is advantageous because it reflects the basic additivity of σ^2 terms; the denominator, unlike in resolution, is a sum of additive terms. In addition, F is directly proportional to the column length.

Evolvement of Resolution During Migration: Length and Plate Requirements

The equations of this section can be used to amplify the qualitative discussion of resolution presented in Section 2.5. In that section it was shown by simple reasoning that during chromatographic migration the gap between zones increases more rapidly (in proportion to L as compared to $\sqrt{L}$) than the outward spreading of the zones. Given sufficient length, the former will always outdistance the latter and separation will be achieved. The development here is parallel although more quantitative. Equation (7.2-4) does indeed show the gap, Δz, increasing with L and equation (7.2-5) shows the zone quarter-width, σ, increasing with $\sqrt{L}$. We thus find resolution, depending on the ratio of the two, increasing with $L/\sqrt{L} = \sqrt{L}$. This shows that separation can be achieved with sufficient column length, but equations (7.2-6), (7.2-13), and (7.2-14) are in this case specific regarding the length needed for given column parameters. In terms of equation (7.2-13), for instance, we obtain on solving for L

$$L = 16H[Rs(V_m + KV_s)]^2/(V_s\,\Delta K)^2 \tag{7.2-19}$$

Thus if we know the column plate height (by measurement or theory), the partition coefficients and the relative volumes, the length can be established for acquiring a desired resolution Rs. Usually $Rs = 1$ is an adequate separation,† and hence Rs may be replaced by unity in the above equation. This method can also be used to show the number of plates required to obtain the desired resolution in view of the fact that $N = L/H$:

$$N = 16[Rs(V_m + KV_s)]^2/(V_s\,\Delta K)^2 \tag{7.2-20}$$

The other equations for resolution, (7.2-6) and (7.2-14), can be used to obtain expressions equivalent to these.

Resolution and the Degree of Retention

By means of various adjustments in temperature, porosity, amount of bulk stationary phase (partition chromatography), and surface activity

† A resolution of unity means, since $Rs = \Delta z/4\sigma$, that four standard deviations or quarter-widths σ may be placed between the zone maxima. Roughly, there is room for two σ's from each zone, the front of one and the back of the other, and that cross-contamination occurs mainly near the midway point and in proportion to the amount of material in the overlapping half of the zone beyond the 2σ distance. This is about 2%. If less cross-contamination is desired a larger resolution must be acquired, e.g., $Rs = 1.5$ (six σ's between zone maxima, 0.1% beyond 3σ). An elegant quantitative discussion of cross-contamination is found in Glueckauf.[6]

(adsorption chromatography), the degree of retention can be varied over a wide range from nearly total retention ($R = 0$) to a state of free migration ($R = 1$). The total retention limit will occur when the sorptive power is so great that all solute is removed from the mobile phase, and the free migration limit will result when there is no sorption. In any practical system a compromise is reached between these limits. Well must this be so, for a chromatographic system cannot operate without its solute occupying both phases to some extent. Solute must occupy the mobile phase or no migration will occur. It must occupy the stationary phase to make use of the selective retardation offered by that phase. The one extreme (lack of solute in the mobile phase or $R = 0$) leads to an infinite time of separation, a problem to be considered generally in Section 7.4. The other extreme ($R = 1$) fails to produce any resolution, a fact demonstrated clearly by equation (7.2-14)—the value of Rs is seen to go to zero as R approaches unity. This occurs even though the relative selectivity $\Delta K/K$ is finite. The value of R, naturally, does not have to be precisely unity to degrade resolution. Any value near unity is injurious because resolution is proportional to $1 - R$. The achievement of separation is thus not consistent with high R values, a fact of some importance in practical work.

Resolution and Theoretical Plates

The very definition of resolution ($Rs = \Delta z/4\sigma$) implies that it is composed of two elements. One, the Δz term, is related to the differential migration of zone centers and the other, σ, is connected with zone spreading. Since plate height is the conventional measure of the spreading of zones, being given by $H = \sigma^2/L$, one element of resolution can be tied directly to theoretical plates. The significance of this fact will be demonstrated here with the help of the previous equations in this section.

If L/H in equation (7.2-6) is replaced by the number of theoretical plates N, it is found that

$$\sqrt{\frac{N}{16}} = \frac{Rs}{\Delta R/R} \tag{7.2-21}$$

Since R is directly proportional to zone velocity (the latter being equal to Rv), $\Delta R/R$ is the *relative velocity difference* of the zones. Thus[7]

$$\sqrt{N/16} = \text{resolution per unit of relative velocity difference} \tag{7.2-22}$$

This equation shows that $\sqrt{N/16}$ is a kind of specific resolution of the column, indicating the resolution that can be achieved for each unit of $\Delta R/R$. Thus N comes closer to indicating the general effectiveness of a column than does resolution itself. (This is actually a result of the fact that N changes only moderately from zone to zone, and thus approaches a true column characteristic, while Rs varies enormously from pair to pair and consequently is not a column characteristic.) For any particular pair, of course, Rs is the desired parameter. The main use of N is in correlating the individual values of Rs. If, for instance, changes occur, or are made, in column parameters which do not affect the relative velocity difference, the change in resolution for each pair may be approximately gaged by noting the resolution change of one pair or by measuring directly the change in N.

In elution chromatography, particularly gas chromatography, each zone is characterized by its elution time (or volume). For close-lying peaks the relative velocity difference $\Delta R/R$ is approximately equal to the *relative retention time difference* $\Delta t/t$. Thus a slightly more convenient relationship between N and Rs is[7]

$$\sqrt{N/16} = \text{resolution per unit of relative retention time (or volume) difference} \tag{7.2-23}$$

In the last two equations resolution is given with respect to the derived (although directly measurable) parameters—zone velocity and retention time. Resolution can be related to a more basic parameter, K, characterizing the distribution between phases, by means of equation (7.2-14). From this is obtained

$$\sqrt{\frac{N}{16}(1 - R)^2} = \frac{Rs}{\Delta K/K} \tag{7.2-24}$$

that is, N multiplied by $(1 - R)^2$ is a measure of resolution per unit of relative selectivity $\Delta K/K$.[7] The quantity $N(1 - R)^2$ is called the number of effective theoretical plates.[8,9] The merit of this parameter is related to the fact that it goes to zero as R approaches one, and it thus vanishes, unlike N, as resolution itself is destroyed by the approach of R toward unity.

Resolution and Free Energy

While a complete study of thermodynamics is beyond the scope of this volume, it is helpful to relate resolution to the most basic of thermodynamic equilibrium properties, the free energy change of phase transfer. We wish particularly to emphasize the limiting practical case of two close-lying zones, for in this case a very simple equation, not generally recognized, is applicable. The treatment here follows an earlier approach by the author.[5]

Of primary concern in regard to the thermodynamics of separation is the difference in distribution coefficients, ΔK, between two solutes. Of particular importance is the relative selectivity $\Delta K/K$ appearing in equation (7.2-14). We shall proceed to relate these quantities to free energy changes.

The quantity ΔK may be written as $K_{\mathrm{I}} - K_{\mathrm{II}}$ (the difference between K values for the two specific peaks of interest). The latter can be rearranged to give

$$(K_{\mathrm{I}} - K_{\mathrm{II}}) = (K_{\mathrm{I}}K_{\mathrm{II}})^{1/2}\left[\left(\frac{K_{\mathrm{I}}}{K_{\mathrm{II}}}\right)^{1/2} - \left(\frac{K_{\mathrm{II}}}{K_{\mathrm{I}}}\right)^{1/2}\right] \qquad (7.2\text{-}25)$$

The individual values of K are given by the usual equation for thermodynamic equilibrium

$$K = \exp\left(-\Delta G^{\circ}/\mathscr{R}T\right) \qquad (7.2\text{-}26)$$

where ΔG° is the standard Gibbs free-energy change for solute in the mobile phase passing into the stationary phase. The ratio of two such terms, as required by equation (7.2-25), is equal to

$$\frac{K_{\mathrm{I}}}{K_{\mathrm{II}}} = \exp\left(\frac{\Delta G_{\mathrm{II}}^{\circ} - \Delta G_{\mathrm{I}}^{\circ}}{\mathscr{R}T}\right) \qquad (7.2\text{-}27)$$

The term in square brackets in (7.2-25) is thus

$$\left[\ \right] = \left[\exp\left(\frac{\Delta G_{\mathrm{II}}^{\circ} - \Delta G_{\mathrm{I}}^{\circ}}{2\mathscr{R}T}\right) - \exp\left(-\frac{\Delta G_{\mathrm{II}}^{\circ} - \Delta G_{\mathrm{I}}^{\circ}}{2\mathscr{R}T}\right)\right] \qquad (7.2\text{-}28)$$

This is in the well-known form $\exp(x) - \exp(-x)$, and is thus twice the hyperbolic sine of x. Consequently

$$\left[\ \right] = 2\sinh\left(\frac{\Delta G_{\mathrm{II}}^{\circ} - \Delta G_{\mathrm{I}}^{\circ}}{2\mathscr{R}T}\right) \qquad (7.2\text{-}29)$$

This is a general expression which can be used in (7.2-25). In our case, since the zones are stipulated to be close to one another, the difference in their ΔG° values cannot be large and the argument of the hyperbolic sine is

$$(\Delta G_{\mathrm{II}}^{\circ} - \Delta G_{\mathrm{I}}^{\circ})/2\mathscr{R}T \ll 1 \qquad (7.2\text{-}30)$$

Under these circumstances the hyperbolic sine may be replaced by its argument† and equation (7.2-29) becomes

$$\left[\ \right] = \frac{\Delta G_{\mathrm{II}}^{\circ} - \Delta G_{\mathrm{I}}^{\circ}}{\mathscr{R}T} \qquad (7.2\text{-}31)$$

If this is substituted back into equation (7.2-25) and the product $K_{\mathrm{I}}K_{\mathrm{II}}$

† When $\Delta G_{\mathrm{II}}^{\circ} - \Delta G_{\mathrm{I}}^{\circ} = \mathscr{R}T$ the error of this step is only about 2%. Chromatographically this is an extremely large increment, much larger than that characterizing pairs difficult to resolve. This can be shown by combining equation (7.2-31) and (7.2-14); from the combination we find that with this increment such zones could be separated with a mere 16 to 32 plates for an R range of 0–0.5. Such zones cannot be regarded as close to one another. When they actually become so, the error is much less than 2%.

replaced by K^2, we obtain the relative selectivity

$$\frac{\Delta K}{K} = \frac{\Delta G_{II}{}^{\circ} - \Delta G_{I}{}^{\circ}}{\mathscr{R}T} = -\frac{\Delta(\Delta G^{\circ})}{\mathscr{R}T} \qquad (7.2\text{-}32)$$

This is the desired limiting expression for close-lying peaks.† It shows, in the range of usual interest, that the ΔK value increases in direct proportion to the difference in the free energy change, and that exponential relationships need not be invoked.

Special Problems

In the introduction to this section we indicated that resolution was only a partial measure of the success of separation. Other criteria such as time, convenience, etc. must sometimes be used to cover special situations. A different kind of limitation on the use of the resolution concept is found when zones are distorted or when two zones are grossly unequal in size. A zone which tails obviously leaves a trail of solute behind that will cross-contaminate with the following zones and thus interfere with purity (if solute collection is a goal), quantitative analysis, or, if the following peak is small, qualitative analysis. A large zone by the side of a small one will have a similar effect; even the relatively low concentrations at the extremes will contaminate the smaller zones to an abnormal extent. In each of these cases the zones must be removed further from one another to gain the desired objective. This means simply that resolution must be increased beyond normal, although usually not much.

These examples do point out the fact that a fixed resolution, adequate under one circumstance, is not always satisfactory. The difficulty with this is usually not serious, and will not be pursued further.

7.3 Optimum Conditions Based on the Resolution Concept

In Section 7.2 equations were obtained showing the dependence of resolution on various column parameters. It is possible, as a next step, to find values of these parameters which will lead to maximum resolution. This procedure may be termed the *optimization with respect to resolution*, and it is based on the concept that resolution is the best criterion of merit for chromatographic runs.‡ This concept is true only to a degree depending on the objectives of the run. The speed of separation is another significant criterion (Section 7.4), the gratification of which will erode away the optimum domain based on the resolution concept. Convenience, practicality

† This expression can also be obtained in its limiting form by taking the derivative of K with respect to ΔG^0 using equation (7.2-26).

‡ This concept, in slightly different terms, has been discussed for gas chromatography by Giddings.[10]

with available equipment, detectability, and sample size are other criteria which must usually be considered along with resolution. In view of these varying shades of complexity our procedure will be to establish optima based on the resolution concept. We shall then indicate how these optima are to be modified in view of some of the other requirements. Our discussion will remain general and rather brief considering the enormous range to be covered. Specific systems will be considered in more detail in the subsequent volumes.

The optimization with respect to resolution is based partly on the form of the plate height equation. As indicated in the last section, we have chosen the simplest equations which represent the elementary facts of zone spreading. The frequent need for other terms is beyond the scope of the present treatment.

In a practical sense the maximization of resolution is significant only for those components not easily separated. Solutes with widely differing structures can be separated by the worst of techniques. In many cases the majority of adjacent pairs may be well separated, and attention will thus be drawn to the one or more overlapping pairs for which resolution is not adequate. In some cases all pairs can be easily separated, and it is wasteful to pursue further optimization. The reason is that once the necessary resolution is acquired, nothing can be gained by its additional increases.

One approach to achieving better separations is connected with searching out more selective systems with larger $\Delta R/R$ or $\Delta K/K$ values. This approach is valuable but it is not, as implied occasionally, the only approach to the problem. The prodigious resolution acquired in some capillary columns (gas chromatography), for instance, is in no way related to the enhancement of selectivity; it is a sole consequence of increasing the number of theoretical plates up to values approaching one million. Also, as pointed out in Section 2.1, the separation of a many-component mixture depends on finding room for the sheer volume of the zones present more than it involves selectivity. Important strides have been made in increasing selectivity in chromatography, but unfortunately little correlation exists between the various developments and they remain scattered throughout a wide literature. Selectivity is primarily a problem of thermodynamic distribution, and for this reason will not be dealt with to any significant degree in this study of dynamics.

Maximum Resolution at Fixed Length and Retention

Equation (7.2-14) shows that resolution can be written as

$$Rs = \frac{\sqrt{L/16}(\Delta K/K)(1 - R)}{\sqrt{H}} \tag{7.3-1}$$

When column length and all retention parameters are fixed, the numerator of this equation remains constant. The sole variation in resolution occurs in plate height H. This quantity, as earlier chapters have made clear, depends on numerous column parameters. We shall discuss these individually later, but the point of interest here is the fact, discussed in Sections 2.11 and 4.8, that even with all parameters optimal the plate height can only be reduced to a certain minimum level, usually equal to one or two particle diameters. This level is fixed by the structural characteristics typical of nearly all chromatographic packing material. Any column exhibiting a plate height this small can be improved very little by any further adjustments. Any column having a significantly larger plate height ($H > 5$ or $10d_p$) is not very efficient from a resolution point of view, and remedial steps should at least be considered.

The choice of particular parameters, discussed below, should be made with the above fact in mind. It is entirely possible to approach the minimum plate height level with some parameters far from optimal. This will occur whenever resolution (or plate height) is not very sensitive to certain parameters. For instance, if mass transfer effects within the stationary phase are negligible for any physically reasonable arrangement of a given system, resolution will not be affected to any marked degree by wide variations in the configuration of the stationary phase.

The foregoing considerations indicate that particle size is of considerable importance in efficient systems. It is thus useful to establish a criterion for the minimal column length required for separation in terms of particle diameter d_p. This is comparable to equation (7.2-19) which gives the required length based on plate height. The particle size criterion, as given here, may be more useful for the initial design of a chromatographic system.

Equation (7.3-1) shows that

$$L/H = 16Rs/[(\Delta K/K)(1 - R)]^2 \qquad (7.3\text{-}2)$$

If plate height is written as $H > 2d_p$ this becomes

$$L/d_p > 32Rs/[(\Delta K/K)(1 - R)]^2 \qquad (7.3\text{-}3)$$

The quantity $1 - R$ is generally near unity, and $Rs = 1$ is usually indicative of an adequate separation. Thus as an approximate criterion we have

$$L/d_p > 32(K/\Delta K)^2 \qquad (7.3\text{-}4)$$

indicating that column length must always be at least $32K/\Delta K$ times as large as d_p. It is easy to show, comparing equations (7.2-6) and (7.2-14), that this criterion may be replaced by

$$L/d_p > 32(R/\Delta R)^2 \qquad (7.3\text{-}5)$$

for cases in which R values are more accessible than K values. The extent of the above inequalities depends on how well the column has been optimized; for well-prepared and well-operated columns an equal sign should be approximately valid in these expressions.

Optimum Parameters

We shall now proceed to obtain optimum parameters for resolution. Our considerations will be based on equation (7.3-1) and the related equations of Section 7.2. A discussion similar to that below but designed specifically for gas chromatography has been published.[12]

Initial Sample Size. If the initial mixture of solutes occupies a large (by chromatographic standards) physical volume, each zone will start out with a considerable width. This, along with a possible nonlinear outgrowth, effectively increases the plate height beyond its normal value and reduces resolution. Thus the optimum sample size, by the resolution criterion, is zero. Practically, of course, a finite sample size must be employed. All things considered our conclusion must be that sample size should be kept to a minimum consistent with detectability, and, if necessary, reproducibility.

Dead Volume. Any large pockets within the column or at its ends will also increase plate height. Theoretically and practically these should be kept very small. In the subsequent discussion these first two effects will be assumed negligible. This is in keeping with good experimental technique.

Column Length *L*. Equation (7.3-1) shows that Rs increases indefinitely with column length. This conclusion is based, of course, on the assumption that other parameters, such as flow velocity, remain constant with increases in length. An increasing pressure drop is needed to maintain constant flow under such circumstances, and eventually a point is reached where such flow can no longer be sustained. We may think of this as an equipment limitation on the optimum. It is a very important limit if the utmost is demanded of the system, and will thus be discussed further in Section 7.5.

There are other considerations which also moderate the length requirement based solely on resolution. First, a long column under given conditions will require a proportionately long separation time; second, long columns are often inconvenient to prepare and use. An optimum for many circumstances will be found as that length which is adequate to separate all critical pairs with a resolution of about unity, but no more. A longer column will increase resolution beyond unity, an increase that is often unnecessary and wasteful.

From here on we will assume that column length is limited and will proceed to find other optimum parameters. In Section 7.5 we will investigate the case in which inlet pressure, not column length, is the chief limit on indefinite increases in resolution.

Flow Velocity v. The only term affected by flow velocity in equation (7.3-1) is the plate height H. Since this is a denominator term, resolution will be a maximum at the velocity which gives a minimum plate height. The plate height passes through a distinct minimum with increasing velocity. This minimum occurs at a compromise velocity for which neither the term B/v, inversely proportional to v, nor those proportional to v (see equations (7.5-15) and (7.5-16)) are excessive.† In Section 2.11 it was shown that the location of the minimum was in the vicinity of $\nu = 1$ for well-designed systems. Since ν, the reduced velocity, equals $d_p v/D_m$, this means that the optimum velocity is approximated by

$$v_{\text{opt}} \sim D_m/d_p \tag{7.3-6}$$

This equation may be used to estimate the general location of the optimum. A few experimental runs will then serve to pinpoint its value if such is deemed necessary. As indicated in Section 2.11 the optimum velocity is very roughly approximated by 10 cm/sec in gas chromatography and 10^{-3} cm/sec in liquid chromatography.

The optimum velocity should be used whenever we wish to obtain the maximum resolution from any given column. However, it is not the fastest way to achieve separation; Section 7.5 will show that separation speed is obtained with velocities above optimal (usually coupled with an increased column length). The speed criterion is especially important in liquid chromatography, which is inherently slow because D_m is small in liquids, and thus v is often set well above the optimum.‡ The value of v need not be located exactly at the optimum for most work because the minimum in plate height (and thus the maximum in resolution) is rather flat over a reasonable interval. This can be shown by writing equations (7.2-15) and (7.2-16) for plate height in the form

$$H = B/v + Cv \tag{7.3-7}$$

The minimum value of H and the optimum velocity which yields this minimum, found by setting the derivative dH/dv equal to zero, are

$$H_{\min} = 2\sqrt{BC} \tag{7.3-8}$$

$$v_{\text{opt}} = \sqrt{B/C} \tag{7.3-9}$$

† As indicated in Section 2.2, the principle of a compromise velocity was first stated by Wilson.[11]

‡ Operating velocities up to 0.17 cm/sec in ion exchange columns have been reported.[12]

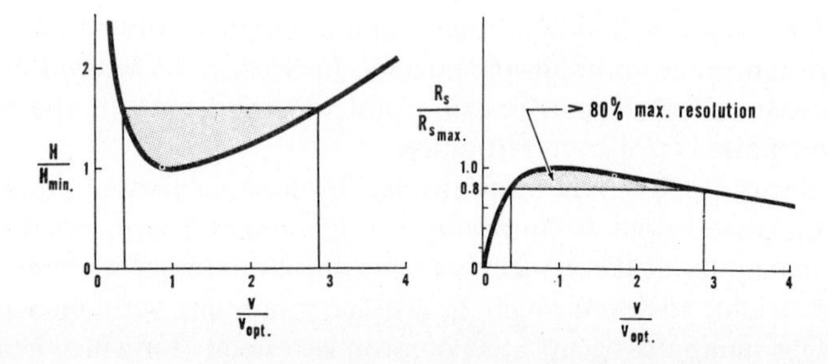

Figure 7.3-1 Plate height relative to its minimum value and resolution relative to its maximum value as a function of the ratio of velocity to the optimum velocity. The maximum in resolution is sufficiently flat that flow velocity need not be fixed exactly. At least 80% of maximum resolution, for instance, may be obtained anywhere along the horizontal lines.

By using these expressions, it can be shown quite easily that

$$\frac{H}{H_{\min}} = \frac{1}{2}\left(\frac{v}{v_{\text{opt}}} + \frac{v_{\text{opt}}}{v}\right) \tag{7.3-10}$$

and since resolution is inversely proportional to the square root of plate height

$$\frac{Rs}{Rs_{\max}} = 1 \Big/ \left[\frac{1}{2}\left(\frac{v}{v_{\text{opt}}} + \frac{v_{\text{opt}}}{v}\right)\right]^{1/2} \tag{7.3-11}$$

A plot of these two equations, shown in Figure 7.3-1, demonstrates that there is some latitude in the choice of velocity without sacrificing resolution. Table 7.3-1 makes this point more explicit by showing the percentage of maximum resolution which can be obtained over a given velocity range. It is seen, for instance, that nearly one-half the maximum resolution can

Table 7.3-1 Resolution Loss as a Function of the Departure from Optimum Flow Velocity

Percentage of maximum resolution	Extremes of (v/v_{opt})
89	$\frac{1}{2}$–2
77	$\frac{1}{3}$–3
69	$\frac{1}{4}$–4
62	$\frac{1}{5}$–5
45	$\frac{1}{10}$–10

be achieved anywhere within a hundredfold range. These figures are based, of course, on the simple form of equation (7.3-7). The coupling mechanism for eddy diffusion (Section 2.10), not allowed for in this simple equation, will extend the velocity range for any given resolution. In those cases where C originates mainly with the mobile phase (especially in liquid chromatography), the upper velocity extreme may be extended considerably, thus making it possible to operate with velocities considerably higher than optimal without great sacrifice.

Mean Particle Diameter d_p. Once again the only effect on resolution, equation (7.3-1), is through the plate height in the denominator. The simple plate height equations of the last section, (7.2-15) and (7.2-16), are adequate to assess the influence of particle size on resolution. The first term in each equation (the longitudinal molecular diffusion term), $2\gamma D_m/v$, is independent of particle size. This term is dominant at low velocities so that when $v \ll v_{opt}$ there is no significant particle size effect. The second term in each equation, representing mobile phase nonequilibrium, increases with the square of particle size—$\omega d_p^2 v/D_m$. The final term of these equations, representing nonequilibrium connected with the stationary phase, is generally independent of particle size. However in some forms of partition chromatography (paper, ion exchange), where the bulk stationary phase is represented by the particle itself (as opposed to cases where this phase occupies separate pores within the particle), the depth d of the stationary phase increases in proportion to d_p. In these cases the final term also increases in proportion to d_p^2. The last two terms, in varying proportions, dominate the plate height at high flow rates, $v \gg v_{opt}$. With rapid flow, then, H invariably increases with particle size. The degree of variation depends on whether the second or third term is dominant and on the nature of the third term. At the extreme H increases with d_p^2 and thus resolution is proportional to $1/d_p$.

The plate height also increases with particle size at optimum flow. The increase is less severe than above because the B/v terms, independent of particle size, has an equal influence with the C terms at the optimum. This is shown by equation (7.3-8). Even at the extreme with C proportional to d_p^2, H_{min} increases only with the first power of d_p.

The foregoing arguments indicate that a zero particle size is the optimum. However very little resolution gain is found beyond a certain reduction in d_p because the term mainly affected, $\omega d_p^2 v/D_m$, rapidly becomes negligible. Thus a point is reached in reducing d_p where there is very little theoretical advantage, and severe practical disadvantages to further reduction. The practical disadvantages stem from the excessive pressure drop needed to force mobile fluid through the column and the difficulty of preparing a uniform packing of extremely fine materials.

Pressure and Pressure Drop. Normally changes in pressure will affect only the plate height component of resolution, so that once again optimization can be based solely on a plate height criterion. (Exceptions exist in gas-solid chromatography where carrier gas sometimes competes with the solute for adsorption sites, and might exist in all forms of chromatography where the pressure is raised enough to change molecular interactions.)

Returning to the plate height expressions of Section 7.2 (equations (7.2-15) and (7.2-16)), it might be observed that no explicit pressure dependence is evident. In fact the plate height in liquid chromatography will not depend on pressure (or mean pressure since a pressure gradient must always be present) unless we approach such values that molecular interactions are affected. This is generally outside the practical range of operation.† For gas chromatography the diffusion coefficient in the mobile (gas) phase, D_m, is inversely proportional to pressure, thus strongly affecting the $2\gamma D_m/v$ and $\omega d_p^2 v/D_m$ terms. For this reason high average pressure is beneficial to resolution when $v \ll v_{\text{opt}}$ and detracts from resolution when $v \gg v_{\text{opt}}$. At the optimum velocity high pressure improves resolution, but the degree of improvement rapidly disappears as the gas phase nonequilibrium term $\omega d_p^2 v/D_m$ equals and then exceeds the nonequilibrium term for the stationary phase.

While high average pressures may improve resolution in many laboratory systems, the gain is not always easy to achieve. High pressures are inconvenient (and sometimes dangerous) to work with, particularly with the many leak-prone connections needed for chromatography. It is a nearly invariable practice of chromatography to maintain the column outlet at atmospheric pressure and to adjust the inlet pressure to suit convenience and to obtain the desired flow.‡ A high average pressure will usually either require an increased pressure drop, with attendant flow changes, or an increase in outlet pressure above atmospheric.

Pressure drop, of course, is related to the mean pressure when outlet conditions are fixed. Its function is different however, being connected with propelling the mobile fluid through the column (this discussion is not relevant, naturally, to paper and thin layer chromatography where capillary forces are mainly responsible for flow). Pressure drop is largely fixed by the parameters discussed earlier. Thus a given column length and particle diameter, coupled with a specified flow velocity, will fix the

† Assuming that the product of diffusivity and viscosity is constant when pressure is applied to liquids we may estimate the change in diffusivity from the pressure dependence of viscosity.[13] This indicates that diffusivity will be reduced about 10% for 100 atmospheres and be cut in half at about 1000 atmospheres.

‡ A rare exception is reported by Scott.[14]

pressure drop within narrow limits. In this sense, pressure drop has already been optimized in the earlier discussions. The real importance of pressure drop is that there are generally equipment limitations which restrict its magnitude. A given choice of column length, particle diameter, and flow velocity may be impossible to achieve because they may require an impossibly high pressure drop. In this case a judicious sacrifice must be made with respect to the intended length or flow velocity, or the particle size must be increased.

In Section 7.5 we will discuss more fully the case in which resolution is limited by the available pressure drop.

Column Radius r_0. The column radius r_0 does not appear anywhere in the resolution expressions of the last section. Nor does it appear in the plate height equations, (7.2-15) and (7.2-16). The reason for the latter omission is that the tube radius effect is apparently small in most columns and was thus excluded from the simple forms of (7.2-15) and (7.2-16). This is not always justified, however, in view of the known resolution loss of large-scale columns and in view of the theoretical treatments of Chapter 2 and 4. As a result of both theoretical and experimental evidence now available, we may tentatively summarize this matter as follows. Tube radius is not highly important in the intermediate range where tube/particle diameter is 5 to 50 and the velocity is near optimum. Large-diameter columns operated at high flow velocities are, on the other hand, subject to serious resolution loss. There are many borderline cases which are simply not well enough understood at present to lead to conclusions. However, as a generally safe and often helpful principle, column diameters should be reduced as much as possible consistent with experimental limitations.

Physical Properties of the Mobile Phase. Very little attention has been given to the physical characteristics of the mobile phase in its relationship with resolution. Nonetheless this matter deserves consideration. In liquid chromatography, of course, the mobile phase strongly influences the relative migration rate, and is thus a big factor in getting a proper range of R values and in controlling selectivity. Aside from this, however, the mobile phase has several significant roles and associated properties. Its viscosity determines the freedom for working with length, velocity, and particle size with the limited pressure drop usually available. Diffusion rates within the mobile phase affect the plate height terms for longitudinal molecular diffusion and mobile phase nonequilibrium. In paper and thin layer chromatography the surface tension (and viscosity) controls the rate at which the mobile phase may advance into the medium.

Temperature T. Temperature influences resolution in a number of ways. Referring to the resolution expression of equation (7.3-1), it is found that changes in temperature strongly affect the $\Delta K/K$ term, the $1 - R$ term, and the plate height H. The only term unaffected is the column length L (neglecting, of course, thermal expansion). We will discuss the influence of temperature on the various terms separately, but at the same time keep in mind that it is the combined relationship of these terms that establishes the role of temperature in the achievement of separation.

Resolution is always proportional to the value of $1 - R$, and, as discussed in the last section, it is desirable to keep R small and absolutely necessary to keep it from approaching unity. The part played by temperature can be seen after using equation (7.2-9) to obtain the expression

$$1 - R = KV_s/(V_m + KV_s) \tag{7.3-12}$$

Aside from minor thermal expansion effects, the only change caused by temperature is in the distribution coefficient K. Furthermore it can be easily shown that $1 - R$ always changes in the same direction as K. Since from equation (7.2-26), $K = \exp(-\Delta G°/\mathcal{R}T)$, and since the free energy change, $\Delta G°$, of solute going from the mobile phase to the stationary phase is generally negative, both K and $1 - R$ may be expected to decrease with temperature.† This is in accord with the usual observation that R increases with temperature. The change in $1 - R$ (and K) is usually quite rapid since it is governed by the exponential form given above, commonly with a substantial energy term in the neighborhood of 10 kcal/mole. However, as $1 - R$ approaches zero, the change is much less rapid.

Practically speaking, it is only necessary that R be kept from unity (usually $R < 0.5$) so that $1 - R$ remains substantial. If R is small (say, 10^{-3}) it may change enormously (say, by a factor of 10^2) without any real effect. However, when R becomes an appreciable fraction of unity, small changes have a noticeable effect on resolution. In this range of R values we may expect, considering the exponential dependence of K, to notice significant changes in resolution due to this factor for temperature changes of roughly 20 C°.

The next factor affected by temperature is the relative selectivity $\Delta K/K$. Equation (7.2-31) gives this quantity as

$$\Delta K/K = -\Delta(\Delta G°)/\mathcal{R}T \tag{7.3-13}$$

where $\Delta G°$ is the standard free energy change for the mobile phase-stationary phase transition and T the absolute temperature. Since $\Delta G° = \Delta H° - T\,\Delta S°$, this becomes

$$\Delta K/K = -\Delta(\Delta H°)/\mathcal{R}T + \Delta(\Delta S°)/\mathcal{R} \tag{7.3-14}$$

The first of these two terms is essentially proportional to $1/T$ (the minus

† Actually it is only the enthalpy component of $\Delta G°$ which is responsible for this decrease.

sign disappears when the actual numbers are substituted in, and thus is not important here), and the second term is essentially independent of T. By "essentially" we mean that changes in the quantities ΔH° and ΔS° themselves are regarded as negligible. Under these circumstances the equation may be written as

$$\Delta K/K = a/T - b \tag{7.3-15}$$

where the constants a and b are normally positive. As long as $a/T > b$, temperature increases have a detrimental effect on $\Delta K/K$. This is the usual situation in practice. When $a/T < b$, temperature increases are beneficial since negative values of ΔK (reversing zone order) are as valid for separation as positive values providing the magnitude is as great. This unusual but sometimes real case is dominated by entropy effects (i.e., by the b term). If the temperature happens to be such that $a/T = b$, then a shift in either direction will restore the finiteness of $\Delta K/K$ and make the separation possible again.

Finally, it is necessary to investigate the effect of temperature on plate height. Since the diffusion coefficient D_m increases with temperature whether the mobile fluid be gas or liquid, the longitudinal molecular diffusion term $2\gamma D_m/v$ increases with T. This term, of course, dominates the low-velocity range $v \ll v_{opt}$. The mobile phase nonequilibrium term $\omega d_p{}^2 v/D_m$ is affected in just the opposite manner because D_m is in the denominator. Thus if stationary phase nonequilibrium terms were absent, leaving only the above two, temperature increases would be detrimental with $v < v_{opt}$, beneficial when $v > v_{opt}$, and unimportant when $v = v_{opt}$.

For partition chromatography, equation (7.2-15), the stationary phase term is $qR(1 - R)d^2v/D_s$. The main temperature dependence arises in the R and D_s quantities. Both increase with temperature but at variable rates. In gas chromatography R, when less than 0.5 or so, is usually more sensitive to temperature than D_s. This is not necessarily so in liquid chromatography. In the former, then, this term usually increases with temperature (assuming a low R value) but in the latter the effect is likely quite variable. This effect is big enough in gas chromatography to make temperature increases generally detrimental by the plate height criterion. The overall effect in liquid chromatography is highly variable, and depends on the particular system considered.

The stationary phase term in adsorption chromatography has not been studied with respect to temperature. It is most probable that this term has no severe temperature dependence, so that the overall effect of temperature may roughly follow the dependence of the first two terms.

When we combine all the above facts, we find that temperature increases are more often than not detrimental to resolution. This would imply, using the resolution criterion, that temperature should be reduced indefinitely. There is one overriding consideration here which mitigates this conclusion; as temperature is lowered and the R value approaches zero, the zone velocity, proportional to R, also becomes very small. The increase in separation time may therefore become prohibitive for practical work. Thus temperature is usually established as a compromise between better resolution (when it is lowered) and higher speed (when it is raised). It should be noted that our conclusions in regard to temperature parallel those made in Section 7.2 on the effect of a changing R value. This is quite naturally the case because one of the main roles of temperature is that of fixing the R value, most other parameters being less affected.

Other Parameters. Parameters other than those considered above generally require a more specific discussion of particular chromatographic systems and methods. The same approach is applicable, but the details are much too numerous to recount here. Such excursions will be reserved for the subsequent volumes of this series.

7.4 Fast Separations

The first and uppermost demand on chromatography is that it provide us with a separation. The last section was designed to show some of the profitable means of improving separability, but this often leads to a system that is overengineered. We frequently find resolution many times larger than needed, particularly for easy separations. Excess resolution can be traded for greater separation speed, a barter that is especially valuable for sluggish or repetitive separations. Chromatography is often regarded as a rapid separation technique when compared to other methods, but it is probably safe to say that only a small fraction of its potential speed is used in most work.

Any complete study of fast separations must be prepared to deal with a wide variety of special circumstances which investigators may encounter. A complete optimization based on the speed of separation (instead of resolution) is possible, but this may require certain operating conditions and column parameters which cannot be easily attained. In this case a less ambitious optimum must be sought, revolving around these special circumstances. An investigator may wish to use a column designed for purposes other than speed, but still may wish to hasten separation as much as possible. Here the column length and radius, the particle size, and the nature of the stationary phase are fixed, but temperature and flow velocity, etc. may be optimized. In other cases limits on temperature, pressure, column length, etc. may restrict the free choice of optima. All of these special cases cannot be covered in a treatment this short. We will try, instead, to introduce the reader to some of the basic concepts and a few of the practical cases. The columns in question will be assumed uniform even though some important effects arise because of gradients (particularly in gas chromatography). Even with gradients, however, the arguments are qualitatively correct.

Increased Speed with a Fixed Column

In order to illustrate our approach with a practical case, let us take the situation in which the column is fixed without regard to fast separations. Let us suppose that the column provides more than adequate resolution when optimized, and that it is desired to trade this excess resolution for speed. The fact that the column has unneeded resolution means, among other things, that it could perform adequately with fewer plates. Thus as a first step we might attempt to trade plates for speed. This can be done easily by increasing the flow velocity v through the column. The effect of this procedure is illustrated in Figure 7.4-1. As we move away from the

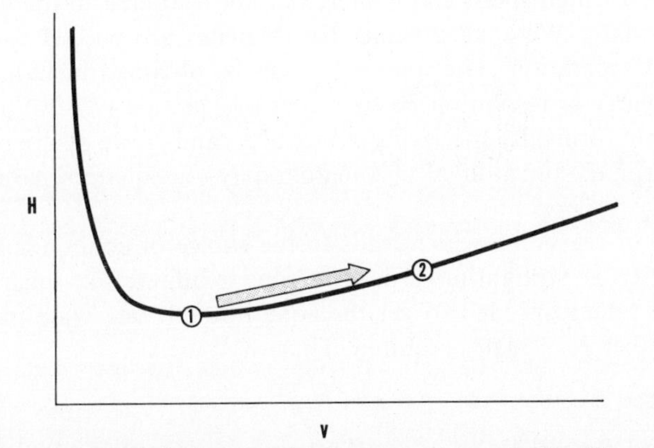

Figure 7.4-1 An increase in flow velocity from point 1 to point 2 will increase analysis time, but only by increasing plate height and thus sacrificing plates. If excess plates exist, the sacrifice is worthwhile.

optimum velocity (following the heavy arrow from point 1 to point 2 of the figure), the plate height increases and there is a corresponding drop in the number of plates. The increase in flow velocity between points 1 and 2 may be considerable, however, and this increase will lead to a proportional increase in separation speed. We may, in general, increase the velocity until a point is reached where no more plates can be sacrificed. This will be the best velocity to achieve the desired goals.

The quantitative basis of the foregoing argument is identical to that considered in Table 7.3-1. There we determined the resolution loss for particular liberties taken with the velocity. In the present case an estimation can first be made for the percentage of optimum resolution required. The permissible velocity, relative to optimum velocity, is then given as the second figure in the second column. For instance, if it is established that a separation will be adequate with only 45% of the maximum resolution, Table 7.3-1 shows that the velocity can be increased by a factor of about 10, and the separation time thus reduced by a factor of 10. In general, we see that rather small reductions in resolution can lead to considerable increases in separation speed.

A second alternative for increasing separation speed is to increase the temperature. An increase in temperature will generally reduce resolution, but will increase the speed of separation. Since some resolution can be sacrificed, the trade is once again profitable.

The Choice of Column Length and Flow Velocity

We will now consider the case in which the investigator wishes to design his column for high speed and is therefore not restricted to the use of an arbitrary length. We shall assume that N plates are needed to achieve the desired separation. The number N can be obtained in terms of the required degree of resolution Rs by means of equation (7.2-20) or one of its equivalent forms. By specifying values of Rs and N, we assure ourselves that the primary function of chromatography—adequate separation—is realized.

The line of reasoning which leads to the choice of column length and flow velocity is straightforward. The time required for separation is simply the time t_p needed to get the zone through one plate multiplied by the number N of plates required. Thus

$$t = Nt_p \tag{7.4-1}$$

The passage time through a single plate t_p, is given by the length (or height) of that plate H divided by the zone velocity Rv: $t_p = H/Rv$. Consequently, the separation time may be written as

$$t = \frac{N}{R}\frac{H}{v} \tag{7.4-2}$$

This shows that the separation time is proportional to H/v, so it is desirable to reduce this ratio as much as possible.† The H/v ratio can be obtained directly from the plate height-velocity plot, thus making this the key to fast separations. Figure 7.4-2 shows how the plate height plot may be

† Recognition of the importance of the H/v ratio in high speed gas chromatography, and its relationship to C, was first made by Purnell.[15]

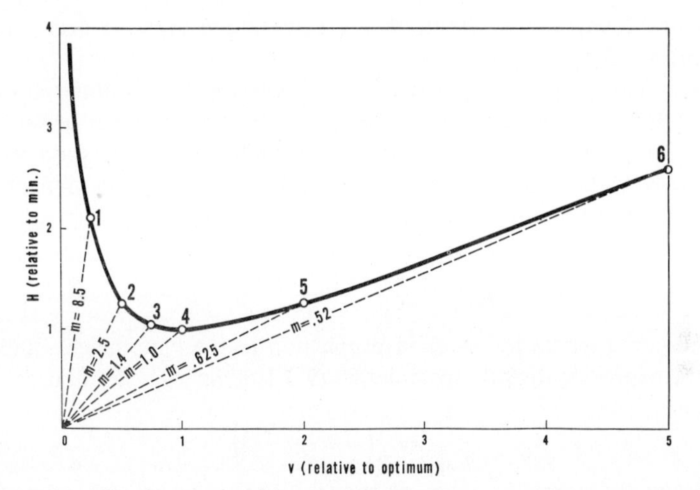

Figure 7.4-2 The shortest possible separation time for any point on the plate height-flow velocity curve is proportional to the slope, *m*, of the line joining that point to the origin. This plot shows that increased separation speed may be attained at high flow velocities.

used to obtain the best conditions for fast analysis. This plot is based on the same simple equation used earlier in this section, $H = B/v + Cv$; in practice we might use an experimental curve or a more elaborate theoretical form.

In the figure a series of hypothetical points 1 through 6 indicate some of the possible conditions for a run. The slope m of the line drawn from the origin to the point is simply the H/v ratio. It is seen that this slope, proportional to separation time, decreases as flow velocity increases. A point of diminishing return is soon reached, however, as seen in the figure. In going from the optimum velocity to twice the optimum velocity (point 4 to point 5), for instance, separation time is cut to 62.5% of its original value. An increase in velocity by a factor of 5 (point 4 to point 6) does little better than this, decreasing separation time only to 52%. Even in the limit of an infinite velocity increase this figure approaches only 50% (this can be seen from equation (7.3-10) by ignoring the last term because $v \to \infty$; the ratio $(H/v)/(H_{min}/v_{opt})$ is observed to approach $\frac{1}{2}$). In actual practice the possible time reduction is probably greater than indicated above. The neglect of further terms (particularly the coupling expression for eddy diffusion) in the simple plate height equation used here minimizes the gains to be made.

Once the velocity has been selected, the column length necessary for the separation is obtained as NH. The value of H can be read directly from the plot for any particular velocity. Since H increases as flow velocity is advanced beyond the optimum, a greater length becomes necessary for separation.

As long as the plate height can be approximated by the simple equation $H = B/v + Cv$, the above observations can be put in mathematical form. Using equation (7.3-10), based on this same plate height expression, we find that separation time can be reduced to the following fraction of that obtained at optimum velocity

$$\frac{H/v}{H_{\min}/v_{\text{opt}}} = \frac{1}{2}\left[1 + \left(\frac{v_{\text{opt}}}{v}\right)^2\right] \tag{7.4-3}$$

The required length increases in proportion to the plate height, and can thus be expressed directly by equation (7.3-10), viz.

$$\frac{L}{L_{\text{opt}}} = \frac{1}{2}\left(\frac{v}{v_{\text{opt}}} + \frac{v_{\text{opt}}}{v}\right) \tag{7.4-4}$$

There is a noted contrast between the gain in speed indicated above and that which was obtained for the "fixed" column. In the latter case it was found that speed could be increased by large amounts—a factor of 10, while preserving adequate resolution, whereas in the present case an increase of 2 is all that is possible. The reason for this contrast is simple: in the fixed column speed is traded for resolution while the variable column, as discussed here, is always held to the minimum acceptable resolution (or plate number). The bulk of the gain in the fixed column is due to this trade; only a smaller portion is due to the proper choice of velocity. This can be seen by noting that a column operated at optimum velocity and having twice the needed resolution (four times the needed plates) is four times longer than necessary. A reduction in length by 4, keeping the same optimal velocity, will reduce the separation time by a factor of 4. On top of this one now increases velocity and length (the latter back to the original), as described above, thus gaining a lesser factor of about 2. These two independent changes are together equivalent to the single change described for fixed columns.

It should not be difficult to exploit the above conclusions for practical work. First, we note that a long column is potentially capable of achieving faster separation than a short column because of the decreasing H/v ratio (see Figure 7.4-2). Thus we start with a column which is as long as practically feasible. Under optimum conditions this will hopefully over-separate the desired components. We may then increase flow velocity until the resolution is barely adequate. This will provide the utmost speed of analysis. If the pressure drop required for such high flow rates and long columns cannot be obtained, the procedure should be followed with a shorter column.

In gas chromatography there is the rather unique situation that the C_m term depends strongly on the location within the column (due to the dependence of gaseous diffusion rates on pressure). Under these circumstances it has been established theoretically that highest speed can be attained with a column of definite (rather than "infinite") length.[16] It is also found that a vacuum outlet leads to increased separation speed.

Other Factors in Fast Separations

Further gains in separation speed are possible through changes in particle size and temperature and in the nature of the mobile and stationary phases. These will be considered briefly here.

It was shown in equation (7.4-2) that minimum separation time is given by $t = NH/Rv$. Subsequently, it was shown that the highest speeds could be obtained with flow velocities well beyond the optimum. At high flow rates the plate height approaches the form $H = Cv$ since the longitudinal molecular diffusion term, B/v, is small. We may therefore replace H/c by C

$$t = (N/R)C \tag{7.4-5}$$

In order to make this expression a more explicit function of the basic parameters, N and C should be written in terms of these parameters. It will be recalled that N is the number of plates needed for the desired resolution, Rs, and is given by equation (7.2-20) as

$$N = 16[Rs(V_m + KV_s)]^2/(V_s\,\Delta K)^2 \tag{7.4-6}$$

An alternative expression, obtained by the same procedure as used on this one except starting with equation (7.2-14), is in the form

$$N = 16(Rs)^2/[(\Delta K/K)(1 - R)]^2 \tag{7.4-7}$$

The problem of writing out an explicit C term is more difficult. Not only does this term vary from technique to technique, but its apparent magnitude is affected by the coupling phenomenon, Section 2.10, and thus depends on the precise velocity range used (see Section 2.11). At very high velocities the mobile phase term C_m disappears completely, thus making the potential separation speed even greater.† For the purposes of the present discussion we shall assume that only a moderately high velocity is used. We shall illustrate the discussion with partition chromatography as an example. Thus the overall C term may be written as (see equation (7.2-15))

$$C = C_m + C_s = \omega\, d_p^{\,2}/D_m + qR(1 - R)\, d^2/D_s \tag{7.4-8}$$

† This condition may be of greatest interest in liquid chromatography where the v/v_{opt} ratio is often quite large.

Upon substituting equation (7.4-7) and (7.4-8) into (7.4-5), we obtain

$$t = \frac{16(Rs)^2}{(\Delta K/K)^2(1 - R)^2 R}\left[\omega \frac{d_p^2}{D_m} + qR(1 - R)\frac{d^2}{D_s}\right] \tag{7.4-9}$$

This shows, among other things, that R cannot be allowed to approach unity and that, quite obviously, the relative selectivity $\Delta K/K$ should be enhanced as much as possible. It also shows that separation time can be reduced by decreasing particle size and by choosing mobile and stationary phases with large diffusivities, providing this is consistent with a high degree of selectivity. The depth d of the units of stationary phase should be kept minimal and the geometry of the units consistent with a low configuration factor q. The relative importance of each of these parameters depends on the case being considered, especially in view of the changing role of the various terms in equation (7.4-9).

If we wished to consider the effect of temperature T or the mobile phase-stationary phase volume ratio V_m/V_s, it would be necessary to use equation (7.4-6) for N and equation (7.2-9) for the remaining R terms of C_s. It would then be necessary to use the exponential temperature equation, (7.2-26), for K, and equation (7.2-31) for $\Delta K/K$. The detailed analysis is quite complex because of the numerous terms and their changing roles. Semiquantitative reasoning shows that increased temperature would generally be desirable in increasing R, D_m, and D_s. The increase would have to be carefully watched to make sure that R was not allowed to approach unity and the critical term $1 - R$ thus kept well above zero.

We spoke earlier of the need to choose parameters for fast separations partly on the basis of special circumstances and limitations inherent to the investigator's goals and equipment. In the field of gas chromatography Knox has reasoned that the ultimate limit on separation speed is determined by the pressure drop available for maintaining column flow.[17,18] This concept is significant for liquid chromatography as well[1] (see Section 7.5). The reason for Knox's conclusion is as follows. The analysis speed, equation (7.4-9), contains a term proportional to d_p^2. As particle size is decreased to gain more speed, the column becomes less permeable to flow. An offsetting pressure increase at the inlet is needed to maintain adequate flow. Eventually a point is reached where inlet pressure cannot be increased further, and this limits the attainable separation speed.

The practical importance of this concept probably varies from case to case. If d is independent of d_p (e.g., gas-liquid chromatography), the beneficial effect of decreasing d_p may vanish as the C_m term (containing d_p) becomes negligible compared to C_s. In other cases (e.g., ion-exchange chromatography), where d is proportional to d_p, the concept is undoubtedly significant. Even here, however, we must assume that columns can be successfully fabricated with the small or even extremely small (in the case of simple separations) particles demanded by theory.

A simple quantitative view of the importance of pressure limitations can be given when d is proportional to d_p, or when the C_s term is altogether negligible. In either case the minimum separation time of equation (7.4-9) is proportional to d_p^2, it being assumed that the other parameters remain fixed by separate criteria. The pressure drop through the column is proportional to Lv/d_p^2 (see Chapter 6). In order to keep corresponding conditions, the v of this ratio must increase in proportion to v_{opt}. The latter varies directly as $(B/C)^{1/2}$, and thus as $1/d_p$. The column length varies in proportion to plate height, Cv, and is consequently proportional to $d_p^2/d_p = d_p$. Thus the pressure drop varies directly as $1/d_p^2$. Since separation time is proportional to d_p^2, it is also inversely proportional to pressure drop

$$t = \text{constant}/\Delta p \tag{7.4-10}$$

The above arguments give a simple view of the role of pressure drop in one limiting case. This matter will be discussed in more detail in Section 7.5.

7.5 Comparison of Gas and Liquid Chromatography

The two main classes of chromatographic analysis—gas and liquid chromatography—will be treated individually in the subsequent volumes of this series. We wish to discuss here some of the similarities and differences of these two broad categories of chromatography.

It is commonly known that gas and liquid chromatography differ in many ways as far as practical analysis is concerned. Gas chromatography, for instance, is obviously limited to the separation of volatile compounds. It is also known that gas chromatography has a general advantage in speed over the liquid methods. There are, moreover, recognized differences in technique, detection, identification, etc. These differences, however, are superficial when regarded from a basic viewpoint. (By "superficial" we do not wish to imply that these differences are any less important in the applications of chromatography.) They are chiefly an outgrowth of secondary differences—differences of degree but not of kind—in the basic chromatographic process. As such, they are of less importance than generally imagined, and do not form an inseparable barrier preventing the transfer of knowledge and data back and forth to the mutual benefit of each.

While the practice of chromatography requires, due to the equipment and technique differences, that the gas and liquid methods be set apart there are very few fundamental reasons for their separation. In fact the underlying processes suggest that the main divisions of chromatography be the partition and adsorption methods. Here, at least, there is a major difference in the mechanism of retention. The partition-adsorption contrast has appeared frequently in this book. Earlier in this chapter, for instance, it was essential to write separate plate height equations, (7.2-15) and (7.2-16), for partition and adsorption chromatography, but not for gas and liquid chromatography. Yet little advantage has ever been taken of the common features of the gas and liquid methods. This failure is largely due to the different paths followed by proponents of each method. It would be advantageous to reunite the two with a common language and theory, and with increased data transfer, in spite of the disparity of laboratory techniques. Some small advancements have recently been made in this direction.[19–21] It is hoped that this book may further catalyze the change.

Nature of the Similarities and Differences

While the similarities of gas and liquid chromatography are much more prominent than the differences, they can be stated much more succinctly; zone migration and separation occur by virtue of the same kinds of thermodynamic, flow, kinetic, and diffusion processes, and are thus subject to the same theoretical laws.

The differences between gas and liquid chromatography are naturally centered around the differences in gases and liquids. As far as chromatography is concerned, the most significant differences are related to some of the physical properties and their immense variation from liquids to gases. These are, in rough order of importance:

1. "density" of intermolecular attraction (~10,000-fold variation)
2. diffusivity (~100,000-fold variation)
3. viscosity (~100-fold variation)
4. surface energy (~10,000-fold variation)
5. density (~1000-fold variation)

Properties of little direct importance to the separation process (such as refractive index, etc.) are naturally excluded. Many properties of liquids and gases, including the above, are interrelated. Among the five properties listed above, the last four can be traced back to the first. Exceedingly slow diffusion in liquids, for example, is due to the immobilizing effect of the surrounding molecular force field on the diffusing species.

Because of the central role of the first property listed above, it deserves a more quantitative basis. We may redefine this term as the negative of either the thermodynamic enthalpy change, ΔH, or energy change, ΔE, attendant to the assemblage of the gas or liquid from infinitely removed molecules. For a liquid, of course, this is very close to the enthalpy or energy of vaporization. The most direct role of intermolecular attraction has to do with retention. A gas, in contrast to a liquid, shows almost no attraction for molecules. Thus a gas has no mechanism for dislodging larger (nonvolatile) molecules from their normal stationary phase environment, within liquids and on solids, where they are anchored by intermolecular forces.† A liquid mobile phase, by contrast, can attract large solute molecules competitively. The attraction, furthermore, depends on the nature of the liquid, thus giving the operator an additional degree of freedom for adjusting retention and selectivity.

The enormous difference, about 10^5 (down to 10^4 on occasion), in diffusivity strongly influences chromatographic resolution and separation speed. At velocities near the optimum, liquid chromatography is likely to show superior resolution for reasons related to diffusivity and outlined in Section 4.8. When we wish to approach the theoretical limit of resolution, as will be discussed shortly, the low diffusivity of liquids is a distinct advantage to liquid chromatography. However the optimum flow velocity and thus the separation speed, is apt to be 10^4–10^5 times slower in liquid chromatography. Sluggish diffusion in the liquid phase probably leads to relatively long separation times in liquid chromatography under any conditions.

The low viscosity of gases is an advantage to gas chromatography, but it is sometimes offset by the diffusivity difference. With a given pressure we should be able to operate a longer column, having more plates, in gas chromatography than liquid chromatography. This would be true were the flow velocities equal. Liquid chromatography, however, is invariably run with slower flow (about 10^4 times slower at the optimum) such that longer columns (and larger plate numbers) are generally possible. Such liquid runs would be slow, however, as compared to gas chromatography. The impressive speed of the latter would not usually be possible without the relatively low viscosity of gases.

Primarily as a result of the relatively strong attractive forces between liquid molecules, liquids possess a significant surface energy (or tension). Surface energy is utterly negligible in gases. The techniques of paper and thin layer chromatography work with liquids but not gases because of this difference in surface energy. First, the main driving force for flow in these systems is capillarity, a phenomenon based on surface (or interfacial) energy. Second, these systems are usually open in the sense that no column wall restrains the escape of mobile phase. This escape is prevented by the surface energy of the mobile liquid.

† Smaller molecules enter the gas, without the inducement of attractive forces, by virtue of their entropy gain. With large molecules "energy" considerations dominate the entropy effects.

The final property on the list is density. The relatively high density of liquids makes it possible to use gravity as the driving force for flow in liquid systems. It also makes the use of centrifugal forces (in paper chromatography) feasible. The reverse effect occurs in ascending paper chromatography where gravity retards the movement of mobile liquid and changes its distribution on the paper. Density also affects the nature of flow as turbulence is approached (Chapter 5).

The Upper Theoretical Limit of Resolution: A Pressure-Drop Criterion[1]

While the theoretical limit may not be easily achieved in either liquid or gas chromatography, a discussion of this limit will show the nature of various factors which, unfortunately, restrain the indefinite increase in separating power. It should also indicate the ultimate potential of liquid as compared to gas chromatography.

The achievement of high resolution (or moderate resolution with pairs difficult to separate) obviously requires the choice of a system with high relatively selectivity $\Delta K/K$ and with an R value well under unity so that the value of $1 - R$ does not go to zero (see equation (7.3-1)). Liquid and gas chromatography have approximately the same potential for filling these requirements. Large values of resolution also require a maximum number of theoretical plates, $N = L/H$, as indicated by the proportionality of Rs to $(L/H)^{1/2}$ shown in equation (7.3-1). We will consider the importance of N below.

An apparent way to increase N without limit is to keep adding to the column length. As indicated in Section 7.3, however, we eventually run into pressure drops of such magnitude that the apparatus fails to operate. In addition, a study of gas chromatography has shown that there is a critical inlet pressure below which separation cannot be achieved no matter how the column length is varied.[16] Thus we must base our prediction of a maximum achievable N on criteria other than length, and with the role of pressure and pressure gradients adequately accounted for.

It was shown in Chapter 5 that the pressure gradient in a column is directly proportional to viscosity and flow velocity, and inversely proportional to the particle diameter squared:

$$\Delta p/L = 2\phi\eta v/d_p^2 \tag{7.5-1}$$

where ϕ is a dimensionless structural constant of order 300. By solving for L, we can write the number of theoretical plates $N = L/H$ as

$$N = d_p^2\,\Delta p/2\phi\eta v H \tag{7.5-2}$$

Thus N appears as a function of the basic column parameters excluding length. The plate height H may be written in the simple form (somewhat abbreviated in its last term) used earlier in this chapter, equation (7.2-15)

$$H = 2\gamma D_m/v + \omega\, d_p^2 v/D_m + C_s v \tag{7.5-3}$$

Thus N becomes

$$N = \Delta p/2\phi\eta(2\gamma D_m/d_p^2 + \omega v^2/D_m + C_s v^2/d_p^2) \tag{7.5-4}$$

Several conclusions are immediately evident. First, the achievable N is directly proportional to the pressure drop Δp available and inversely proportional to viscosity η. Its value increases with particle size d_p, in contrast to the case for high speed separations. Its value increases as the velocity decreases (with Δp fixed a velocity reduction is equivalent to an increase in length). In fact the limiting achievable N, obtained as v approaches zero, is

$$N_{\text{lim}} = \frac{d_p^{\,2}\Delta p}{4\phi\gamma\eta D_m} \tag{7.5-6}$$

This expression is also valid for more complex forms of the plate height equation, especially those allowing for eddy diffusion. A modification is needed, however, if longitudinal molecular diffusion in the stationary phase is important.

In comparing gas and liquid chromatography, we may regard $d_p^2/\phi\gamma$ as a constant. We can also assume that the obtainable pressure drop Δp is comparable for the two methods. We find, then, for the ratio of theoretical maximum N values

$$\frac{N_{\text{lim}}(\text{LC})}{N_{\text{lim}}(\text{GC})} = \frac{\eta_g D_g}{\eta_l D_l} \tag{7.5-7}$$

where LC and GC stand for liquid and gas chromatography, respectively, and the subscripts g and l stand for gas and liquid values. Since ordinarily $\eta_g \sim 10^{-2}\eta_l$ and $D_g \sim 10^5 D_l$, a thousandfold more plates would be attainable by liquid chromatography than gas chromatography. This advantage is lessened by accounting for the fact that D_g is strongly pressure dependent, and we are considering fairly high pressures. Thus a more realistic comparison of the two methods would entail the equation (see Chapter 5)

$$D_g = D_g'/\bar{p} \tag{7.5-8}$$

where D_g' is the diffusion coefficient at one atmosphere, $\sim 10^5 D_l$. It can be shown that the mean column pressure $\bar{p}$ is equal to $\frac{2}{3}$ times the inlet pressure p_i, when the inlet to outlet pressure ratio is large. Also under these conditions $p_i \sim \Delta p$. Thus $\bar{p} \cong 2\,\Delta p/3$ and

$$D_g \cong 3D_g'/2\,\Delta p \tag{7.5-9}$$

Hence the ratio of obtainable plates, equation (7.5-7), becomes

$$\frac{N_{\text{lim}}(\text{LC})}{N_{\text{lim}}(\text{GC})} = \frac{3\eta_g D_g'}{2\eta_l D_l \,\Delta p} \sim 10^3/\Delta p \tag{7.5-10}$$

where Δp is expressed in atmospheres. This shows that gas chromatography would approach the liquid method in potential only if $\sim 10^3$ atmospheres were used for the inlet pressure. This is far beyond present equipment limitations.

It is interesting to calculate the magnitude of N_{lim} for liquid systems with reasonable parameters. We may assume $d_p^2 \sim 10^{-3}$ cm^2 (40–60 mesh), $\Delta p \sim 10$ atm (10^7 dynes/cm^2), $4\phi\gamma \sim 10^3$, $\eta \sim 10^{-2}$ poise, and $D_m \sim 10^{-5}$ cm^2/sec. We thus find

$$N_{\text{lim}} \sim 10^8 \text{ plates} \tag{7.5-11}$$

a value much larger than has ever been reported experimentally.

The foregoing discussion indicates the theoretical maximum to be obtained from a single column without regard to its length. The number of plates (and thus resolution) should be capable of even further increases, however, by linking columns together with pumps. In that case each column would provide up to N_{lim} plates. The pumps, however, would be required to have a dead volume near zero.

Equation (7.5-6) gives the maximum number of plates that can be obtained as a function of the pressure drop for fixed values of d_p, η, and D_m. We may turn this around and regard Δp as the lowest possible inlet pressure (outlet pressure negligible) from which N plates can be gotten. In this case Δp is a critical inlet pressure p_c for the achievement of N plates, the other factors being constant. Thus

$$p_c = 4\phi\gamma\eta D_m N/d_p^2 \tag{7.5-12}$$

For gas chromatography $D_m = D_g \sim 3D_g'/2\,\Delta p$, as shown by equation (7.5-9). On substituting this above and writing $\Delta p = p_c$, each being the inlet pressure at negligible outlet pressure, we obtain

$$p_c' \cong (6\phi\gamma\eta D_g' N/d_p^2)^{1/2} \tag{7.5-13}$$

This agrees in form with a more rigorous derivation[16] which accounted exactly for gas compressibility, but is too small by a numerical factor of $(\frac{2}{3})^{1/2}$ or 0.82.

Comparison of Potential Separation Speed[1]

The above comparison of gas and liquid chromatography is based on maximum resolution (or plates) only, and does not account for the speed of separation. In fact maximum resolution occurs as the flow velocity and thus separation speed approach zero (instead of the previously determined optimum value, v_{opt}) when pressure drop is limiting. The longitudinal molecular diffusion term B is thus critical in this case, whereas the nonequilibrium or C terms are more important for obtaining fast separations.

An equation for the theoretical limit of separation speed, based on a limiting pressure drop, can be readily obtained. As mentioned previously, Knox first suggested the use of a pressure drop limit for fast separations.[17,18] The development of this concept is similar to that of the last section.

The separation time, as indicated by equation (7.4-2), is given by $t = NH/Rv$. The flow velocity v is related to the pressure drop Δp by means of the basic pressure-gradient equation, (7.5-1)

$$v = d_p^{\,2}\,\Delta p/2\phi\eta L \tag{7.5-14}$$

On substituting this into $t = NH/Rv$, we obtain

$$t = 2\phi\eta N^2H^2/R\,d_p^{\,2}\,\Delta p \tag{7.5-15}$$

where L has been replaced by NH. The value of H is dependent, of course, on the flow velocity v in the approximate form of $H = B/v + Cv$. As long as L is written as NH in equation (7.5-14), this provides another relationship between H and v. The value of H (and also v), fixed by these two simultaneous equations, can be obtained as

$$H = d_p^{\,2}\,\Delta p/(2\phi\eta N\,d_p^{\,2}\,\Delta p/C - 4\phi^2\eta^2N^2B/C)^{1/2} \tag{7.5-16}$$

When this is substituted back into equation (7.5-15), we obtain

$$t = \frac{NC}{R}\left[\frac{1}{1 - 2\phi\eta NB/d_p^{\,2}\,\Delta p}\right] \tag{7.5-17}$$

For very large Δp values the separation time approaches the same limit, $t = NC/R$, as discussed in Section 7.4. Otherwise this equation shows how a finite Δp will reduce the latter theoretical limit. As Δp is reduced, a point is reached where the denominator is zero—the separation time goes to infinity. This is the minimum pressure drop able to provide N plates. By writing $B = 2\gamma D_m$, we see that this is the same as the critical pressure (or pressure drop) p_c of equation (7.5-12). The equation for t may be most simply written in terms of this parameter,

$$t = \frac{NC}{R}\left[\frac{\Delta p/p_c}{(\Delta p/p_c) - 1}\right] \tag{7.5-18}$$

The relative merits of gas and liquid chromatography depend critically on the magnitude of, first, C and, second, p_c. For simple separations, requiring few plates, p_c is small (since p_c increases with N) and the ratio $\Delta p/p_c$ is thus large for any reasonable value of Δp. In this case t

approaches the limiting separation time NC/R. In these cases gas chromatography has an enormous advantage because C may be 10^3–10^5 times smaller than in liquid chromatography. As more difficult separations are considered and p_c becomes larger, the quantity in square brackets dwindles in value. This decrease is much more rapid for gas than liquid chromatography since p_c is 10^2–10^3 times larger in the former. A point is reached in gas chromatography, with a sufficient increase in the number of plates required, where p_c becomes as large as the maximum pressure drop Δp. At this point the separation can no longer be achieved in a finite time. With liquid chromatography, requiring the same number of plates, the separation time is still only slightly larger than the theoretical limit NC/R, because of the smallness of p_c. This method continues to achieve separation until 10^2 to 10^3 times more plates are required. There is no doubt then, in this region of extremely difficult separations, that liquid chromatography has the advantage by virtue of the fact that gas chromatography cannot make such separations at all.

The foregoing discussion of separation time in gas and liquid chromatography is illustrated in Figure 7.5-1. This figure demonstrates the

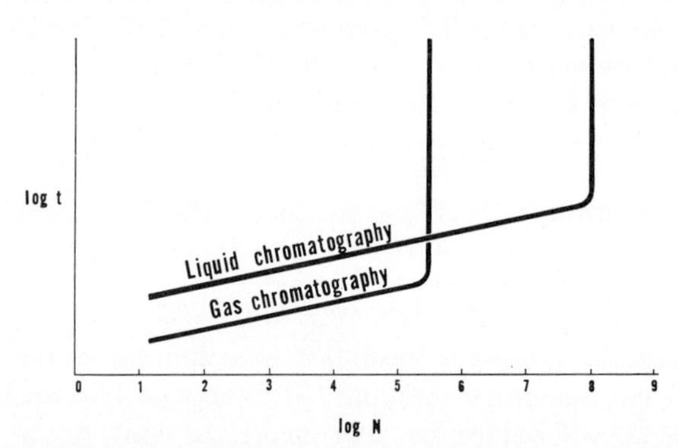

Figure 7.5-1 An illustration of the theoretical limit of minimum separation time for gas and liquid chromatography. For most separations gas chromatography has a distinct advantage, but for extremely difficult cases liquid chromatography, only, can make the separation. Note that the scale is logarithmic. The abscissa values have order of magnitude significance at most.

initial advantage of gas chromatography, and shows that this advantage is abruptly lost at a given number of plates. The transition occurs as C loses its importance compared to the critical pressure p_c.

As was the case with the theoretical limit of resolution, the present consideration of minimum separation time requires special attention for gas chromatography. This is a result of the strong dependence of the diffusion coefficient D_g on pressure.

If we assume that the outlet pressure is negligible, Δp may be written as the inlet pressure p_i. The diffusion coefficient for the gas may, as before, be approximated by $D_m = D_g = 3D_g'/2p_i$, where D_g' is the value of D_g at unit pressure. With this the ratio $\Delta p/p_c$ in equation (7.5-18) becomes $p_i/p_c = p_i^2/p_c'^2$. The critical pressure for gas chromatography, p_c', is given explicitly by equation (7.5-13). If we abbreviate p_i/p_c' by ρ, equation (7.5-18) becomes

$$t = (NC/R)\rho^2/(\rho^2 - 1) \tag{7.5-19}$$

In the simplest case we may assume that C is a sum of stationary phase and mobile phase contribution, $C_s + C_m$. The latter, usually written as $\omega\, d_p^2 v/D_m$, also depends on the diffusion coefficient $D_m = D_g$ and thus on pressure. If we write $C_m = C_m'\bar{p}$ to account for this fact, and write the mean pressure $\bar{p}$ as $2p_i/3$ (see equation (7.5-8) and the material following), we have

$$C_m = 2C_m'p_i/3 \tag{7.5-20}$$

In this case t can be expressed as

$$t = (N/R)(C_l + 2C_m'p_i/3)\rho^2/(\rho^2 - 1) \tag{7.5-21}$$

This equation agrees almost exactly with one derived earlier† in which pressure gradients were rigorously accounted for. Total correspondence is achieved if the ratio $\frac{2}{3}$ appearing with C_m' is replaced by $\frac{3}{4}$. This provides further assurance that strong gradients, as found here, are not very important by themselves.

Equation (7.5-21) shows an important property unique to gas chromatography; namely, t is not necessarily reduced by operating the inlet at its maximum possible pressure. While the ratio $\rho^2/(\rho^2 - 1)$ always decreases with an increase p_i, the appearance of p_i in the numerator, as found only with gas chromatography, will reverse this effect. Thus there is often a distinct optimum‡ for the inlet pressure below its experimental limit.§ This matter will be considered in detail in the subsequent volume.

7.6 References

1. J. C. Giddings, *Anal. Chem.*, **36,** 741 (1964).
2. J. C. Giddings, in *Gas Chromatography*, N. Brenner et al., ed., Academic Press, New York, 1962, Chap. V.
3. L. Hagdahl in *Chromatography*, E. Heftmann, ed., Reinhold, New York, 1961.

† Equation 15 in reference 16.

‡ This optimum is described in reference 16.

§ This does not nullify the Knox concept of pressure-limited separation speed. If d_p is lowered so as to decrease C_m' and possibly C_s, the critical pressure p_c becomes larger. The optimum inlet pressure $p_i > p_c'$ will thus eventually be unattainable.

4. R. W. Ohline and D. D. DeFord, *Anal. Chem.*, **35,** 227 (1963).
5. J. C. Giddings, *Anal. Chem.*, **32,** 1707 (1960).
6. E. Glueckauf, *Trans. Faraday Soc.*, **51,** 34 (1955).
7. J. C. Giddings, *J. Gas Cnromatog.*, **2,** 167 (1964).
8. D. H. Desty, A. Goldup, and W. T. Swanton in *Gas Chromatography*, N. Brenner et al., ed., Academic Press, New York, 1962, Chap. VIII.
9. I. Halász in *Gas Chromatography*, N. Brenner et al., ed., Academic Press, New York, 1962, Chap. VIIIa.
10. J. C. Giddings in *Advances in Analytical Chemistry and Instrumentation*, Vol. 3, C. N. Reilley, ed., Interscience, New York, 1964, p. 315.
11. J. N. Wilson, *J. Am. Chem. Soc.*, **62,** 1583 (1940).
12. P. B. Hamilton, *Anal. Chem.*, **32,** 1779 (1960).
13. F. A. Moelwyn-Hughes, *Physical Chemistry*, 2nd Ed., Pergamon, Oxford, 1961, Chap. XVI.
14. R. P. W. Scott in *Gas Chromatography, 1958*, ed. D. H. Desty, Academic Press, New York, 1958, p. 189.
15. J. H. Purnell, *Ann. N.Y. Acad. Sci.*, **72,** 592 (1959).
16. J. C. Giddings, *Anal. Chem.*, **34,** 314 (1962).
17. J. H. Knox, *J. Chem. Soc.*, **1961,** 433.
18. J. H. Knox, *Gas Chromatography*, Methuen, London, 1962, pp. 34 ff.
19. J. C. Giddings, *Anal. Chem.*, **35,** 1338 (1963); **35,** 2215 (1963).
20. C. Karr, Jr., E. E. Childers, and W. C. Warner, *Anal. Chem.*, **35,** 1290 (1963).
21. J. C. Giddings, *J. Chromatog.*, **13,** 301 (1964).

APPENDIX A

Principal Symbols

General symbols, used throughout book, are indicated by ■. For those symbols used mainly in one or two sections, the section numbers are given at the end of the definition. Minor symbols (e.g., those used only once or used as lesser constants) are not shown.

A, A_i	■	Eddy diffusion contribution to plate height.
A	■	Area of adsorbing surface per unit column volume (specific surface).
A_j		Component of above area due to type j sites (6.5).
A		Cross sectional area of single channel (6.4).
A_i		Fraction of cross sectional area occupied by phase i (6.3).
A_m		Fraction of cross sectional area occupied by mobile phase (6.3).
A_w		Surface area per unit mass (gram) (6.5).
a		$= k_1 t$, stochastic theory (2.13).
a_0	■	Constant in equation, $a = a_0 x^n$, for pore area in tapered pores.
B	■	Longitudinal molecular diffusion term in plate height equation.
b		$= k_2 t$, stochastic theory (2.13).
C	■	Nonequilibrium or mass transfer term.
C_m	■	C term for mobile phase diffusion.
C_s	■	C term for diffusion in bulk stationary phase.
C_k	■	C term for kinetic effects (e.g., adsorption-desorption).
c	■	Overall concentration (amount per unit column volume).
$c_{\max}$	■	Value of c at zone center.
c_s, c_m	■	Overall concentration for stationary phase solute and mobile phase solute, respectively.
c_s^*, c_m^*	■	Equilibrium values of c_s and c_m.
c_s', c_m'	■	Local concentrations (amount per unit volume of stationary phase and mobile phase, respectively).
$c_s'^*, c_m'^*$	■	Equilibrium values of c_s' and c_m'.
c_0		Plateau concentration in frontal analysis (2.14).

$\bar{c}$		Mean molecular velocity (4.8,6.5).
D	■	Diffusion coefficient or effective diffusion coefficient.
D_s,D_m	■	Molecular diffusion coefficients for stationary and mobile phases, respectively.
D_l,D_g	■	Molecular diffusion coefficients for liquids and gases, respectively.
D_i	■	Molecular diffusion coefficient in phase *i*.
d_p	■	Mean particle diameter.
d	■	Thickness (or depth) of the units of bulk stationary phase.
d_c	■	Column diameter.
ΔE_i		Energy difference of type 1 and type *i* sites (6.5).
ΔE_{21}		Energy difference of type 1 and type 2 sites (6.5).
F		Separation function (7.2).
F		Force on fluid element (5.3).
f_2,f_4		Fraction of partially attached molecules which complete their adsorption (4.2).
f	■	Porosity.
f_0	■	Interparticle porosity, $\sim$0.4 for most granular matter.
G_n		Coefficients in velocity expansion equation (4.5).
g_0,g_1	■	Integration constants in nonequilibrium theory.
H	■	Plate height or its components.
$\hat{H}$	■	Apparent (i.e., experimental) plate height.
H_f	■	Plate height for flow exchange mechanism in the mobile phase.
H_D	■	Plate height for diffusive exchange mechanism in the mobile phase.
H_i		Plate height in *i*th segment (2.14).
H_t		Limiting *H* due to turbulence (5.4).
$\Delta H^{\ddagger}$		Activation enthalpy (6.5).
h	■	Reduced plate height, H/d_p.
h		Planck's constant (6.5).
I_0,I_1		Bessel functions (2.13).
J	■	Solute flux (unit area, unit time).
J^*	■	*J* when equilibrium exists between phases.
ΔJ	■	$= J - J^*$.
J_c		Solute flux due to diffusion in single channel (6.4).
K	■	Distribution of partition coefficient (stat. phase conc./mobile phase conc.).
$K_{\text{I}},K_{\text{II}}$	■	*K* value for specific components.
K_0		Specific permeability (5.3).
k		Ratio of solute in stationary and mobile phases (1.2).

k ■ General symbol for rate constants.

k_a ■ Apparent first order rate constant for adsorption.

k_d ■ First order rate constant for desorption.

k_i, k_1, k_2 ■ Various rate constants.

L ■ Column length or length migrated if the latter is less.

L_i Length of ith segment (2.14).

l ■ Random walk step length.

l, l_t Length of normal and tortuous paths, respectively (6.4).

M Constant (in Gaussian) proportional to solute mass (6.2).

M_2 Molecular weight of solvent in diffusion studies (6.3).

M_1 Molecular weight of solute in diffusion studies (6.3).

m ■ Column diameter/particle diameter (d_c/d_p).

N Number of flow passages per unit cross section (5.3).

N Avogadro's number (6.3).

n ■ Taper factor describing pore shape (see equation (4.4-49)).

n ■ Number of random walk steps.

n Molecules per unit volume (6.3,6.5).

$\dot{n}$ ■ n/t, no. random walk steps per unit time.

n_f, n_0 No. steps (exchanges) due to flow and diffusion, respectively (2.10).

P ■ Ratio of inlet to outlet pressure, p_i/p_0.

P ■ Probability distribution (or density), proportional to the zone concentration profile.

P_c, P_e P for column only and external regions only, respectively (2.14).

p ■ Pressure.

p_i, p_o ■ Inlet and outlet pressure, respectively.

$\bar{p}$ ■ Mean column pressure (length average).

p_c ■ Critical inlet pressure for achieving a given separation.

p_c' Critical pressure, explicitly for gas chromatography (7.5).

Q Volumetric fluid flux through single capillary (5.3).

q Fluid flux per unit column area (5.3).

R ■ Rentention ratio, i.e., passage time of nonsorbing zone/passage time of given component; or, equivalently, the fraction of solute in the mobile phase.

$\mathscr{R}$ ■ Gas constant.

R_f ■ Distance to zone/distance to mobile phase front; a parameter similar but not identical to R.

R_0	■	Coil radius.
Rs	■	Resolution.
Re		Reynold's number, a criterion of turbulence (5.4).
$Re_{1/2}$		Re value for partially developed turbulence (5.4).
r	■	Radial coordinate, i.e., distance from center.
r		Number of sorptions during column migration (1.4, 2.13).
r_c, r_p	■	Column radius and particle radius, respectively.
r_1		Radius of solute molecule (6.3).
S	■	Persistence of path distance for molecular migration (see Chapter 2).
$\Delta S^{\ddagger}$		Entropy of activation (6.5).
s	■	Accumulation rate (nonequilibrium theory).
s		Saturation (5.2).
T	■	Absolute temperature (usually degrees Kelvin).
$\mathscr{T}$		Tortuosity (6.4).
T_c		Critical temperature (6.3).
t	■	The time variable.
t	■	Retention time (these two main uses are easily distinguished by contex).
t_i		Retention time in segment i (2.14).
t_p		Passage time through one plate (7.4).
t_m		Time spent by molecule in mobile phase, L/v, during migration.
t_s	■	Time spent by molecule in stationary phase during migration.
t_a, t_d	■	Mean adsorption and desorption times, $1/k_a$ and $1/k_d$, respectively.
$\bar{t}_d$	■	Mean desorption time from nonuniform as well as uniform surfaces.
t_e	■	Exchange (or escape) time between flow paths.
t_e		Time spent by zone external to column (2.14).
t_D	■	Time needed for diffusion over specified distance.
t_A, t_B		Time spent in A and B forms in the general scheme, $A \rightleftarrows B$ (2.13).
Δt_s		Departure of t_s from $\bar{t}_s$ (2.13).
$\mathscr{V}$	■	Velocity of zone center.
$\dot{V}$	■	Volumetric flow rate, usually cm^3/sec.
V_m	■	Volume of mobile phase per unit volume of column.
V_s	■	Volume of stationary phase per unit volume of column.
V_i	■	Volume of phase i phase per unit volume of column.

V_j Volume of unit j of stationary phase (gen. combination law) (4.4).

V_0 Port or detector volume (2.14).

V_1 Molar volume of solute (6.3).

V_c Critical volume (6.3).

v ■ Regional velocity of mobile phase, identical to the downstream velocity of a nonsorbing zone in that region (this is the most significant flow parameter, and is often termed simply "velocity" or "flow velocity").

v_i ■ Regional velocity in state i or phase i (nonequilibrium theory).

v_i ■ Regional velocity at column inlet.

v_i' ■ Local (point) velocity in state i or phase i.

Δv Difference between mean and extreme velocity (2.8).

$v_{1/2}$ ■ Transition velocity in coupling theory.

$\tilde{v}$ ■ Cross sectional average of v (important in large columns).

v_0 Interparticle velocity (5.4).

W ■ Activation energy.

W_l Activation energy for liquid diffusion (6.3).

W_s Activation energy for surface diffusion (6.3)

w Distance along coordinate normal to surface (4.3).

X $= (4k_a k_d t_m t_s)^{1/2}$, stochastic theory (2.13).

X_i ■ Fraction of solute in state i or phase i.

X_i^* ■ Fraction of solute in state i at equilibrium.

x ■ Lateral coordinate; particularly distance from bottom of pore.

x Relative time spent in B form in scheme $A \rightleftarrows B$ (2.13).

Y ■ Number of moles of solute in zone (6.5).

y $= t/t_m$, a reduced coordinate used to describe tailing (2.13).

Z Number of molecules striking unit area in unit time (6.5).

Z Polymer chain length (6.3).

z ■ Distance along column axis.

z_f ■ Distance to liquid front in paper and thin layer chromatography.

Δz Distance separating neighboring zones (2.5,7.2).

Δz Distance from zone center (4.7).

α $= \Delta t_s/\tau$, stochastic theory (2.13).

α Probability that initial state is A in scheme $A \rightleftarrows B$ (2.13).

α		Dimensionless ratio as defined by equation (4.7-19) (4.7).
α	■	Accommodation or sticking coefficient.
α_0		Preexponential constant in exponential form of accommodation coefficient (6.5).
α		Constant in Stefan-Maxwell equation (6.3).
$\alpha(\delta)$		Pore size distribution function (5.2).
β		Probability that initial state of $A \rightleftarrows B$ is B (2.13).
β_s		Reduced longitudinal diffusion term for the stationary phase (2.11).
γ	■	Obstruction factor for diffusion through granular materials.
γ_p		Obstructive factor within solid particles (4.5).
γ		surface tension (5.3).
Δ	■	= increment, e.g., ΔK = increment in distribution coefficient from one solute to the next.
δ		Pore diameter (5.2).
$\delta(x)$	■	Delta function (infinitely thin spike of unit area).
ϵ	■	Fractional departure from equilibrium concentration.
ϵ_m, ϵ_s	■	Overall departure term (in given region) for mobile and stationary phases, respectively.
ϵ_m', ϵ_s'	■	Local departure term for mobile and stationary phases, respectively.
η	■	Viscosity.
θ, θ_0		Nonequilibrium parameters of general combination law (4.4).
$\bar{\theta}$		Average of θ over stationary phase mass (4.4).
θ	■	Fractional surface coverage.
θ_i, θ_1		Fractional surface coverage of type i and type 1 sites, respectively (6.5).
κ		Coefficient in $z_f^2 = \kappa t$ describing flow in paper and thin layers (5.3).
Λ		Heterogeneity factor, giving increase in plate height due to surface nonuniformities (6.5,4.8).
λ, λ_i	■	Eddy diffusion coefficients, e.g., $A = 2\lambda\, d_p$.
ν	■	Reduced velocity, $d_p v / D_m$.
ν		Frequency factor in reaction rates (6.5).
ν		Ratio of outside to inside velocity extremes (4.5,5.4).
ν_L, ν_T		Ratio of velocity extremes in laminar and turbulent flow, respectively (5.4).
ρ	■	Density.

σ	■	Standard deviation ("quarter-width") of zone, in length units.
σ_1, σ_2	■	Components of standard deviation.
σ_{12}		Collision diameter in kinetics theory of gases (6.3).
τ	■	Standard deviation of zone, in time units (applicable to elution chromatography).
Φ	■	Ratio of interparticle free volume to total (interparticle plus intraparticle) free volume.
ϕ		Fraction of retained solute which is adsorbed (4.6).
ϕ	■	Basic flow resistance parameter, defined by equation (5.8-8), ranging from 300 to 600.
ϕ'		$= \phi\Phi$, a number of roughly 300 in most columns (5.3).
χ		Fractional increase in τ due to external effects (2.14).
Ψ_r		Chance that r adsorptions occur during run, used in stochastic theory (2.13).
Ψ_2		Association factor for liquid diffusion, equation (6.3-5) (6.3).
Ω	■	Dimensionless nonequilibrium term, $\Omega = CD_m/d_p^2$.
Ω_s, Ω_k		Values of Ω for bulk stationary phase diffusion and adsorption-desorption, respectively (2.11).
Ω_{12}		Collision integral, gas kinetic theory (6.3).
ω	■	Coefficient of mobile phase plate height contribution, $H = \omega d_p^2 v / D_m$.
ω_i, ω_1 etc.	■	Values of ω for particular processes in the mobile phase.
ω_α		Ratio of exchange distance (i.e., between velocity extremes) to d_p (2.8,2.10).
ω_β		Ratio of velocity difference to mean velocity (2.8,2.9, 2.10).
ω_λ		Ratio of persistence of path, S, to d_p (2.9,2.10).

APPENDIX B

Significant Equations

A summary of the most fundamental and useful equations applicable in the general study of chromatography. The listing is by equation number. Symbols not defined are given in Appendix A.

(1.4-3)	$R = \dfrac{V_m}{V_m + KV_s}$	The Martin-Synge equation, relating R to the distribution coefficient K and the volumes of mobile and stationary phases.
(1.4-4)	$R = \dfrac{V_m}{V_m + \sum K_i V_{si}}$	The generalization of above giving R when more than one retentive mechanism is active.
(2.3-1)	$H = \sigma^2/L$	Definition of plate height in uniform column.
(2.3-2)	$H = d\sigma^2/dL$	Definition of plate height applicable to any column.
(2.3-2a)	$H = L\tau^2/t^2$	Plate height for elution method.
(2.5-1)	$\sigma = l\sqrt{n}$	Zone "quarter width" in terms of random walk parameters (step length and number).
(2.5-5)	$\sigma^2 = \sigma_1{}^2 + \sigma_2{}^2 + \sigma_3{}^2 + \ldots$	Summation rule for variance (σ^2) components of zone.
(2.5-7)	$\sigma^2 = 2Dt_D$	Einstein's equation relating variance to diffusion coefficient and time.
(2.6-1)	$H = \dfrac{2\gamma D_m}{v}$	Plate height for longitudinal molecular diffusion in the mobile phase.
(2.7-6)	$H = 2R(1 - R)vt_d$	Plate height relationship to mean desorption time, t_d, from uniform surface (adsorption chromatography).

(2.8-2)	$H = qR(1 - R)d^2v/D_s$	General plate height expression for diffusion through bulk stationary phase (partition chromatography) of depth d.
(2.8-9)	$H = \omega_i d_p^2 v/D_m$	H for the ith process of mass transfer in the mobile phase.
(2.9-3)	$H = 2\lambda_i d_p$	Classical eddy diffusion expression for plate height.
(2.10-9)	$H = \sum_i \frac{1}{\frac{1}{2}\lambda_i d_p + D_m/\omega_i v d_p^2}$	Coupling expression for eddy diffusion (combining the plate height effects of eddy diffusion and mobile phase mass transfer).
(2.10-12)	$\nu = d_p v/D_m$	Definition of reduced velocity.
(2.14-1)	$\hat{H} = L\tau^2/t^2$	Apparent plate height in elution chromatography.
(2.14-14)	$P(t) = \int_{\infty}^{\infty} P_c(t - t')P_e(t')\,dt'$	General elution profile resulting from the combination of column and external effects.
(3.2-1,2)	$c = c^*(1 \times \epsilon)$	Basic nature (by definition) of equilibrium departure term, ϵ.
(3.2-36)	$H = 2R(1 - R)v/k_d$	Plate height as function of desorption rate constant, k_d, from uniform surface.
(3.2-41)	$H = \frac{-2\epsilon_m}{\partial \ln c/\partial z}$	General relationship of H to mobile phase departure term and gross concentration gradient for stationary phase processes.
(3.3-35)	$H = \frac{2}{3}R(1 - R)\frac{d^2v}{D_s}$	Plate height for diffusion through a uniform film of stationary phase with depth d.
(4.2-31)	$H = -\frac{2\sum X_i^*\epsilon_i v_i}{\mathscr{V}\,\partial \ln c/\partial z}$	General plate height expression (as a function of ϵ terms) for kinetic nonequilibrium.
(4.2-45)	$H = 2R(1 - R)v\bar{t}_d$	Relationship of plate height to the mean desorption time, $\bar{t}_d$, for a nonuniform surface.

(4.3-24) $$H = -\frac{2\sum X_i^*\left(\frac{1}{A_i}\right)\int \epsilon_i' v_i' dA_i}{\mathscr{V}\,\partial \ln c/\partial z}$$

General plate height expression for diffusional nonequilibrium.

(4.4-25) $$H = qR(1 - R)d^2v/D_s$$

H for diffusion in any geometry of bulk stationary phase. The configuration factor is q.

(4.4-26) $$H = \sum (V_j/V_s)H_j$$

General combination law showing how the plate for bulk stationary phase is composed of the constituent plate heights, H_j, of its separate units, weighted by the volume, V_j, of stationary phase in each unit.

(4.4-56) $$H = \frac{2}{(n+1)(n+3)} \times R(1 - R)\frac{d^2v}{D_s}$$

H for "narrow pores" with taper factor n. Of particular interest as special cases we have $n = 0$, uniform film; $n = 1$, rod; $n = 2$, sphere; $n = 3$, bead contact point.

(4.5-4) $$H = \omega d_p^2 v/D_m$$

H contribution from mobile phase nonequilibrium.

(4.5-17) $$H = (6R^2 - 16R + 11) \times r_c^2 v/24D_m$$

Plate height contribution from flow and diffusion in the mobile phase of a capillary column.

(4.6-18) $$H = H(k_1, \infty) + H(k_2, \infty) + \ldots + H(D_1, \infty) + H(D_2, \infty) + \ldots$$

Additive law for the plate height contribution of different processes.

(5.3-8) $$-\frac{dp}{dz} = \frac{2\phi\eta v}{d_p^2}$$

Pressure gradient related to viscosity η, particle diameter d_p and flow velocity v. The flow resistance parameter, ϕ, is usually 300–600.

(5.3-10) $$K_0 = \frac{d_p^2}{180}\frac{f_0^3}{(1 - f_0)^2}$$

Kozeny-Carman equation for specific permeability.

(5.4-1) $Re = \rho v d_p / \eta$

Turbulence develops when the Reynolds number, as defined here, passes through the range from 1 to 100.

(6.2-3) $$\frac{\partial c}{\partial t} = D \frac{\partial^2 c}{\partial z^2}$$

Ficks second law of diffusion.

(6.2-7) $$\frac{\partial c}{\partial t} = -Rv \frac{\partial c}{\partial z} + D \frac{\partial^2 c}{\partial z^2}$$

Ficks second law modified for a zone moving at velocity Rv.

(6.3-1) $D_l = D_0 \exp(-W_l / \mathscr{R}T)$

Equation showing liquid diffusion as an activation process (activation energy = W_l).

(6.3-5) $$D_l = \frac{7.4 \times 10^{-8} (\Psi_2 M_2)^{0.5} T}{\eta V_1^{0.6}}$$

Wilke-Chang equation for predicting diffusion rates in liquids.

(6.3-9) $$D_g = \frac{0.0043 T^{3/2} \left(\frac{1}{M_1} + \frac{1}{M_2} \right)^{1/2}}{p (V^{1/3} + V^{1/3})^2}$$

Gilliland equation (approximate), a practical modification of the Stefan-Maxwell hard sphere model for gaseous diffusion.

(6.3-11) $$D_g = \frac{0.00186 T^{1/2} \left(\frac{1}{M_1} + \frac{1}{M_2} \right)^{1/2}}{p \sigma_{12}^2 \Omega_{12}}$$

Hirschfelder-Bird-Spotz equation (rigorous) for gaseous diffusion.

(6.3-12a) $$D_g = \frac{0.00125 T^{1.75} \left(\frac{1}{M_1} + \frac{1}{M_2} \right)^{1/2}}{p [(\sum v_i)^{1/3} + (\sum v_i)_2^{1/3}]^2}$$

Fuller-Schettler-Giddings equation for gaseous diffusion.

(6.4-13) $\gamma = [\mathscr{T}^2 \bar{A} \overline{(1/A)}]^{-1}$

Obstructive factor as a function of tortuosity and constriction effects.

(6.5-1) $k = \nu \exp(-W / \mathscr{R}T)$

Arhennius equation for the rate constant. Applicable to adsorption and desorption.

(6.5-38) $$H = \frac{8 f (1 - R)^2 v}{\alpha_0 \bar{c} \rho A_w} \times \exp(W / \mathscr{R}T) \Lambda$$

Plate height for surface adsorption and desorption; generalized to incorporate nonuniform surfaces.

(6.5-39)	$\Lambda = \sum x_i{}^* \rho_i$	Heterogeneity factor giving the increase in plate height due to surface nonuniformities.
(7.2-1)	$Rs = \Delta z/4\sigma$	Resolution expressed in terms of peak separation, Δz, and mean "quarter width," σ.
(7.2-14)	$Rs = \sqrt{\dfrac{L}{16H}} \dfrac{\Delta K}{K} (1 - R)$	Resolution as a function of relative selectivity, $\Delta K/K$, and other parameters.
(7.2-23)	$\sqrt{N/16}$ = resolution per unit of relative retention time (or volume) difference.	Relationship of resolution to number of theoretical plates, N.
(7.2-32)	$\dfrac{\Delta K}{K} = -\dfrac{\Delta(\Delta G^0)}{\mathcal{R}T}$	Relative selectivity as a function of the difference in the free energies of phase change.
(7.4-2)	$t = \dfrac{N}{R} \dfrac{H}{v}$	Analysis time in terms of the number of plates, N, needed for separation, the plate height-velocity ratio and R.
(7.5-12)	$p_c = 4\phi\gamma\eta D_m N/d_p{}^2$	The critical pressure, p_c, is the minimum inlet pressure below which it is not possible to achieve a separation requiring N plates.
(7.5-18)	$t = \dfrac{NC}{R} \left[\dfrac{\Delta p/p_c}{(\Delta p/p_c) - 1} \right]$	Analysis time is increased above the theoretical minimum, NC/R, for a finite pressure drop as shown here.

AUTHOR INDEX

Numbers in parentheses are reference numbers and indicate that an author's work is referred to although his name is not cited in the text. Numbers in italics show the page on which the complete reference is listed.

A

B

C

D

E

K

L

M

O

P

R

S

T

V

W

Y

Z

SUBJECT INDEX

F

G

H

I

K

L

M

N

O

P

Z